CHEMICAL
REACTOR
DESIGN

CHEMICAL INDUSTRIES

A Series of Reference Books and Textbooks

Consulting Editor

HEINZ HEINEMANN

1. *Fluid Catalytic Cracking with Zeolite Catalysts,* Paul B. Venuto and E. Thomas Habib, Jr.
2. *Ethylene: Keystone to the Petrochemical Industry,* Ludwig Kniel, Olaf Winter, and Karl Stork
3. *The Chemistry and Technology of Petroleum,* James G. Speight
4. *The Desulfurization of Heavy Oils and Residua,* James G. Speight
5. *Catalysis of Organic Reactions,* edited by William R. Moser
6. *Acetylene-Based Chemicals from Coal and Other Natural Resources,* Robert J. Tedeschi
7. *Chemically Resistant Masonry,* Walter Lee Sheppard, Jr.
8. *Compressors and Expanders: Selection and Application for the Process Industry,* Heinz P. Bloch, Joseph A. Cameron, Frank M. Danowski, Jr., Ralph James, Jr., Judson S. Swearingen, and Marilyn E. Weightman
9. *Metering Pumps: Selection and Application,* James P. Poynton
10. *Hydrocarbons from Methanol,* Clarence D. Chang
11. *Form Flotation: Theory and Applications,* Ann N. Clarke and David J. Wilson
12. *The Chemistry and Technology of Coal,* James G. Speight
13. *Pneumatic and Hydraulic Conveying of Solids,* O. A. Williams
14. *Catalyst Manufacture: Laboratory and Commercial Preparations,* Alvin B. Stiles
15. *Characterization of Heterogeneous Catalysts,* edited by Francis Delannay
16. *BASIC Programs for Chemical Engineering Design,* James H. Weber
17. *Catalyst Poisoning,* L. Louis Hegedus and Robert W. McCabe
18. *Catalysis of Organic Reactions,* edited by John R. Kosak
19. *Adsorption Technology: A Step-by-Step Approach to Process Evaluation and Application,* edited by Frank L. Slejko
20. *Deactivation and Poisoning of Catalysts,* edited by Jacques Oudar and Henry Wise
21. *Catalysis and Surface Science: Developments in Chemicals from Methanol, Hydrotreating of Hydrocarbons, Catalyst Preparation, Monomers and Polymers, Photocatalysis and Photovoltaics,* edited by Heinz Heinemann and Gabor A. Somorjai
22. *Catalysis of Organic Reactions,* edited by Robert L. Augustine

78. *The Desulfurization of Heavy Oils and Residua, Second Edition, Revised and Expanded,* James G. Speight
79. *Reaction Kinetics and Reactor Design: Second Edition, Revised and Expanded,* John B. Butt
80. *Regulatory Chemicals Handbook,* Jennifer M. Spero, Bella Devito, and Louis Theodore
81. *Applied Parameter Estimation for Chemical Engineers,* Peter Englezos and Nicolas Kalogerakis
82. *Catalysis of Organic Reactions,* edited by Michael E. Ford
83. *The Chemical Process Industries Infrastructure: Function and Economics,* James R. Couper, O. Thomas Beasley, and W. Roy Penney
84. *Transport Phenomena Fundamentals,* Joel L. Plawsky
85. *Petroleum Refining Processes,* James G. Speight and Baki Özüm
86. *Health, Safety, and Accident Management in the Chemical Process Industries,* Ann Marie Flynn and Louis Theodore
87. *Plantwide Dynamic Simulators in Chemical Processing and Control,* William L. Luyben
88. *Chemicial Reactor Design,* Peter Harriott
89. *Catalysis of Organic Reactions,* edited by Dennis Morrell

ADDITIONAL VOLUMES IN PREPARATION

Handbook of Fluidization and Fluid-Particle Systems, edited by Wen-Ching Yang

Fundamentals of Polymer Engineering, Anil Kumar and Rakesh K. Gupta

Conservation Equations and Modeling of Chemical and Biochemical Processes, Said S. E. H. Elnashaie and Parag Garhyan

Batch Fermentation: Modeling, Monitoring, and Control, Ali Cinar, Satish J. Parulekar, Cenk Ündey, and Gülnur Birol

Lubricant Additives: Chemistry and Applications, edited by Leslie R. Rudnick

CHEMICAL REACTOR DESIGN

Peter Harriott

Cornell University
Ithaca, New York, U.S.A

MARCEL DEKKER, INC.

NEW YORK · BASEL

Library of Congress Cataloging-in-Publication Data
A catalog record for this book is available from the Library of Congress.

ISBN: 0-8247-0881-4

This book is printed on acid-free paper.

Headquarters
Marcel Dekker, Inc.
270 Madison Avenue, New York, NY 10016
tel: 212-696-9000; fax: 212-685-4540

Eastern Hemisphere Distribution
Marcel Dekker AG
Hutgasse 4, Postfach 812, CH-4001 Basel, Switzerland
tel: 41-61-260-6300; fax: 41-61-260-6333

World Wide Web
http://www.dekker.com

The publisher offers discounts on this book when ordered in bulk quantities. For more information, write to Special Sales/Professional Marketing at the headquarters address above.

Preface

This book deals with the design and scaleup of reactors that are used for the production of industrial chemicals or fuels or for the removal of pollutants from process streams. Readers are assumed to have some knowledge of kinetics from courses in physical chemistry or chemical engineering and to be familiar with fundamental concepts of heat transfer, fluid flow, and mass transfer. The first chapter reviews the definitions of reaction rate, reaction order, and activation energy and shows how these kinetic parameters can be obtained from laboratory studies. Data for elementary and complex homogeneous reactions are used as examples. Chapter 2 reviews some of the simple models for heterogeneous reactions, and the analysis is extended to complex systems in which the catalyst structure changes or in which none of the several steps in the process is rate controlling.

Chapter 3 presents design equations for ideal reactors — *ideal* meaning that the effects of heat transfer, mass transfer, and partial mixing can be neglected. Ideal reactors are either perfectly mixed tanks or packed bed and pipeline reactors with no mixing. The changes in conversion with reaction time or reactor length are described and the advantages and problems of batch, semibatch, and continuous operation are discussed. Examples and problems are given that deal with the optimal feed ratio, the optimal temperature, and the effect of reactor design on selectivity. The design of adiabatic reactors for reversible reactions presents many optimization problems, that are illustrated using temperature-conversion diagrams.

The major part of the book deals with nonideal reactors. Chapter 4 on pore diffusion plus reaction includes a new method for analyzing laboratory data and has a more complete treatment of the effects of complex kinetics, particle shape, and pore structure than most other texts. Catalyst design to minimize pore diffusion effects is emphasized. In Chapter 5 heat transfer correlations for tanks, particles, and packed beds, are reviewed, and the conditions required for reactor stability are discussed. Examples of unstable systems are included. The effects of imperfect mixing in stirred tanks and partial mixing in pipeline reactors are discussed in Chapter 6 with examples from the literature. Recommendations for scaleup or scaledown are presented.

Chapters 7 and 8 present models and data for mass transfer and reaction in gas–liquid and gas–liquid–solid systems. Many diagrams are used to illustrate the concentration profiles for gas absorption plus reaction and to explain the controlling steps for different cases. Published correlations for mass transfer in bubble columns and stirred tanks are reviewed, with recommendations for design or interpretation of laboratory results. The data for slurry reactors and trickle-bed reactors are also reviewed and shown to fit relatively simple models. However, scaleup can be a problem because of changes in gas velocity and uncertainty in the mass transfer coefficients. The advantages of a scaledown approach are discussed.

Chapter 9 covers the treatment of fluidized-bed reactors, based on two-phase models and new empirical correlations for the gas interchange parameter and axial diffusivity. These models are more useful at conditions typical of industrial practice than models based on theories for single bubbles. The last chapter describes some novel types of reactors including riser reactors, catalyst monoliths, wire screen reactors, and reactive distillation systems. Examples feature the use of mass and heat transfer correlations to help predict reactor performance.

I am greatly indebted to Robert Kline, who volunteered to type the manuscript and gave many helpful suggestions. Thanks are also extended to A. M. Center, W. B. Earl, and I. A. Pla, who reviewed sections of the manuscript, and to D. M. Hackworth and J. S. Jorgensen for skilled professional services. Dr. Peter Klugherz deserves special credit for giving detailed comments on every chapter.

Peter Harriott

Contents

Contents

Appendix Diffusion Coefficients for Binary Gas Mixtures

System	T, K	$D_{AB}P$ (cm^2/s) atm
Air		
ammonia	273	0.198
benzene	298	0.0962
carbon dioxide	273	0.136
	317	0.181
	1000	1.32
chlorine	273	0.124
ethanol	298	0.132
helium	276	0.640
	346	0.926
n-hexane	294	0.082
	328	0.095
naphthalene	303	0.088
sulfur dioxide	273	0.122
toluene	298	0.0844
water	298	0.260
	1273	3.25
Carbon dioxide		
ethanol	273	0.0693
hydrogen	273	0.550
methane	273	0.153
propane	298	0.0863
water	307	0.203

Helium
benzene	273	0.317
hydrogen	293	1.64
methanol	432	1.060
nitrogen	352	1.151

Hydrogen
ammonia	298	0.783
benzene	273	0.317
methane	288	0.694
nitrogen	294	0.783
	573	2.481

Nitrogen
ammonia	298	0.230
carbon monoxide	288	0.192
ethylene	298	0.163
oxygen	273	0.181

Oxygen
benzene	311	0.103
ethylene	298	0.182
water	352	0.362

Water
hydrogen	307	1.02
helium	307	0.902
methane	308	0.292
nitrogen	308	0.256
oxygen	352	0.352

1

Homogeneous Kinetics

DEFINITIONS AND REVIEW OF KINETICS FOR HOMOGENEOUS REACTIONS

Reaction Rate

When analyzing kinetic data or designing a chemical reactor, it is important to state clearly the definitions of *reaction rate*, *conversion*, *yield*, and *selectivity*. For a homogeneous reaction, the *reaction rate* is defined either as the amount of product formed or the amount of reactant consumed per unit volume of the gas or liquid phase per unit time. We generally use moles (g mol, kg mol, or lb mol) rather than mass to define the rate, since this simplifies the material balance calculations.

$$r \equiv \frac{\text{moles consumed or produced}}{\text{reactor volume} \times \text{time}} \tag{1.1}$$

For solid-catalyzed reactions, the rate is based on the moles of reactant consumed or product produced per unit mass of catalyst per unit time. The rate could be given per unit surface area, but that might introduce some uncertainty, since the surface area is not as easily or accurately determined as the mass of the catalyst.

$$r \equiv \frac{\text{moles consumed or produced}}{\text{mass of catalyst} \times \text{time}} \tag{1.2}$$

For fluid–solid reactions, such as the combustion of coal or the dissolution of limestone particles in acid solution, the reaction rate is based on the mass of solid or, for some fundamental studies, on the estimated external surface area of the solid. The mass and the area change as the reaction proceeds, and the rates are sometimes based on the initial amount of solid.

Whether the reaction rate is based on the product formed or on one of the reactants is an arbitrary decision guided by some commonsense rules. When there are two or more reactants, the rate can be based on the most valuable reactant or on the limiting reactant if the feed is not a stoichiometric mixture. For example, consider the catalytic oxidation of carbon monoxide in a gas stream containing excess oxygen:

$$CO + \frac{1}{2}O_2 \xrightarrow{\text{cat}} CO_2$$

$$r_{CO} = \frac{\text{moles CO oxidized}}{s, \; g_{cat}}$$

The rate of reaction of oxygen is half that of carbon monoxide, if there are no other reactions using oxygen, and the rate of carbon dioxide is equal to that for carbon monoxide:

$$r_{O_2} = \frac{\text{moles } O_2 \text{ used}}{s, \; g_{cat}} = \frac{1}{2}r_{CO}$$

$$r_{CO_2} = \frac{\text{moles } CO_2 \text{ formed}}{s, \; g_{cat}} = r_{CO}$$

If the goal is to remove carbon monoxide from the gas stream, the correlation of kinetic data and the reactor design equations should be expressed using r_{CO} rather than r_{O_2} or r_{CO_2}.

For synthesis reactions, the rate is usually given in terms of product formation. For example, methanol is produced from synthesis gas by complex reactions over a solid catalyst. Both CO and CO_2 are consumed, and the reaction rate is given as the total rate of product formation.

$$CO + 2H_2 \leftrightarrow CH_3OH$$

$$CO_2 + 3H_2 \leftrightarrow CH_3OH + H_2O$$

$$r = \frac{\text{moles } CH_3OH \text{ formed}}{s, \; g_{cat}}$$

In the definitions given for homogeneous and heterogeneous reactions, all the rates are defined to be positive, even though the amounts of reactants are decreasing. In some texts, the rate is defined to be negative for materials that are consumed and positive for products formed, but this distinction is generally unnecessary. It is simpler to think of all rates as positive and to use material balances to show increases or decreases in the amount of each species.

With a complex reaction system, the *reaction rate* may refer to the rate of an individual reaction or a step in that reaction or to the overall rate of reactant consumption. The partial oxidation of hydrocarbons is often accompanied by the formation of less desirable organic byproducts or by complete oxidation. In the following example, B is the desired product and C, CO_2, and H_2O are byproducts; the equations are not balanced, but this example is used later to demonstrate yield and selectivity.

$$A + O_2 \xrightarrow{1} B$$

$$A + O_2 \xrightarrow{2} C$$

$$A + O_2 \xrightarrow{3} CO_2 + H_2O$$

If only the concentrations of A and B are monitored, the reaction rate could be based on either the formation of B or on the total rate of reaction of A, which would generally be different.

If a complete analysis of the products permits the rate of each step to be determined, the individual rates could be expressed as r_1, r_2, r_3, and combined to give the overall rate for A:

$$r_A = r_1 + r_2 + r_3$$

$$r_B = r_1$$

The reaction rate should *not* be defined as the rate of change of concentration, as is sometimes shown in chemistry texts, since, for gas-phase reactions, the concentration can change with temperature, pressure, or the total number of moles as well as with chemical reaction. For a reaction such as the oxidation of carbon monoxide in a flow system, the moles of product formed are less than the moles of reactant used, and the reactant concentration at 50% conversion is greater than half the initial concentration. Using just the change in concentration of CO would give too low a value for the reaction rate.

For other reactions, there may be a large increase in total moles, as in the cracking of hydrocarbons. Test data for thermal cracking of *n*-hexadecane show 3 to 5 moles of product formed for each mole cracked [1]:

$$C_{16}H_{34} \rightarrow \text{olefins} + \text{paraffins} + H_2$$

The concentration of hexadecane falls much more rapidly than the number of moles of reactant. If the change in total moles is not allowed for, it can lead to errors in determination of reaction order and in reactor scaleup.

For liquid-phase reactions, the densities of reactants and products are often nearly the same, and the slight change in volume of the solution is usually neglected.* Then for a *batch reaction* in a perfectly mixed tank, the reaction rate is the same as the rate of change of reactant or product concentration. To prove this, consider a stirred batch reactor with V liters of solution and a reactant concentration C_A mol/L. The amount reacted in time dt is $V(-dC_A)$, and the reaction rate is $-dC_A/dt$, a positive term:

$$r_A \equiv \frac{\text{moles A reacted}}{\text{L, sec}} = V(-dC_A)\frac{1}{V}\frac{1}{dt} = -\frac{dC_A}{dt} \qquad (1.3)$$

If a reaction is carried out at steady state in a continuously stirred tank reactor, the reactant and product concentrations are constant, and it wouldn't make sense to define the rate as a concentration change. The rate should always be defined as given by Eqs. (1.1) and (1.2).

For reactions with two fluid phases, the definition of *reaction rate* is arbitrary. When a reactant gas is bubbled through a liquid in a tank or column, the rate could be expressed per unit volume of clear liquid or per unit volume of gas–liquid mixture, and these volumes may differ by 5–30%. Unless the reactor is made of glass or has several measuring probes, the froth height is unknown, and the original or clear liquid volume may have to be used to express the rate. Unfortunately, many literature sources do not state the basis for calculation when reporting kinetic data for gas–liquid systems.

When dealing with a reaction in a liquid–liquid suspension or emulsion, the rate is usually based on the total liquid volume, even though the reaction may take place in only one phase. Of course, the rate would then vary with the volume ratio of the phases.

Gas–liquid reactions are sometimes carried out in packed columns. Although the reaction takes place in the liquid phase, the holdup of liquid is not measured, and the reaction rate is given per unit volume of the packed column. The rate is then a function of packing characteristics, liquid rate, and physical properties that affect the holdup as well as kinetic factors.

---

*For a polymerization reaction, the decrease in volume can be as much as 20% and the kinetics can be studied by following the change in volume in a special laboratory reactor called a *dilatometer* [2].

Conversion, Yield, and Selectivity

The *conversion*, x, is defined as the fraction (or percentage) of the more important or limiting reactant that is consumed. With two reactants A and B and a nearly stoichiometric feed, conversions based on each reactant could be calculated and designated x_A and x_B. In most cases, this is not necessary, and only one conversion is calculated based on A, the limiting reactant, and no subscript is needed for x.

$$x \equiv \frac{\text{mole A reacted}}{\text{moles A fed}} \tag{1.4}$$

For a continuous-flow catalytic reactor with W grams of catalyst and F_A moles of A fed per hour, the average reaction rate is calculated from the conversion

$$r_{\text{ave}} = F_A \frac{x}{W} \tag{1.5}$$

The differential form of this equation is used later for analysis of plug-flow reactors:

$$F_A \, dx = r \, dW \tag{1.6}$$

The *yield*, Y, is the amount of desired product produced relative to the amount that would have been formed if there were no byproducts and the main reaction went to completion:

$$Y \equiv \frac{\text{moles of product formed}}{\text{maximum moles of product, } x = 1.0} \tag{1.7}$$

For a system where n moles of A are needed to produce 1 mole of product B but A also gives some byproducts, the yield can be expressed in terms of F_A, the feed rate of A, and the rate of product formation, F_B, both in moles/hr:

$$nA \rightarrow B$$

$$Y = \frac{F_B}{F_A/n}$$

The *selectivity* is the amount of desired product divided by the amount of reactant consumed. This ratio often changes as the reaction progresses, and the selectivity based on the final mixture composition should be called an average selectivity. For $nA \rightarrow B$,

$$S_{\text{ave}} = \frac{\text{B formed}}{\text{A used}} = \frac{F_B}{F_A x/n} = \frac{Y}{x} \tag{1.8}$$

The local selectivity, S, is the net rate of product formation relative to the rate of reactant consumption. The difference between S_{ave} and S can be illustrated with a partial-oxidation example (Fig. 1.1). These equations are not balanced, but 1 mole of A is consumed to make 1 mole of desired product B:

$$A + O_2 \xrightarrow{1} B$$

$$B + O_2 \xrightarrow{2} C$$

$$A + O_2 \xrightarrow{3} CO_2 + H_2O$$

$$S = \frac{r_1 - r_2}{r_1 + r_3}$$

At the start, no B is present, $r_2 = 0$ and $S = r_1/(r_1 + r_2)$. As B accumulates and r_2 increases, S decreases and may even become negative, which would mean B is being destroyed by reaction 2 faster than it is formed by reaction 1. The average selectivity also decreases with increasing conversion but at a lower rate.

The selectivity is a very important parameter for many reaction systems. On scaleup from laboratory reactors to pilot-plant units to industrial reactors, slight decreases in selectivity often occur, and these are generally more important than changes in conversion. Decreases in conversion on scaleup may be corrected for by small changes in reaction time or temperature. However, it is not easy to correct for greater byproduct formation, which may mean more difficult product purification as well as greater raw material cost. A few percent decrease in selectivity may be enough to make the process uneconomic. Factors affecting selectivity changes, such as heat transfer, mass transfer, and mixing patterns, are discussed in later chapters.

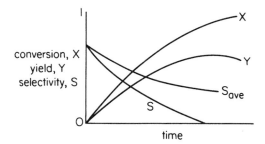

FIGURE 1.1 Changes in conversion, yield, and selectivity for a partial oxidation.

Reaction Order and Activation Energy

Kinetic data are often presented as simple empirical correlations of the following type:

$$r = kC_A^n \quad \text{or} \quad r = kP_A^n P_B^m$$

The reaction order is the exponent in the rate equation or the power to which the concentration or partial pressure must be raised to fit the data. When the exponents are integers or half-integer values, such as $1/2$, 1, $1\frac{1}{2}$, 2, they may offer clues about the mechanism of the reaction. For example, if the gas-phase reaction of A with B appears to be first order to A and first order to B, this is consistent with the collision theory. The number of collisions per unit volume per unit time depends on the product of the reactant concentrations, and a certain fraction of the collisions will have enough energy to cause reaction. This leads to the following equation:

$$A + B \rightarrow C$$

$$r = kC_A C_B$$

If the rate data fit this expression, the reaction is described as first order to A and first order to B. Calling the reaction second order is ambiguous, since a total order of 2 could mean $r = kC_A^{1.5}C_B^{0.5}$ or $kC_A^0 C_B^2$.

Many unimolecular reactions (only one reactant) appear first order over a wide range of concentrations, though second order might seem more logical. Molecules acquire the energy needed to break chemical bonds by collision with other molecules; and if only type A molecules are present, the rate of collisions would vary as C_A^2. The Lindemann theory [3] of unimolecular reactions explains first-order behavior and shows that the order may change with concentration. For the reaction $A \rightarrow B + C$, high-energy molecules A^* are created by collision, but this process is reversible:

$$A + A \xrightarrow{1} A^* + A$$

$$A + A^* \xrightarrow{2} A + A$$

Some of the A^* molecules decompose to $B + C$ before the energy is reduced by step 2:

$$A^* \xrightarrow{3} B + C$$

If steps 1 and 2 are very rapid relative to step 3, so that $r_1 \cong r_2$, an equilibrium concentration of A^* is established, and the reaction to produce B and C appears first order to A:

$$k_1 C_A^2 \cong k_2 C_A^* C_A$$

$$C_A^* = \left(\frac{k_1}{k_2}\right) C_A$$

$$r = k_3 C_A^* = \left(\frac{k_3 k_1}{k_2}\right) C_A$$

In the more general case, C_A^* is assumed to reach a pseudo-equilibrium value, where the rate of formation of A^* is equal to the sum of the rates of the steps removing A^*:

$$\text{For } \frac{dC_A^*}{dt} = 0, \qquad k_1 C_A^2 = k_2 C_A^* C_A + k_3 C_A^*$$

$$C_A^* = \frac{k_1 C_A^2}{k_2 C_A + k_3}$$

At moderate or high pressures, $k_2 C_A \gg k_3$ *and* C_A^* is proportional to C_A, giving first-order kinetics. At very low pressures, the reaction rate might appear second order:

$$\text{if } k_2 C_A \ll k_3, \qquad \text{then} \qquad C_A^* \cong \frac{k_1}{k_3} C_A^2$$

$$r = k_3 C_A^* = k_1 C_A^2$$

At intermediate pressures, a unimolecular reaction might appear to have a noninteger order, such as 1.3 or 1.75, but such values have no physical significance, and the order is likely to change when the concentration is varied over a wider range.

A reaction order of ½ is often found when dealing with molecules that dissociate before reacting. For example, the initial rate of nitric oxide formation reaction in air at high temperature is first order to nitrogen and half order to oxygen:

$$N_2 + O_2 \rightleftharpoons 2NO$$

$$r_i = k P_{O_2}^{1/2} P_{N_2}$$

The half order indicates that the slow step of the reaction involves oxygen atoms, which are nearly in equilibrium with oxygen molecules. Nitric oxide formation is an example of a chain reaction that was first explained by Zeldovitch [4] and is treated in more detail later in this chapter.

Catalytic hydrogenation can also appear half order when H_2 dissociates on the catalyst:

$$H_2 \xrightarrow{1} 2H$$

$$2H \xrightarrow{2} H_2$$

At steady state,

$$k_1 C_{H_2} = k_2 (C_H)^2$$

$$C_H = \left(\frac{k_1}{k_2} C_{H_2}\right)^{1/2}$$

If the reaction order is zero for one reactant, it means that the rate is independent of the reactant concentration, at least for the range of concentrations covered in the tests. It does not mean that the reaction can take place at zero reactant concentration. Zero order to A may indicate that the overall reaction requires several steps, and the rate-limiting step does not involve A. However, at very low values of C_A, some step involving A will become important or controlling, and the reaction order for A will change to a positive value. For a two-phase reaction system, such as A + B(gas) → C, mass transfer of B could be the rate-limiting step, making the reaction appear zero order to A over a wide range of concentrations.

Negative reaction orders are sometimes observed for bimolecular reactions on solid catalysts. Increasing the partial pressure of one reactant, A, which is strongly adsorbed, can lead to a surface mostly covered with adsorbed A, leaving little space for adsorption of reactant B. However, the negative order for A would change to zero order and then to a positive order as the partial pressure of A is reduced to very low values. Reactions that show negative order because of competitive adsorption are discussed in Chapter 2.

Why is it worthwhile to determine the reaction order when analyzing kinetic data or scaling up laboratory results? Finding the reaction order usually does not verify a proposed mechanism, since different models may lead to the same reaction order. The first benefit is that the reaction order is a convenient way of referring to the effect of concentration on the reaction rate, and it permits quick comparisons of alternate reactor designs or specifications. For example, if a first-order reaction in a plug-flow reactor achieves a certain conversion for a given residence time, doubling the residence time will result in the same percent conversion of the remaining reactant. If 50% conversion is measured and the reaction is first order, then doubling the residence time will result in 50% conversion of the material remaining, for an overall conversion of 75%. For a zero-order reaction, doubling the residence time would double the conversion. For a second-order reaction, more than twice the time would be needed to go from 50% to 75% conversion.

The reaction order is also useful when comparing a continuous-flow mixed reactor (CSTR) with a plug-flow reactor (PFR) or a batch reactor. The ratio of reactor volumes, VCSTR/VPFR, increases with reaction order and with the required conversion. For a first-order reaction this ratio is

$$\frac{V_{CSTR}}{V_{PFR}} = \frac{\dfrac{x}{1-x}}{\ln\left(\dfrac{1}{1-x}\right)} = 3.91 \qquad \text{for } x = 0.9$$

For a fractional-order reaction, this volume ratio is smaller than that for first-order kinetics; for second order, the ratio is much larger. Some examples are given in Chapter 3.

Effect of Temperature

For most reactions, the rate expression can be written as the product of a rate constant, which is temperature dependent, and a concentration term:

$$r = k(T)f(C_A, C_B, C_C, \ldots) \qquad (1.9)$$

The rate constant often follows the Arrhenius relationship:

$$k = k_0 e^{-E/RT} \qquad (1.10)$$

where

k_0 = frequency factor (different units)

E = activation energy, J/mol or cal/mol

R = gas constant, 8.314 J/mol K or 1.987 cal/mol K

T = absolute temperature, K

The activation energy has been equated to the energy needed by colliding molecules for reaction to occur. For an endothermic reaction, E is at least somewhat greater than the heat of reaction. For a reversible exothermic reaction, the difference in activation energies of the forward and reverse steps is the heat of reaction, as shown in Figure 1.2.

The variation of k with temperature is often shown using the logarithmic form of Eq. (1.10). For a temperature change from T_1 to T_2, the change in k is

$$\ln\left(\frac{k_2}{k_1}\right) = -\frac{E}{R}\left(\frac{1}{T_2} - \frac{1}{T_1}\right) \qquad (1.11)$$

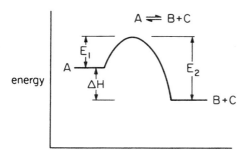

FIGURE 1.2 Activation energies and heat of reaction for a reversible exothermic reaction.

The activation energy can be calculated from two values of k using Eq. (1.11), but it is better to use several data points and make a plot of $\ln(k)$ versus $1/T$, which will have a slope of $-E/R$ if the Arrhenius equation holds.

The derivative of the logarithmic form of Eq. (1.10) is another way to bring out the strongly nonlinear temperature dependence:

$$\frac{d\ln(k)}{dT} = \frac{E}{RT^2} \tag{1.12}$$

If $E/R = 10^4\,\text{K}$ ($E = 20\,\text{kcal/mol}$), a $1°\text{C}$ increase in temperature at 300 K will increase k by 12%. A $1°\text{C}$ increase at 600 K will increase k by only 3% for the same value of E.

SCALEUP AND DESIGN PROCEDURES

The design of large-scale chemical reactors is usually based on conversion and yield data from laboratory reactors and pilot-plant units or on results from similar commercial reactors. A reactor is hardly ever designed using only fundamental rate constants from the literature, because of the complexity of most reaction systems, possible changes in catalyst selectivity, and the effects of heat transfer, mass transfer, and mixing patterns. By contrast, heat exchangers, distillation columns, and other separation equipment can be designed directly from the physical properties of the system and empirical correlations for transport rates.

The normal procedure for a new reaction product or a major process change is to make laboratory tests over a range of conditions to determine the reaction rate, selectivity, and catalyst life. After favorable conditions have been tentatively determined, there are two approaches to scaleup or design of a production unit.

The first method is to scale up in stages using the same type of reactor, the same inlet conditions, and the same reaction time. Batch tests in a 2-liter stirred vessel might be followed by tests in a 5-gallon pilot-plant reactor and then a 50-gallon demonstration unit, operated batchwise or continuously. Data from these tests would be used to estimate the performance and cost of a several-thousand-gallon reactor for the plant. This approach is costly and time consuming, but it is often necessary because the reaction rate and selectivity may change on scaleup. Even with three or four stages in the scaleup procedure, it is often difficult to predict the exact performance of the large reactor, as illustrated in the following example.

Example 1.1

Runs to make a new product were carried out in lab and pilot-plant equipment using both batch and continuous operations. For the tests shown in Table 1.1, the temperature, initial concentrations, and reaction time were the same. How accurately can the performance of the large reactor be predicted?

Solution. The slight decrease in conversion on going from 2 to 30 liters and the further decrease on going from batch to continuous might not be very important. By increasing the residence time, adding more catalyst, or using two reactors in series, the conversion in the plant reactor could probably be raised to 85% to match the original lab tests. However, the gradual decrease in selectivity is a serious problem and could make the process uneconomical, particularly if there is a still further loss in selectivity on going to the full-scale reactor. More tests are needed to study byproduct formation and to see if it is sensitive to factors such as agitation conditions and heat transfer rate.

Stirred reactors are sometimes scaled up keeping the power per unit volume constant; but in other cases, constant mixing time or constant maximum shear rate is recommended. It is impossible to keep all these parameters constant on scaleup and maintain geometric similarity, so tests are

TABLE 1.1   Scaleup Tests with Stirred Reactors

Volume, liters	2	30	30	10,000?
Mode of operation	Batch	Batch	Continuous	Continuous
Conversion	0.85	0.83	0.75	0.750.85?
Yield	0.80	0.76	0.67	?
Selectivity	0.94	0.92	0.89	?

needed to show which parameters are most important. Then it may be necessary to consider a tentative, practical design for the large reactor and *scale down* to a laboratory reactor that can be tested at the same parameters that are achievable in the large unit.

Similar problems arise in scaleup of tubular reactors. For a solid-catalyzed gas-phase exothermic reaction, initial tests might be carried out in a small-diameter jacketed tube packed with crushed catalyst. Suppose that the reactor is 1-cm diameter × 45 cm long with 1-mm catalyst particles and that satisfactory conversion is obtained with a nominal residence time of 1.5 seconds. A reactor with many thousand 1-cm tubes would be impractical, so 5-cm-diameter tubes 4.5 m long are considered for the large reactor (see Fig. 1.3). With a gas velocity 10 times greater, the residence time would be the same, but the pressure drop would be very large, so the particle size might be increased to 5 mm. The D_p/D_t ratio is the same, but the particle Reynolds number and the heat and mass transfer parameters are quite different. One solution to the scaleup problem is to build a pilot plant with a single-jacketed tube, 5 cm × 4.5 m, packed with the 5-mm catalyst pellets. The scaleup to a multitube reactor would be straightforward for boiling fluid in the jacket, but could still pose some problems if a liquid coolant is used, because of temperature gradients in the jacket.

The second scaleup method is to determine the intrinsic kinetics from laboratory tests carried out under ideal conditions, that is, conditions where only kinetic parameters influence the results. If this is not possible, the test data should be corrected for the effects of diffusion, heat transfer, and

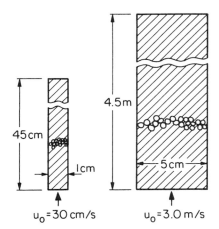

FIGURE 1.3 Scaleup of a tubular reactor.

mixing to determine the intrinsic kinetics. The corrected data are used to determine the reaction order, the rate constant, and the activation energy for the main reaction and the principal byproduct reactions. Overall reaction rates for a larger reactor are predicted by combining the intrinsic kinetics with coefficients for mass transfer and heat transfer and correlations for partial mixing effects.

One advantage of the second method is that the design need not be limited to the same type of reactor. Data taken in a stirred reactor and manipulated to get intrinsic kinetic parameters could be used to estimate the performance of a tubular reactor, a packed bed, or perhaps a new type of contactor for the same reaction. Fundamental kinetic parameters obtained from a small fixed-bed reactor might lead to consideration of a fluidized-bed reactor for the large unit. Of course, pilot-plant tests of the alternate reactor type would be advised.

INTERPRETATION OF KINETIC DATA

There are two main types of laboratory tests used to get kinetic data: batch or integral reactor studies, and tests in a differential reactor. Batch tests are discussed first, since they are more common and often more difficult to interpret. Differential reactors are used primarily for reactions over solid catalysts, which are discussed in Chapter 2.

In a *batch reactor*, all the reactants are charged to a stirred vessel, and the contents are sampled at intervals to determine how the conversion changes with time. If the reactor is a sealed vessel, such as a shaker tube or reaction bomb, the conversion is measured at the end of the test, and other runs are made to show how the conversion varies with time. The semibatch reactor is a variation in which one reactant is charged at the start and the second is added continuously or as frequent pulses as the reaction proceeds. If the second reactant is a gas such as air, it may be fed in large excess and unreacted gas vented from the reactor while products accumulate in the solution.

A type of continuous reactor with performance similar to a batch reactor is the *plug-flow reactor*, a tubular or pipeline reactor with continuous feed at one end and product removal at the other end. The conversion is a function of the residence time, which depends on the flow rate and the reactor volume. The data for plug-flow reactors are analyzed in the same way as for batch reactors. The conversion is compared with that predicted from an integrated form of an assumed rate expression. A trial-and-error procedure may be needed to determine the appropriate rate equations.

To determine the reaction order from batch tests or plug-flow reactor tests, the data are compared with conversion trends predicted for different assumed orders to see which, if any, give a satisfactory fit. There are several steps in this procedure.

1. Plot the data as conversion versus time (x vs. t) for a homogeneous reaction or as x vs. W/F for a catalytic reaction, where W is the mass of catalyst and F is the feed rate. Note the shape of the plot, and consider whether some data points have large deviations from the trend and should perhaps be omitted.

2. Based on the shape of the plot, guess the reaction order, and integrate the corresponding rate equation, allowing for any change in the total number of moles for a gas-phase reaction. If the arithmetic plot shows a gradual decrease in slope with increasing conversion, a first-order reaction is a logical guess. If the decrease in rate is obvious from the tabulated data, step 1 can be omitted and the data presented directly on a first-order plot, such as $\ln(1/1 - x))$ versus t.

3. Rearrange the integrated equation so that a function of x is a linear function of t, and replot the data in this form. If this plot shows definite curvature, guess another order and repeat steps 2 and 3. Use common sense in selecting another order or rate expression rather than making an arbitrary choice. For example, if a first-order plot of $\ln(1/(1 - x))$ versus t shows a decrease in slope at high x, it means that the reaction has slowed down more than expected for a first-order reaction. Therefore a higher order, such as 1.5 or 2, should be tried. There would be no point in guessing a lower order, such as $1/2$.

4. When the data give a reasonably good straight line for the assumed order, check to see if some other order would also fit the data. Scatter in the data may make it difficult to determine the correct reaction order, particularly if the highest conversion is only about 50%.

5. From the plot that best fits the data, determine the rate constant and calculate the predicted conversion for each time. The average error should be close to zero, but the average absolute error is calculated as a way to compare the fit with that for other possible rate expressions. However, a slightly better fit should not be taken as proof of the assumed order. It might be better to say, for example, "The reaction appears to be first order in A, but almost as good a fit is obtained for an order of 1.5. Tests at higher conversions are needed to check the order."

The reaction order determined from batch tests can be checked by varying the initial concentration and comparing initial reaction rates. Sometimes a reaction appears to be first order using initial rate data but higher order by fitting conversion-versus-time data. A possible explanation for such behavior is inhibition by one of the reaction products, which can be checked by runs with some product present at the start.

When the data are accurate enough to clearly show that no simple reaction order gives a satisfactory fit, more complex reactions schemes can be considered. There may be two reactions in parallel that have different reaction orders, which would make the apparent order change with concentration. For a combination of first- and second-order equations, the data can be arranged to determine the rate constants from a linear plot:

$$r = k_1 C_A + k_2 C_A^2$$

$$\frac{r}{C_A} = k_1 + k_2 C_A$$

Example 1.2

Determine the reaction order for the data in Table 1.2 from the air oxidation of compound A in a semibatch reactor:

$$A + O_2 \rightarrow B$$

Solution. Try first order, since the rate seems to be decreasing with time:

$$-\frac{dC_A}{dt} = k_1 C_A$$

$$-\int \frac{dC_A}{C_A} = \ln\left(\frac{C_{A_0}}{C_A}\right) = k_1 t$$

TABLE 1.2 Data for Example 1.2

Time, min	Conversion, x
15	0.06
25	0.11
30	0.21
40	0.25
50	0.36
70	0.44

or, since $C_A = (1 - x)C_{A_0}$,

$$\ln\left(\frac{1}{1 - x}\right) = k_1 t$$

A semilog plot is used for a plot of $1 - x$ versus t, as shown in Figure 1.4(a). A pretty good straight line can be fitted to the data, but the line does not go to 1.0 at $t = 0$. Taking the rate constant from the slope of this line is not correct. The dashed line through $(1.0, 0)$ could be used to get an average value for k.

A slightly curved line could be drawn through the data points, including 1.0 at $t = 0$. Since this line would curve downward, indicating a higher conversion with increasing time than expected for first-order kinetics, a half-order reaction is assumed for the next trial:

$$\frac{dx}{dt} = k_{1/2}(1 - x)^{1/2}$$

$$\int_0^x \frac{dx}{(1 - x)^{1/2}} = k_{1/2}t = 2\big[(1 - x)^{1/2}\big]_x^0 = 2\big(1 - (1 - x)^{1/2}\big)$$

$$1 - (1 - x)^{1/2} = k_{1/2}\frac{t}{2}$$

A plot of $1 - (1 - x)^{1/2}$ vs. t is shown in Figure 1.4(b). A reasonable fit is obtained, but again the straight line does not have the proper intercept.

A third plot is used to test for second-order kinetics:

$$\frac{dx}{dt} = k_2(1 - x)^2$$

$$\int \frac{dx}{(1 - x)^2} = k_2 t$$

$$\frac{1}{1 - x} - 1 = k_2 t \qquad \text{or} \qquad \frac{x}{1 - x} = k_2 t$$

Figure 1.4(c) shows a good straight-line fit, but again the intercept is not at the origin and the fit is not satisfactory.

The order of reaction can't be determined from these results, since assumed orders of $\frac{1}{2}$, 1, and 2 give reasonable straight-line fits to the data, but all have incorrect intercepts. If the run had been extended to conversions of 70–80%, the difference between first and second order would probably be clear, but it might still be hard to decide between closer orders, such as 1 and 1.5.

The data indicate that there may be an induction period of several minutes before significant reaction occurs. This could be checked by taking several samples in the first 10 minutes. An induction period might result

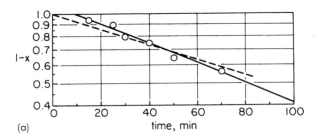

(a)

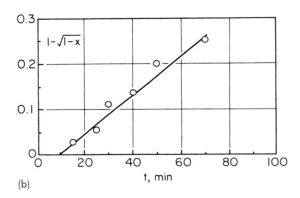

(b)

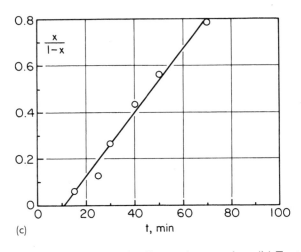

(c)

FIGURE 1.4 (a) Test for first-order reaction. (b) Test for half-order kinetics. (c) Test for second-order kinetics.

from inhibition of the reaction by an impurity that is gradually oxidized or from a delay in reaching the desired temperature. The slow initial reaction could also be caused by a complex reaction scheme with an autocatalytic effect. Determining the cause of this behavior is at least as important as deciding on the reaction order once the reaction proceeds.

Another way of analyzing batch data is to determine the reaction rate for different concentrations from the slope of the plot of C vs. t or x vs. t. Then a log-log plot of the rate versus concentration is made, and the order is the slope of the plot. However, this method works only when there is a continuous record of conversion versus time or when there are many very accurate measurements of the conversion during the run. For data such as those in Example 1.2, taking the rate as $\Delta x/\Delta t$ gives values with a lot of fluctuation. Fitting a smooth curve to the points and measuring the slope is not as accurate as using an integrated form of the rate expression.

Example 1.3

Data for gas-phase cracking of a normal paraffin in a tubular reactor are given in Table 1.3. For moderate conversion, about 4 moles of product are formed for each mole cracked.

   a. Is the reaction first order?
   b. If the change in moles is neglected, would the apparent order be different?

Solution.

   a. $A \rightarrow nP$

   At x fraction converted, the total moles per mole of A fed are

   $$1 - x + nx = 1 + (n - 1)x$$

   Neglecting any changes in temperature and pressure,

TABLE 1.3 Data for Example 1.3

L/u_0, sec	x
0.62	0.212
1.35	0.351
3.05	0.488
4.84	0.602
8.60	0.748
12.2	0.830

$$C_A = C_{A_0} \frac{1-x}{1+(n-1)x}$$

For a tubular reactor with cross section S and length L,

$$F_A \, dx = r \, dV$$

$$u_0 S C_{A_0} \, dx = k_1 C_A \, dV = \frac{k_1 C_{A_0}(1-x) \, dL \, S}{1+(n-1)x}$$

$$\int dx \frac{(1+(n-1)x)}{1-x} = \int k_1 \frac{dL}{u_0}$$

For $n = 4$,

$$\ln\left(\frac{1}{1-x}\right) - 3x + 3\ln\left(\frac{1}{1-x}\right) = \frac{k_1 L}{u_0}$$

$$f(x) = 4\ln\left(\frac{1}{1-x}\right) - 3x = \frac{k_1 L}{u_0}$$

Values of $f(x)$ are given plotted against L/u_0 in Figure 1.5. A straight line through the origin gives a pretty good fit, so first order is probably satisfactory for design purposes.

b. If the change in moles is neglected, a plot of $\ln(1/(1-x))$ vs. L/u_0 is the test for a first-order reaction. The plot in Figure 1.6(a) shows a curve with a pronounced decrease in slope as conversion increases. This might suggest a second-order reaction, but the plot of $x/(1-x)$ vs. L/u_0, Figure 1.6(b), shows increasing slope.

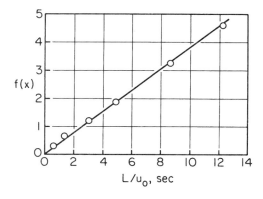

FIGURE **1.5** Test of first-order reaction for Example 1.3: $f(x) = 4 \ln(1/1 - x)$ $-3x$.

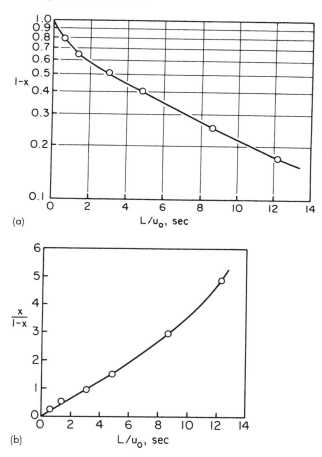

FIGURE **1.6** (a) Test of first-order reaction for Example 1.3 ignoring volume change. (b) Test of second-order reaction for Example 1.3 ignoring volume change.

An assumed order of between 1.0 and 2.0—say, 1.5—might appear satisfactory, but it would be incorrect and could lead to errors in design.

Adiabatic Reactors

Although most kinetic tests are carried out at constant temperature and runs at different temperatures are used to get the activation energy, it is possible to get the rate constant and activation energy from one test in an adiabatic reactor if the reaction is moderately exothermic. For a batch reaction in a well-insulated stirred reactor, the heat released is stored as sensible heat of the fluid and the reactor wall:

$$VC_{A_0}x(-\Delta H) = M_{cp}(T - T_0) + M_w c_{pw}(T - T_0) \tag{1.13}$$

The increase in temperature as the reaction proceeds raises the rate constant; at first this more than offsets the decrease in reactant concentration, and the reaction accelerates. A plot of temperature versus time is then S-shaped, as shown by the solid line in Figure 1.7. If the heat capacities are constant, which can often be assumed over a moderate temperature range, the temperature rise is proportional to the conversion. If the reaction is irreversible and goes to completion, x can be calculated from the relative temperature change without knowing the heat capacities:

$$x = \left(\frac{T - T_0}{T_{\max} - T_0}\right) \tag{1.14}$$

The reaction rate, $C_{A_0}(dx/dt)$ can be obtained from the slope of the temperature plot or from the change in temperature for a short time interval:

$$r = C_{A_0}\frac{dx}{dt} \cong C_{A_0}\left(\frac{\Delta x}{\Delta t}\right) \tag{1.15}$$

If the reaction is assumed to be first order, the rate constant is calculated from $(1 - x)$ and dx/dt.

$$k_1 C_{A_0}(1 - x) = C_{A_0}\frac{dx}{dt} \tag{1.16}$$

It might seem possible to confirm the assumed order by trying other reaction orders and seeing which order gives rate constants that follow the

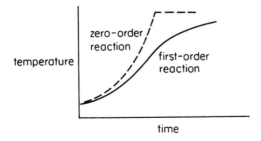

FIGURE 1.7 Temperature change for a batch adiabatic reaction.

Arrhenius equation. If the reaction was zero order, the slope of the temperature–time plot would continue to increase until the reaction was complete, as shown by the dashed line in Figure 1.7. However, for orders of $\frac{1}{2}$, 1, or 2, the curves are quite similar in shape, and very accurate data would be needed to distinguish between different orders. Furthermore, many reactions do not follow the Arrhenius equation exactly, and slight differences in activation energy should not be used to decide between possible orders or mechanisms. The correlation of adiabatic kinetic data is discussed by Rodriguez [5].

Many fixed-bed industrial reactors operate adiabatically, and the temperature profiles can be used to follow changes in catalytic activity and to optimize reactor performance. The temperature profile for an exothermic reaction is similar to the temperature–time curve for a batch reaction. The energy released by reaction is carried out by the fluid, since, except for startup, there is no accumulation in the catalyst. The conversion at any distance from the inlet can be calculated from the temperature rise relative to that for complete conversion.

Example 1.4

The cracking of furfural to furan and carbon monoxide was carried out in an adiabatic reactor using a pelleted catalyst. Data from a large reactor operating at 1.5 atmospheres are given in Tables 1.4 and 1.5. Six moles of steam were used per mole of furfural to decrease the temperature rise in the bed. Analysis of the exit stream showed less than 0.01% of the furfural was unreacted.

TABLE 1.4 Data for
Example 1.4

Bed length, ft	Temperature, °C
0	330
1	338
2	348
3	361
4	380
5	415
6	447
7	454
8	459
9	458

TABLE 1.5   Material Balance
for Example 1.4

For 1 mol feed	Feed	Product
A	1	$1 - x$
B	0	x
CO	0	x
H_2O	6	6
	Total	$7 + x$

a. Assuming first-order reaction and allowing for the increase in number of moles, determine the relative rate constants for each section of the bed, and estimate the activation energy.

b. Repeat the calculation for an assumed order of 2 and compare the estimated values of E.

Solution.

$$\begin{array}{ccc} A & & B \\ C_5H_4O_2 & \rightarrow & C_4H_4O + CO \end{array}$$

Heat balance : $\quad F_A x(-\Delta H) = \sum nc_p(T - T_0)$

The drop in temperature in the last part of the bed may be due to heat loss, and complete conversion probably corresponds to a temperature change of $130°C$. Intermediate conversions are based on the fractional temperature rise, except for the last 3 feet of bed, where the temperature is too close to the final value for an accurate estimate of x.

$$x = \frac{T - 330}{130}$$

The reaction rate depends on the partial pressure, P_A:

$$P_A = \frac{(1 - x)}{(7 + x)} P$$

$$F_A \, dx = r \, dw = \frac{k_1(1 - x)P}{(7 + x)} \rho_b S \, dL$$

Since $(7 + x)$ doesn't change much, an average value is used for each section of bed, and the rate expression is integrated to get k for each section:

$$\int_{x_1}^{x_2} \frac{dx}{1-x} = \ln\left(\frac{1-x_1}{1-x_2}\right) = \frac{k_1 P \rho_b S \, \Delta L}{F_A (7+x)_{ave}}$$

Let

$$k_1' = \frac{k_1 P \rho_b S}{F_A} = \ln\left(\frac{1-x_1}{1-x_2}\right) \frac{(7+x)_{ave}}{\Delta L}$$

Values of k_1' for each 1-ft section up to 6 ft are given in Table 1.6. The last 3 feet are treated as one section using the analysis result to get $x \cong 0.9999$.

A plot of $\ln(k_1)$ vs. $1/T$ is linear (Fig. 1.8), and from the slope, $E = 27\,kcal/mol$. A similar calculation for second-order kinetics gives a steeper plot of $\ln(k_2')$ vs. $1/T$, which is linear up to about 50% conversion, with $E = 34\,kcal/mol$.

COMPLEX KINETICS

Many reactions occur in a series of steps, and the overall rate may not be described by a simple equation with a constant reaction order. Some of the many types of complex rate expressions for heterogeneous catalysts are discussed in Chapter 2. For homogeneous reactions, two examples of complex kinetics are enzyme reactions and chain reactions.

Enzyme Kinetics

Most biological reactions and some important industrial reactions in aqueous media are catalyzed by enzymes, which are macromolecules ($M = 10^4 - 10^6$) composed mainly of proteins. Enzymes are very specific catalysts for particular reactions or for certain classes of reactions. Examples include isomerases, hydrolases, and oxidases, which catalyze isomerizations, hydrolysis, and oxidation reactions, respectively. There are many subtypes

TABLE **1.6** Values of k_1

L, ft	T,°C	T_{ave}, °C	x	$(7+x)_{ave}$	k_1'
0	330		0		
1	338	334	0.062	7.03	0.45
2	348	343	0.138	7.10	0.60
3	361	354	0.238	7.19	0.89
4	380	370	0.385	7.31	1.57
5	415	397	0.654	7.52	4.33
6	447	431	0.90	7.78	9.66
9	458	455	0.9999	7.9	18.2

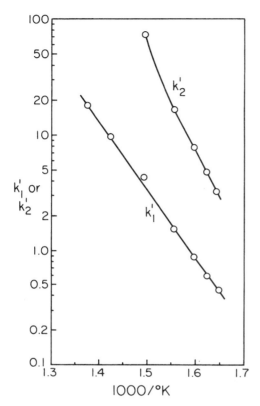

FIGURE 1.8   Rate constants from Example 1.4.

in each of these and other classes. An enzyme acts by binding reversibly to a substrate or reactant and lowering the activation energy for the reaction. The free-energy change and heat of reaction are not affected, but the lower activation energy often increases the rate by several orders of magnitude and permits fairly rapid reaction at ambient conditions.

The rates of enzyme-catalyzed reactions do not fit simple models for first- or second-order kinetics. Typically, the rate is a nonlinear function of concentration, as shown in Figure 1.9. At low substrate concentrations, the reaction appears first order, but the rate changes more slowly at more moderate concentrations, and the reaction is nearly zero order at high concentrations. A model to explain this behavior was developed in 1913 by L. Michaelis and M. L. Menton [6], and their names are still associated with this type of kinetics. The model presented here is for the simple case of a

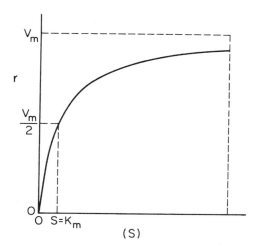

FIGURE **1.9** Effects of substrate concentration on the rate of an enzyme-catalyzed reaction.

single-substrate reaction that is irreversible. Models for reversible reactions, inhibited reactions, or reactions involving multiple substrates are given in specialized texts [7,8].

The first step is the formation of an enzyme–substrate complex, ES. The complex is held together by van der Waals forces or hydrogen bonds, and the rates of formation and dissociation of the complex are very rapid. A near-equilibrium concentration of ES is quickly established, since the rate of product formation is relatively slow:

$$E + S \underset{2}{\overset{1}{\rightleftharpoons}} ES \overset{3}{\longrightarrow} P + E$$

Products are formed when the complex decomposes in such a way that chemical bonds in the substrate break or new bonds are formed, and the enzyme molecule is freed. This step may require another reactant, such as water, but water is usually not included in the kinetic equations. Both the dissociation of the complex to S and E and the product-formation step are assumed to be first order in ES.

A material balance for ES includes the formation rate and the rates of the two reactions removing the complex. Following conventional notation, [S], [E], and [ES] refer to the molar concentrations of substrate, enzyme, and complex, respectively. For a batch reaction,

$$\frac{d[ES]}{dt} = k_1[E][S] - k_2[ES] - k_3[ES] \qquad (1.18)$$

The total enzyme in the system, $[E_0]$, is the sum of the free enzyme and the complex:

$$[E_0] = [E] + [ES] \qquad (1.19)$$

A similar equation is not written for the substrate, since the initial substrate concentration, $[S_0]$, is nearly always much greater than $[E_0]$, and the amount of substrate in the complex is a negligible fraction of the total. Combining Eqs. (1.18) and (1.19), we get

$$\frac{d[ES]}{dt} = k_1[S][E_0] - k_1[S][ES] - (k_2 + k_3)[ES] \qquad (1.20)$$

Since steps 1 and 2 are usually very rapid compared to step 3, ES can be assumed to reach a pseudo-steady-state concentration, and the derivative is set to zero. Solving for [ES] gives

$$[ES] = \frac{k_1[S][E_0]}{k_1[S] + k_2 + k_3} \qquad (1.21)$$

The product formation rate is $k_3[ES]$, and the rate equations can be given without k_1 in the numerator:

$$r = k_3[ES] = \frac{k_3[S][E_0]}{[S] + \dfrac{(k_2 + k_3)}{k_1}} \qquad (1.22)$$

In the literature on enzyme kinetics, the product rate is written in the following form:

$$r = \frac{V_m[S]}{K_m + [S]} \qquad (1.23)$$

where

$$K_m = \frac{k_2 + k_3}{k_1}, \text{ the Michaelis-Menton constant}$$

$$V_m = k_3[E_0], \text{ the maximum reaction rate}$$

The maximum rate is achieved when all the enzyme is present as complex ES, which occurs at high substrate concentration. The reaction then appears zero order to substrate. At very low [S], the rate is $(V_m/K_m)[S]$, and first-order kinetics are observed. The value of K_m, which has units of concentration, can be interpreted as the substrate concentration that gives half the maximum reaction rate.

The model for an enzyme-catalyzed reaction is similar to that for a first-order reaction of a gaseous molecule adsorbed on a solid catalyst, which has a certain number of sites (uniformly active) per unit mass. The surface reaction goes from approximately first order at low partial pressure, when a small fraction of sites are covered, to nearly zero order at high partial pressure and high coverage. Derivations and examples for more complex surface reactions are given in Chapter 2.

Graphical methods can be used to verify the form of the rate equation and to determine K_m and V_m. Inverting Eq. (1.23), we get

$$\frac{1}{r} = \frac{K_m + [S]}{V_m[S]} = \frac{K_m}{V_m}\frac{1}{[S]} + \frac{1}{V_m} \tag{1.24}$$

A plot of $1/r$ versus $1/[S]$ should be a straight line with slope K_m/V_m and a positive intercept, $1/V_m$. Another approach is to multiply both sides of Eq. (1.24) by [S] and plot $[S]/r$ versus [S]:

$$\frac{[S]}{r} = \frac{K_m}{V_m} + \frac{[S]}{V_m} \tag{1.25}$$

The intercept is then K_m/V_m and the slope $1/V_m$.

The two methods of plotting may give different values of K_m and V_m when there is scatter in the data. Very low values of [S] and r have more effect on the best-fit line where their reciprocals are plotted, as in Eq. (1.24). However, the variables are separated in the first method, and the second plot has [S] in both terms.

Example 1.5

The hydration of CO_2 is catalyzed by the enzyme carbonic anhydrase:

$$CO_2 + H_2O \rightleftharpoons HCO_3^- + H^+$$

The initial velocity was measured for both forward and reverse reactions at 0.5°C and pH 7.1 with bovine carbonic anhydrase. Data for the reverse reaction are given in Table 1.7 as reciprocal rates for different substrate concentrations. Plot the results in two ways to determine K_m and V_m.

TABLE 1.7 Data for Example 1.5

$[S] = [HCO_3] \times 10^3$	$\frac{1}{r}$ L-sec/mol $\times 10^{-3}$
2	95
5	45
10	29
15	25

Solution:. A plot of $1/r$ versus $1/[S]$ is given in Figure 1.10(a). From the intercept, $1/V_m = 14.5 \times 10^3$, $V_m = 6.9 \times 10^{-5}$ mol/L-sec. The slope is 153, and the slope is divided by the intercept to get K_m:

$$K_m = \frac{153}{14.5 \times 10^3} = 1.06 \times 10^{-2} \text{ M}$$

The alternate plot of $[S]/r$ versus $[S]$, Figure 1.10(b) seems about as good a fit but gives slightly different values of V_m and K_m:

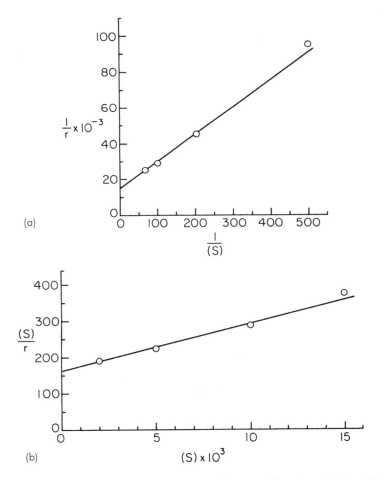

(a)

(b)

FIGURE **1.10** Testing for Michaelis–Menton kinetics: (a) Eq. (1.24); (b) Eq. (1.25).

$$\frac{1}{V_m} = \text{slope} = 1.33 \times 10^4, \qquad V_m = 7.5 \times 10^{-5} \, \text{mol/L-sec}$$

$$K_m = \frac{\text{intercept}}{\text{slope}} = \frac{153}{1.33 \times 10^4} = 1.22 \times 10^{-2} \, \text{M}$$

The data in Example 1.5 are initial rate data obtained at different substrate concentrations. If only batch data are available, an integrated form of the rate equation is needed. For a batch reaction with initial concentration $[S_0]$ and $[E_0]$, the reaction rate is the rate of change of concentration:

$$r = -\frac{d[S]}{dt} = \frac{V_m[S]}{K_m + [S]} \tag{1.26}$$

$$\int \frac{-d[S](K_m + [S])}{[S]} = \int V_m \, dt \tag{1.27}$$

$$K_m \ln \frac{[S_0]}{[S]} + [S_0] - [S] = V_m t \tag{1.28}$$

Dividing by t and K_m gives a convenient form for plotting:

$$\frac{1}{t} \ln \frac{[S_0]}{[S]} + \frac{[S_0] - [S]}{K_m t} = \frac{V_m}{K_m} \tag{1.29}$$

$$\frac{1}{t} \ln \left(\frac{1}{1 - x} \right) = \frac{V_m}{K_m} - \frac{S_0}{K_m t} \tag{1.30}$$

Chain Reactions

Chain reactions take place via a series of steps involving intermediates that are continually reacting and being regenerated in a cyclical process. In most cases, the sequence of steps involves three types of reactions:

1. *Initiation*: One of the reactants or an added initiator decomposes to produce active intermediates.
2. *Propagation*: The active intermediates combine with one or more reactants to produce products and regenerate the intermediates.
3. *Termination*: The active intermediates are removed from the system by recombination, adsorption on the wall, or other mechanisms.

   In a typical chain reaction, the concentration of active intermediates or chain carriers is orders of magnitude lower that the concentration of the main reactants. The chain carrier concentration can be calculated from the initiator and termination rates. The chain length is the number of times a chain carrier goes through the propagation sequence before termination occurs. The chain length may be as high as several thousand.

Nitric Oxide Formation

In the chain reaction producing NO, the initiation step is the dissociation of O_2, which requires a high-energy collision between O_2 and another molecule M (it could be O_2 or N_2):

$$O_2 + M \underset{2}{\overset{1}{\rightleftharpoons}} M + 20 \qquad \Delta H = 118 \, kcal$$

The oxygen atoms produced are free radicals because of the unpaired electron, and they are extremely reactive. The propagation step involves two reactions, which produce NO and regenerate the oxygen atom:

$$O + N_2 \underset{4}{\overset{3}{\rightleftharpoons}} NO + N \qquad \Delta H = 75 \, kcal \quad slow$$

$$N + O_2 \underset{6}{\overset{5}{\rightleftharpoons}} NO + O \qquad \Delta H = -32 \, kcal \quad fast$$

All of the reactions shown are reversible, but to simplify the analysis we focus on the initial rate of NO formation, far from equilibrium, where steps 4 and 6 can be neglected. Then, for each oxygen atom that reacts in step 3, two molecules of NO are formed in the propagation sequence, and the oxygen atom is regenerated. Reaction 3, which is endothermic, is the slow step of this sequence, because of the high activation needed to break the $N \equiv N$ bond. Step 5 is relatively rapid, which makes the concentration of nitrogen atoms very much lower than the concentration of oxygen atoms. The initial rate of NO formation is therefore twice the rate of step 3:

$$r_{i,NO} = 2k_3[O][N_2] \tag{1.31}$$

The termination step is the recombination of oxygen atoms, which is step 2. The rate of recombination of nitrogen atoms and the reaction of nitrogen and oxygen atoms are negligible because of the very low concentration of nitrogen atoms. The initiation and termination are assumed to be very rapid compared to the propagation step, so a pseudo-steady-state concentration is reached:

$$k_1[M][O_2] = k_2[M][O]^2 \tag{1.32}$$

$$[O] = \sqrt{\frac{k_1}{k_2}[O_2]} \tag{1.33}$$

Combining Eqs. (1.31) and (1.33) gives

$$r_{i,NO} = 2k_3 \left(\frac{k_1}{k_2}\right)^{1/2} [O_2]^{1/2} [N_2] \qquad (1.34)$$

Experiments show that Eq. (1.34) gives a good fit to the initial rate of NO formation. An equation allowing for the reverse steps in the propagation sequence has the following form [9]:

$$r_{NO} = k_f [N_2][O_2]^{1/2} - k_r [NO]^2 [O_2]^{-1/2} \qquad (1.35)$$

Note that the reverse reaction term is $-\frac{1}{2}$ order to oxygen, which is necessary to satisfy the equilibrium relationship where $r_{NO} = 0$.

Hydrocarbon Cracking

The thermal cracking of hydrocarbons proceeds by a chain mechanism involving hydrocarbon and hydrogen free radicals. The initiation step is the decomposition of the hydrocarbon to form two free radicals. For ethane the radicals are identical, but for higher paraffins different radicals may be formed.

INITIATION.

$$C_2H_6 \xrightarrow{1} 2CH_3\cdot$$

$$C_nH_{2n+2} \xrightarrow{1} C_xH_{2x+1}\cdot + C_yH_{2y+1}\cdot$$

Radicals can abstract hydrogen from stable molecules to form new radicals, and ethyl or larger radicals can decompose to form an olefin and a smaller radical.

PROPAGATION.

$$CH_3\cdot + C_2H_6 \xrightarrow{2} CH_4 + C_2H_5\cdot$$

$$H\cdot + C_2H_6 \xrightarrow{2} H_2 + C_2H_5\cdot$$

$$C_2H_5\cdot \xrightarrow{3} C_2H_4 + H\cdot$$

$$C_4H_9\cdot \xrightarrow{3} C_3H_6 + CH_3\cdot$$

Several termination steps are possible with ethane cracking, and there are many more for cracking of higher hydrocarbons.

TERMINATION.

$$2C_2H_5 \cdot \xrightarrow{4} C_4H_{10}$$

$$C_2H_5 \cdot + H \cdot \xrightarrow{4} C_2H_6$$

$$CH_3 \cdot + H \cdot \xrightarrow{4} CH_4$$

$$2H \cdot \xrightarrow{4} H_2$$

The overall order of the reaction depends on the relative importance of the different termination steps and whether the initiation step is first or second order. For ethane cracking, experiments indicate an order of 1 or 1.5, depending on the pressure. For larger hydrocarbons, a great many species and over 100 reactions are used in developing models for thermal cracking [10], and apparent orders of 1–1.25 are reported.

Polymerization

Chain reactions also occur in the liquid phase, and many synthetic polymers are produced by free-radical chain polymerizations. The initiation step is the decomposition of an added initiator, an unstable molecule such as a peroxide or persulfate:

$$I_2 \xrightarrow{k_i} 2I \cdot$$

The initiator radical adds to a monomer molecule to form a new radical, starting the propagation step

$$I \cdot + M \xrightarrow{k_p} R_1 \cdot$$

The radical continues to add monomer at a rapid rate, forming larger and larger free radicals:

$$R_n \cdot + M \xrightarrow{k_p} R_m \cdot$$

Since radicals of different length appear to have the same reactivity to monomer, subscripts are not needed and [R·] stands for the total concentration of free radicals.

The termination step is the recombination of radicals to form dead polymer:

$$R_n \cdot + R_m \cdot \xrightarrow{k_t} P_{n+m}$$

Dead polymer molecules can also be formed by a chain transfer reaction between a polymer radical and a molecule of solvent, monomer, or chain transfer agent S:

$$R_n \cdot + S \xrightarrow{k_{t_r}} P_n + S \cdot$$

If the new radical is reactive, chain transfer does not affect the rate of polymerization, but it does decrease the average molecular weight of the polymer formed.

The rate of polymerization is proportional to the steady-state radical concentration, which depends on the rates of initiation and termination:

$$\frac{d[R\cdot]}{dt} = 0 = 2k_i[I_2] - 2k_t[R\cdot]^2$$

$$[R\cdot] = \sqrt{\frac{k_i}{k_t}[I_2]}$$

The monomer conversion reaction is first order to monomer and half order to initiator:

$$r_p = k_p[R\cdot][M] = k_p\sqrt{\frac{k_i}{k_t}}(I_2)^{1/2}M$$

An important feature of free-radical polymerization is that the reaction mixture contains some high-molecular-weight polymer (with a distribution of molecular weights) and some unreacted monomer. There are no dimers, trimers, or low-molecular-weight oligomers, because once a chain is initiated, it adds a great many monomer units before termination occurs. This is in contrast to a stepwise, or condensation, polymerization, where the average chain length grows slowly as chains combine to form larger chains.

NOMENCLATURE

Symbols

C	molar concentration
c_p	heat capacity
c_{pw}	heat capacity of reactor wall
D_p	particle diameter
D_t	tube diameter
E	activation energy, enzyme concentration
ES	concentration of enzyme–substrate complex
F	volumetric feed rate
F_A	molar feed rate of reactant A
K_m	Michaelis–Menton constant, Eq. (1.23)
k	reaction rate constant
L	length

M	any molecule, molecular weight
n	number of moles, reaction order
P_A	partial pressure of A
R	gas constant
r	reaction rate
r_i	initial reaction rate
S	selectivity
S_{ave}	average selectivity
S	cross section, substrate concentration
T	absolute temperature, K
T_0	initial temperature
T_{max}	maximum temperature
t	time
u_0	superficial velocity
V	volume of reactor
V_m	maximum reaction rate for enzyme reaction
W	mass of catalyst
x	fraction converted
Y	yield of desired product

Greek Letters

ρ_b	bed density
ΔH	heat of reaction

PROBLEMS

1.1 Trioxane, the cyclical trimer of formaldehyde, depolymerizes in the presence of acid catalysts. Batch test with $8\,N\,H_2SO_4$ gave the conversions at different temperatures shown in Table 1.8 [11].

    a.   Calculate the rate constants assuming first-order or second-order kinetics, and plot k_1 and k_2 on an Arrhenius plot. Can the order and activation energy be determined from this plot? Would it

TABLE 1.8 Data for
Problem 1.1

T, °C	t, min	x, %
20	300	4.5
40	60	17.5
70	10	62

have been better to determine the time for a constant percent decomposition—say, 50%—at different temperatures?

b. Determine the reaction order and activation energy from the more complete data in Table 1.9.

c. What is the order with respect to sulfuric acid?

1.2 The thermal decomposition of plutonium hexafluoride was studied in a batch reactor made of nickel [12]. The reactor was pretreated with fluorine before use:

$$PuF_{6(g)} \rightarrow PuF_{4(s)} + F_{2(g)}$$

The results shown in Table 1.10 were reported for a reaction time of 90 min at 161°C.

a. Assume concurrent zero-order and first-order reactions (as the authors did), integrate the rate equation, and determine the best values of the rate constants.

b. Repeat the analysis assuming a half-order reaction, and compare the fit to the data.

c. Would any other model give a better or as good a fit?

1.3 Acrylamide was polymerized in aqueous solution using ammonium persulfate and sodium metabisulfite for redox initiation [13]. Typical batch data are in Table 1.11.

a. Determine the reaction order by trial from the conversion–time data in Table 1.11.

b. Determine the reaction order from the initial reaction rates.

c. What type of equation is needed to fit these data? What is the significance of the different orders?

TABLE 1.9 Data for Problem 1.1

T, °C	Normality H_2SO_4	10% Time	50% Time	75% Time	99% Time
20	8	12.3 hr	3.4 days	6.8 days	22.5 days
20	12	55 min	6.1 hr	12.2 hr	40 hr
20	16	4.4 min	29 min	58 min	190 min
20	20	20 sec	22 min	4.4 min	14.5 min
40	8	31 min	3.5 hr	7.0 hr	23 hr
70	8	66 sec	7.3 min	14.6 min	48 min
95	8	11.6 sec	77 sec	2.6 min	8.5 min

TABLE 1.10 Data for Problem
1.2: Partial Pressure of
PuF_6, cmHg

Initial	Final
98.1	70.4
59.6	37.9
51.2	31.5
32.6	16.5
16.8	5.9

1.4   Holbrook and Marsh showed that the gas-phase decomposition of ethyl chloride is apparently first order, but the rate constant at 521°C decreases with decreasing initial pressure (Table 1.12) [14]. How well can the decrease in rate constant be accounted for using the simple activation theory of Lindemann? Determine this value of k_∞, and compare with the author's value of $8.88 \times 10^{-3}\ \text{sec}^{-1}$.

$$C_2H_5Cl \rightarrow C_2H_4 + HCl$$

1.5   The thermal reaction $H_2 + Br_2 \rightarrow 2HBr$ is thought to proceed by the chain mechanism given here, with appropriate rate constants at 800 K.

TABLE 1.11   Data for Problem 1.3

t, sec	M^a	M	M
0	0.40	0.70	1.10
60	0.358	0.625	1.01
180	0.325	0.550	0.91
300	0.301	0.531	0.806
480	0.276	0.478	0.687
600	0.251	0.429	0.644
900	0.223	0.395	0.541
1200	0.201	0.374	0.543
Temperature:	30°C		
Bisulfite:	0.095 g		
Persulfate:	0.114 g		

^{a}M = monomer concentration.

TABLE **1.12** Data for Problem 1.4

P_0, mmHg	$k \times 10^3 \, sec^{-1}$ at 521°C
134	8.20
130	8.56
27.6	6.41
18.1	6.34
12.0	6.03
8.1	5.11
6.1	4.21
4.1	4.17
2.82	4.58
1.74	3.83
1.16	3.14
0.97	2.99
0.41	2.07

	at 800 K
$M + Br_2 \xrightarrow{1} 2Br + M$	$k_1 = 1.6 \, L\text{-mole}^{-1}sec^{-1}$
$Br + H_2 \xrightarrow{2} HBr + H$	$k_2 = 1.2 \times 10^6 \, L\text{-mole}^{-1}sec^{-1}$
$H + Br_2 \xrightarrow{3} HBr + Br$	$k_3 = 7.1 \times 10^{10} \, L\text{-mole}^{-1}sec^{-1}$
$H + HBr \xrightarrow{4} H_2 + Br$	$k_4 = 8.5 \times 10^9 \, L\text{-mole}^{-1}sec^{-1}$
$M + Br + Br \xrightarrow{5} Br_2 + M$	$k_5 = 10^9 \, L^2\text{-mole}^{-2}sec^{-1}$

    a. For an equal-molar mixture of H_2 and Br_2 at 1 atm and 800 K, calculate the steady-state concentration of bromine atoms and compare with the concentration of bromine molecules.

    b. Estimate the concentration of hydrogen atoms and the contribution of H + H and H + Br to the termination step.

    1.6 The enzymatic hydrolysis of *n*-benzoyl L-arginine ethyl ester (BAEE) was carried out in a packed bed using trypsin bound to particles of porous glass. The glass was 200–400 mesh with 355-Å pore diameter, and the bed had a 0.9-cm diameter and was 2.2 cm high (see Table 1.13).

    Assuming that the Michaelis–Menton equation applies, show that a plot of $\frac{1}{x}\ln\left(\frac{1}{1-x}\right)$ vs. $1/x(V/F)$ can be used to get the two constants in the rate expression. Are the constants independent of S_0?

TABLE **1.13** Rate Constants for Problem 1.6: Conversion in a Fixed
Bed for Various Substrate Concentrations

$S_0 = 0.5$ mM		$S_0 = 0.8$ mM		$S_0 = 1.0$ mM	
x	V/F, min $\times 10^2$	x	V/F, min $\times 10^2$	x	V/F, min $\times 10^2$
0.438	5.90	0.372	7.66	0.268	5.0
0.590	8.03	0.410	8.89	0.328	8.3
0.670	9.58	0.496	10.48	0.430	10.72
0.687	9.46	0.602	12.90	0.625	15.80
0.815	11.30	0.680	14.00	0.670	16.80
0.910	14.72	0.792	17.10	0.768	21.21
0.972	18.00	0.844	19.50	0.823	23.30
		0.905	22.04	0.925	29.0
				0.948	31.8

REFERENCES

1. BM Fabuss, JO Smith, RI Lait, AS Borsanyi, CN Satterfield. Ind Eng Chem Proc Des Dev 1:293–299, 1962.
2. F Rodriguez. Principles of Polymer Systems. 4th ed. New York: McGraw-Hill, 1996, p161.
3. FA Lindemann. Trans Faraday Soc 17:598, 1922.
4. YB Zeldovich, PY Sadovnikov, DA Frank-Kamenetskii. Oxidation of Nitrogen in Combustion. M. Shelef (trans.). Academics of Sciences of the USSR. Moscow: Institute of Chemical Physics, 1947.
5. F Rodriguez. Polymer 23:1473, 1982.
6. L Michaelis, ML Menton. Biochem Z 48:333, 1913.
7. ML Shuler, F Kargi. Bioprocess Engineering. Englewood Cliffs, NJ: Prentice-Hall, 1992.
8. KM Plowman. Enzyme Kinetics. New York: McGraw-Hill, 1972.
9. JH Seinfeld. Air Pollution: Physical and Chemical Fundamentals. New York: McGraw-Hill, 1975.
10. GF Froment, KB Bischoff. Chemical Reactor Analyses and Design. 2nd ed. New York: Wiley, 1990.
11. JF Walker, AF Chadwick. Ind Eng Chem 39:974, 1947.
12. J Fischer, LE Trevorrow, GJ Vogel, WA Shinn. Ind Eng Chem Proc Des Dev 1:47, 1962.
13. F Rodriguez, RD Givey. J Polymer Sci 55:713, 1961.
14. KA Holbrook, ARW Marsh. Trans Faraday Soc 63:643, 1967.

2

Kinetic Models for Heterogeneous Reactions

When studying the kinetics of heterogeneous reactions or when designing a large catalytic reactor, there are more factors to consider than when dealing with homogeneous reactions. For a solid-catalyzed reaction, the rate depends on the reactant concentrations at the catalyst surface, but these are not the same as the bulk concentrations, because some driving force is needed for mass transfer to the surface. If the catalyst is porous, as is usually the case, there are further differences in the concentration between the fluid at the external surface and the fluid in the catalyst pores. Models must be developed to predict the surface concentrations as functions of the partial pressures or concentration in the gas or liquid, and the rate expression can then be written in terms of the fluid concentrations.

When the reaction is between a solid and a reactant in the gas or liquid, one must consider not only the foregoing factors but also the problem of changing particle size or shape as the solid is consumed. In this chapter we deal mainly with reactions on a solid catalyst. The equations are derived for gaseous reactants, but they apply to liquids as well when partial pressures are replaced with concentrations.

BASIC STEPS FOR SOLID-CATALYZED REACTIONS

There are five steps in the sequence of mass transfer and reaction over a solid catalyst:

1. Diffusion of reactants to the external surface of the catalyst and into the pores
2. Adsorption of one or both reactants on active sites
3. Reaction on the surface between adsorbed species or between surface species and a reactant in the fluid phase
4. Desorption of the products
5. Diffusion of products out of the pores and into the external fluid

When the system is at steady state, all the steps in the sequence take place at the same rate. However, the overall rate is often controlled by one step, and knowing which step limits the rate is key to understanding the system and developing a sound correlation of the kinetic data. If one step really controls, it means that changing the rate constant or coefficient for that step produces a proportional change in the overall rate and that changing the rate constant for any other step has a very small or negligible effect.

EXTERNAL MASS TRANSFER CONTROL

Catalytic reactions that are controlled by the rate of mass transfer to the external surface are relatively rare, since this requires an extremely active catalyst or very high temperatures. One example is found in the manufacture of nitric acid; a key step in this process is the oxidation of NH_3 to NO on fine platinum wires, and the reaction is controlled by the rate of mass transfer of NH_3 to the catalyst surface. Other examples are the catalytic incineration of hydrocarbon vapors and the oxidation of CO in an automobile catalytic converter. These examples will be discussed in Chapter 10. Here, a simple example is used to illustrate the significance of external mass transfer control.

For the reaction $A + B \rightarrow C$, if there is excess B and mass transfer of A is the controlling step, the external gradients might be as shown in Figure 2.1. In this example, if $P_A = 0.2$ atm and $P_{A_s} = 0.004$ atm, the driving force for mass transfer is $P_A - P_{A_s} = 0.196$ atm. If the mass transfer coefficient is doubled—say, by increasing the velocity—the rate of mass transfer of A and the other rates would almost double. Assuming the surface reaction is first order in A, P_{A_s} would increase to about 0.008 atm, and the driving force for mass transfer would decrease slightly to 0.192 atm Thus, the rate would increase by a factor of $2 \times 0.192/0.196 = 1.96$. If, on the other hand, the kinetic constant for the surface reaction was doubled, the overall rate would

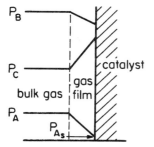

P_B

P_C

bulk gas

P_A

P_{A_s}

catalyst

gas
film

FIGURE 2.1 Gradients for mass transfer of A controlling.

be essentially unchanged, since P_{A_s} would go to about 0.002 atm, and the driving force for mass transfer would go from 0.196 atm to 0.198 atm, a negligible change. No matter how active the catalyst is, the rate can't exceed the maximum possible rate of mass transfer to the catalyst surface.

If external mass transfer is the controlling step or is slow enough to have some effect on the overall rate, steps 1 and 5 should be considered together, since product diffuses outward through the same boundary layer that forms the resistance to reactant diffusion. When the reactant concentration at the surface is much lower than in the bulk gas, the product concentration at the surface will be much higher, and the effect of gas composition on diffusivity may have to be accounted for. There may also be a net flux of molecules to or away from the surface, which must be considered when external mass transfer controls.

Diffusion of reactants into the pores is listed in step 1 as part of the mass transfer process, but a rigorous treatment must consider simultaneous pore diffusion and reaction rather than steps in sequence. This topic is covered in detail in Chapter 4. In many cases, external mass transfer and pore diffusion are rapid enough so that concentrations in the catalyst are almost the same as those in the surrounding gas. The reaction rate is then controlled by step 2, 3, or 4 or some combination of these steps.

MODELS FOR SURFACE REACTION

Before discussing the rate of adsorption as a possible controlling step, we will consider various models for reaction on the surface when the surface is at equilibrium with the gas phase. The concentrations of reactants and products on the surface are given in terms of adsorption isotherms, where the amount adsorbed is expressed as the fraction of surface (or sites) covered, rather than in concentration units such as moles/m^2.

Langmuir Isotherm

The Langmuir isotherm is widely used for reactants that reversibly chemisorb on the catalyst surface. The surface is assumed to be completely uniform, with all sites having equal reactivity. Adsorption occurs when molecules with sufficient energy strike vacant sites or uncovered parts of the surface. The process is described as a reaction between a molecule from the gas phase and an unoccupied site, s.

$$A + s \rightarrow A_s \tag{2.1}$$

The frequency of collisions is proportional to the partial pressure, and the probability of adsorption is incorporated in the rate constant k_1, which has an exponential dependence on temperature. The concentration of vacant sites is expressed as $(1 - \theta)$, where θ is the fraction of occupied sites, and the total site concentration is included in the rate constant k_1. Molecules already adsorbed are assumed to have no effect on the rate of adsorption for nearby vacant sites:

$$r_{ads,A} = k_{1A}P_A(1 - \theta)$$
$$k_{1A} = ae^{-E_{ads}/RT} \tag{2.2}$$

The rate of desorption of A is proportional to the amount of A on the surface, and the rate constant for desorption is also an exponential function of temperature:

$$r_{des,A} = k_{2A}\theta_A$$
$$k_{2A} = be^{-E_{des}/RT} \tag{2.3}$$

The difference between E_{ads} and E_{des} is the heat of adsorption, as shown in Figure 2.2.

If only A is adsorbed, then $\theta = \theta_A$, and at equilibrium the rates of adsorption and desorption are equal:

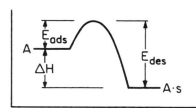

FIGURE 2.2 Chemisorption energy diagram.

$$k_{1A}P_A(1-\theta_A) = k_{2A}\theta_A \tag{2.4}$$

$$\theta_A = \frac{k_{1A}P_A}{k_{2A}+k_{1A}P_A} \tag{2.5}$$

The ratio of adsorption to desorption rate constants is an equilibrium constant with the units of reciprocal pressure.

$$K_A = \frac{k_{1A}}{k_{2A}} \tag{2.6}$$

Dividing both terms of Eq. (2.5) by k_{2A} and introducing K_A gives

$$\theta_A = \frac{K_A P_A}{1 + K_A P_A} \tag{2.7}$$

Equation (2.7) is the Langmuir isotherm [1] for adsorption of a single species, and it shows that the amount adsorbed is nearly proportional to the pressure at low values of $K_A P_A$ and nearly independent of pressure when $K_A P_A$ is much larger than 1.0. Large values of $K_A P_A$ correspond to nearly complete or monolayer coverage. The adsorption constant K_A is a measure of the strength of adsorption of A on that surface. When $P_A = 1/K_A$, the surface is half covered with adsorbed A.

If adsorption follows the Langmuir isotherm, an arithmetic plot of the amount adsorbed versus the pressure will have the shape shown in Figure 2.3. The amount adsorbed per unit mass of solid can be expressed in volume, mass, or mole units, but volume units such as cm^3 STP/g are most common. In Figure 2.3, V_m is the volume corresponding to a monolayer on the surface or to all sites occupied. To test the fit of data to a Langmuir isotherm, a rearranged form of the equation is used. From Eq. (2.7) and the definition of V_m,

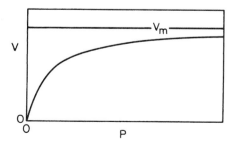

FIGURE 2.3   Typical Langmuir isotherm.

$$V = V_m \frac{K_A P_A}{1 + K_A P_A} \tag{2.8}$$

$$\frac{1}{V} = \frac{1 + K_A P_A}{V_m K_A P_A} = \frac{1}{V_m K_A}\left(\frac{1}{P_A}\right) + \frac{1}{V_m} \tag{2.9}$$

Plotting $1/V$ against $1/P_A$ should give a straight line with intercept $1/V_m$ and slope $1/V_m K_A$, as shown in Figure 2.4.

A great many studies of chemisorption have been published, and in most cases the Langmuir isotherm does not give a very good fit to the data. This may be due to interactions between adsorbed species or to heterogeneity of the surface, factors ignored in the derivation. Although other correlations, including the Freundlich isotherm, $\theta = aP^m$, and the Temkin isotherm, $\theta = RT \ln(A_0 P)$, may give a better fit for single gas adsorption, the Langmuir isotherm is more easily adapted for competitive adsorption, and it often leads to a satisfactory equation for correlating kinetic data.

When two or more types of molecules can adsorb on the same type of sites, they compete for places on the surface. For a binary mixture where both A and B adsorb as molecules, the equations are

$$A + s \rightarrow A_s$$
$$B + s \rightarrow B_s$$
$$r_{ads,A} = k_{1A} P_A (1 - \theta) = k_{1A} P_A (1 - \theta_A - \theta_B)$$
$$r_{des,A} = k_{2A} \theta_A$$
$$r_{ads,B} = k_{1B} P_B (1 - \theta) = k_{1B} P_B (1 - \theta_A - \theta_B)$$
$$r_{des,B} = k_{2B} \theta_B$$

By equating the rates of adsorption and desorption for both gases, the equilibrium coverages are obtained:

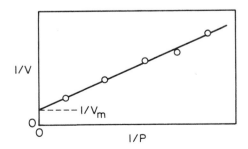

FIGURE 2.4   Test of Langmuir isotherm.

$$\theta_A = \frac{K_A P_A}{1 + K_A P_A + K_B P_B} \tag{2.10}$$

$$\theta_B = \frac{K_B P_B}{1 + K_A P_A + K_B P_B} \tag{2.11}$$

Extending the derivation to multicomponent mixtures gives extra terms in the denominator, such as $K_C P_C$ for a reaction product, or $K_P P_P$ for a trace impurity or poison that is chemisorbed.

$$\theta_A = \frac{K_A P_A}{1 + K_A P_A + K_B P_B + K_C P_C + K_P P_P + \cdots} \tag{2.12}$$

Dissociating Gases

Diatomic gases often dissociate when adsorbing on metal catalysts. Using H_2 as an example, an isotherm can be derived assuming that two adjacent sites are needed for adsorption, and the probability for this varies with $(1 - \theta)^2$:

$$H_2 + 2s \rightarrow 2H_s \tag{2.13}$$

$$r_{ads} = k_1 P_{H_2}(1 - \theta)^2 \tag{2.14}$$

Desorption requires reaction of two adjacent atoms to form a molecule that then desorbs. The atoms may move from site to site by surface diffusion, and the rate of collisions is proportional to the square of the surface concentration:

$$2H_s \rightarrow H_2 + 2s \tag{2.15}$$

$$r_{des} = k_2 \theta^2 \tag{2.16}$$

To get the equilibrium coverage, the rates of adsorption and desorption are set equal, and the square root of both sides is taken to solve for θ.

$$k_1 P_{H_2}(1 - \theta)^2 = k_2 \theta^2 \tag{2.17}$$

$$\frac{\theta}{1 - \theta} = \left(\frac{k_1 P_{H_2}}{k_2}\right)^{0.5} = K_{H_2}^{0.5} P_{H_2}^{0.5} \tag{2.18}$$

$$\theta = \frac{K_{H_2}^{0.5} P_{H_2}^{0.5}}{1 + K_{H_2}^{0.5} P_{H_2}^{0.5}} \tag{2.19}$$

When H_2 is present along with other gases that compete for the same sites, the isotherms are similar to Eq. (2.10), with $(K_{H_2} P_{H_2})^{0.5}$ included:

$$\theta_H = \frac{\left(K_{H_2}P_{H_2}\right)^{0.5}}{1 + \left(K_{H_2}P_{H_2}\right)^{0.5} + K_A P_A + K_B P_B} \tag{2.20}$$

$$\theta_A = \frac{K_A P_A}{1 + (K_H P_H)^{0.5} + K_A P_A + K_B P_B} \tag{2.21}$$

Langmuir–Hinshelwood Kinetics

The use of Langmuir isotherms to interpret kinetic data was proposed by Hinshelwood [2] and discussed at length by Hougen and Watson [3]. Surface reaction rates are assumed to depend on the fraction of sites covered by different species. If the surface is at adsorption-desorption equilibrium, the equations for θ_A, θ_B, θ_H, etc. are used in the rate expressions, and the surface reaction is assumed to be the rate-controlling step. The more complex cases, where adsorption equilibrium is not attained or where mass transfer has some effect, are treated later.

Consider first a unimolecular decomposition reaction where the products are not adsorbed or are very weakly adsorbed.

$$A_s \rightarrow B + C + s$$
$$r = k\theta_A = \frac{kK_A P_A}{1 + K_A P_A} \tag{2.22}$$

The reaction would appear first order to A at low pressures, fractional order at intermediate pressures, and zero order at high pressure, where nearly all the sites are covered with A. Note the similarity to Eq. (1.23) for an enzyme-catalyzed reaction, where the reaction order changes from about 1 to nearly 0 as the substrate concentration is changed from low to high values.

For an irreversible bimolecular reaction between molecules that are competitively adsorbed on the same type of sites, the reaction rate depends on the probability that the molecules are on adjacent sites. This probability is approximately proportional to the product of the fractional coverages:

$$A_s + B_s \rightarrow C_s + D_s$$
$$r = k\theta_A\theta_B \tag{2.23}$$

Equation (2.23) can also be obtained by considering surface migration of adsorbed species. The reaction rate then depends on the rate of collisions, which is proportional to the product of the surface concentrations. Adding the Langmuir isotherms to Eq. (2.23) gives

$$r = \frac{kK_A P_A K_B P_B}{(1 + K_A P_A + K_B P_B + K_C P_C + K_D P_D)^2} \tag{2.24}$$

If the reaction follows this model, the rate is proportional to P_A (or P_B) at low partial pressure, when only a small fraction of the sites are occupied, and the $K_A P_A$ term in the denominator is negligible. As P_A increases, the rate becomes less dependent on P_A, as shown in Figure 2.5, and the reaction order becomes zero at medium values of $K_A P_A$, where the reaction rate goes through a maximum. At very high pressures, the rate decreases as P_A increases. The negative reaction order is caused by competitive adsorption of reactants, with a high surface concentration of A leaving little room for adsorption of B. The possibility of negative-order behavior is a major feature of Langmuir–Hinshelwood models for surface reactions.

Note that there are five constants to be determined in Eq. (2.24) if there is significant adsorption of both products as well as reactants. It would be very difficult to determine these constants from tests in an integral reactor, since P_C and P_D would be increasing with conversion as P_A and P_B were decreasing. Tests should be carried out in a differential reactor where the inlet concentration can be varied independently and there is little change across the reactor. Products can be added to the feed mixture to test for product inhibition. There is no definite limit on the permissible conversion in a differential reactor, but 5–20% would be reasonable. Very low conversion makes it difficult to get an accurate value for the reaction rate, and with high conversion there is some error using the arithmetic average pressure in the rate equation. For conversions greater than 30%, the data should be analyzed using integrated rate equations.

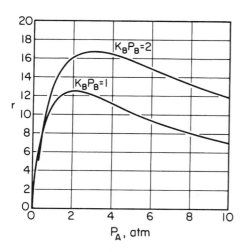

FIGURE 2.5 Reaction rate for Langmuir–Hinshelwood kinetics with competitive adsorption, $K_A = 1$ atm^{-1}.

Noncompetitive Adsorption

If a catalyst has two different types of sites, a reaction might take place between reactant A adsorbed on one type of site and reactant B adsorbed on an adjacent but different type of site. The reaction rate would then depend on the fractional coverage of each type of site. In the simple case where only A adsorbs on type 1 sites and only B on type 2 sites, the rate expression is given in Eq. (2.25). The primed variables refer to type 2 sites.

$$A + s_1 \rightarrow A_{s_1}$$

$$B + s_2 \rightarrow B_{s_2}$$

$$A_{s_1} + B_{s_2} \rightarrow C \tag{2.25}$$

$$r = k\theta_A\theta_B' = \frac{kK_AP_A}{(1 + K_AP_A)}\frac{K_B'P_B}{(1 + K_B'P_B)}$$

If the product is adsorbed on either type of site, term K_CP_C or $K_C'P_C$ could be added to the denominator terms.

The main difference between Eq. (2.25) and Eq. (2.24) is that with noncompetitive adsorption, the rate becomes independent of P_A and P_B at high partial pressures, whereas the rate eventually decreases with P_A or P_B if A and B are competitively adsorbed.

If experimental data show a definite maximum in the plot of rate versus pressure, competitive adsorption is the likely cause, and Eq. (2.24) or a similar equation is needed to fit the results. However, the absence of a rate maximum does not eliminate the competitive adsorption mechanism, since the experiments may not have been conducted over a range of pressures wide enough to show a maximum.

Eley-Rideal Mechanism

Another model to consider is the reaction of adsorbed molecules of A with energetic molecules of B from the gas phase. The reaction rate is assumed proportional to the fraction of the surface covered by A and the collision frequency for B, which is proportional to the partial pressure of B:

$$A_s + B \rightarrow C_s \rightarrow C + s$$

$$r = k\theta_AP_B = \frac{kP_AK_AP_B}{1 + P_AK_A + P_CK_C} \tag{2.26}$$

This model was proposed by Rideal and Eley [4,5] as an alternative to the Langmuir–Hinshelwood models. It predicts that the reaction is first order in B, the reactant from the gas phase, and varying order to A, as shown in Figure 2.6. However, the same type of behavior might be found for

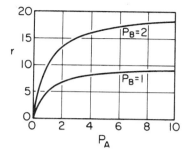

FIGURE 2.6   Reaction rates for Eley–Rideal kinetics.

a Langmuir–Hinshelwood reaction between adsorbed reactants. Going back to Eq. (2.24), if B and D are very weakly adsorbed and A and C are moderately adsorbed, the rate equation becomes

$$r \cong \frac{kK_A P_A K_B P_B}{(1 + K_A P_A + K_C P_C)^2} \qquad (2.27)$$

If the values of P_A are not large enough to make the rate go through a maximum, the kinetic data could probably be fitted with either Eq. (2.27) or Eq. (2.26). Additional special tests, such as chemisorption measurements for the pure gases or tests of the transient response to feed changes, would be needed to discriminate between these mechanisms. Langmuir–Hinshelwood models have been more widely used for correlating kinetics data than the Eley–Rideal model, but a good fit to any model should not be taken as proof of the proposed mechanism [6].

Complex Reactions

Surface reactions often take place in several steps, but one step may control the rate, and a Langmuir–Hinshelwood model may still be appropriate. Consider a two-step surface reaction involving a nonvolatile intermediate, C, that reacts irreversibly with B on the surface to form the product D:

$$A + 2B \longrightarrow D \qquad \text{overall reaction}$$

$$A_s + B_s \xrightarrow{1} C_s \qquad \text{slow step}$$

$$C_s + B_s \xrightarrow{2} D_s \qquad \text{fast}$$

If the first step controls the rate, there is little C on the surface, since it reacts almost immediately with B to form D. The kinetics of step 2 are therefore unimportant, and the overall rate equation is the same as for a single-step bimolecular reaction:

$$r = k\theta_A\theta_B = \frac{kK_AP_AK_BP_B}{(1 + K_AP_A + K_BP_B + K_DP_D)^2} \tag{2.28}$$

If the first step is at equilibrium and step 2 controls, the fraction covered by C becomes important:

$$A_s + B_s \underset{-1}{\overset{1}{\rightleftharpoons}} C_s \qquad \text{at equilibrium}$$

$$C_s + B_s \overset{2}{\longrightarrow} D_s \qquad \text{slow step}$$

$$D_s \rightleftharpoons D_{\text{gas}} \qquad \text{at equilibrium}$$

At pseudo steady-state,

$$k_1\theta_A\theta_B \cong k_{-1}\theta_C$$

$$\theta_C = \left(\frac{k_1}{k_{-1}}\right)\theta_A\theta_B$$

Equilibrium coverage is assumed for A and B:

$$r = k_2\theta_C\theta_B = \left(\frac{k_1k_2}{k_{-1}}\right)\theta_A\theta_B^2$$

$$\tag{2.29}$$

$$r = \frac{kK_AP_A(K_BP_B)^2}{(1 + K_AP_A + K_BP_B + K_DP_D)^3}$$

Although Eqs. (2.28) and (2.29) have different exponents, they both predict that the rate would eventually go through a maximum with increasing values of P_A or P_B. Data over a moderate range of pressures might be fitted with either equation.

If one of the reactants is H_2 or another gas that may dissociate on adsorption, a great many models can be formulated based on the reaction of H_2, H, or 2H in the individual steps. Typical data for hydrogenation of olefins or aromatics can often be fitted quite well by several different models. When data are available from a differential reactor, the models can be tested using reciprocal plots similar to those used for Langmuir isotherms.

Example 2.1

The kinetics of benzene hydrogenation to cyclohexane over a Ni/SiO$_2$ catalyst were studied using a small fixed-bed reactor [7]. The bed had 0.87 g of 0.22-mm catalyst particles diluted with 19.2 g of glass beads.

Data for a bed temperature of 67.6°C are given in Table 2.1. What reaction models give good fit to these data?

Solution. The data show that a 2.85-fold increase in P_{H_2} increases the rate only 18%, and a four-fold increase in P_B increases the rate about 40%. This suggests that both benzene and hydrogen are strongly adsorbed on the catalyst and cover a large fraction of the available sites. Several models can be formulated based on reaction of adsorbed benzene with one or two hydrogen atoms as the slow step and with either competitive adsorption or adsorption on two types of sites.

For competitive adsorption of benzene and hydrogen, simultaneous reaction of two hydrogen atoms with adsorbed benzene is assumed the slow step:

$$B + s \leftrightarrow B_s$$
$$H_2 + 2_s \leftrightarrow 2H_s$$
$$B_s + 2H_s \rightarrow BH_{2_s} \qquad \text{slow step}$$
$$BH_{2_s} + 2H_s \rightarrow BH_{4_s} \qquad \text{fast}$$
$$BH_{4_s} + 2H_s \rightarrow C_s \qquad \text{fast}$$
$$C_s \leftrightarrow C + s \qquad \text{(weak adsorption of C assumed)}$$

$$r = k\theta_B \theta_H^2 = \frac{kK_B P_B \left(\sqrt{K_{H_s} P_{H_2}}\right)^2}{\left(1 + K_B P_B + \sqrt{K_{H_s} P_{H_2}}\right)^3}$$

$$Y = \left(\frac{P_B P_{H_2}}{r}\right)^{1/3} = \frac{1 + K_B P_B \sqrt{K_{H_s} P_{H_2}}}{\left(kK_B K_{H_s}\right)^{1/3}}$$

A plot of Y vs. P_B for $P_{H_2} = 2110$, Figure 2.7(a) gives a straight line with a positive slope and intercept:

TABLE 2.1   Data for Example 2.1

P_{H2}, torr	P_B, torr	F_B, mol/hr	X_B, %	$10^3 r$ mol/hr, g
1050	70	0.1034	3.2	3.81
2105	70	0.0498	7.47	4.27
2988	70	0.0337	11.6	4.50
2110	185	0.1038	4.53	5.40
2110	286	0.01563	3.41	6.12

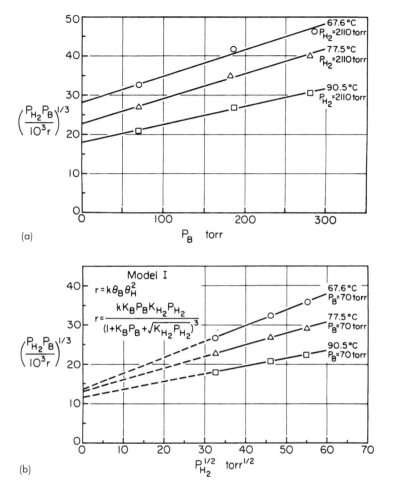

FIGURE 2.7 Test of Model I for benzene hydrogenation: (a) Variable benzene pressure. (b) Variable hydrogen pressure.

$$\text{slope} = \frac{K_B^{2/3}}{k^{1/3}K_{H_2}^{1/3}} = 0.635$$

$$\text{intercept} = \frac{1 + \sqrt{K_{H_2}P_{H_2}}}{\left(kK_BK_{H_2}\right)^{1/3}} = 287$$

A similar plot of Y vs. $P_{H_2}^{1/2}$ for $P_B = 70$, Figure 2.7(b), gives

$$\text{slope} = \frac{K_{H_2}}{k^{1/3}K_B^{1/3}} = 4.12$$

$$\text{intercept} = \frac{1 + K_B P_B}{\left(k K_B K_{H_2}\right)^{1/3}} = 135$$

Solving for the three parameters gives

$$K_{H_2} = 0.00243 \text{ torr}^{-1}$$
$$K_B = 0.0076 \text{ torr}^{-1}$$
$$k = 0.0861$$

The final equation for $T = 67.6°C$ is

$$r = \frac{(1.59 \times 10^{-6}) P_B P_{H_2}}{\left(1 + 0.0075 P_B + 0.0493 P_{H_2}^{1/2}\right)^3}$$

The equation predicts the rate with an average error of 3%.

If the hydrogen and benzene are adsorbed on different types of sites and adsorption is noncompetitive, the rate expression is

$$r = k\theta_B \theta_H^2 = \frac{k K_B K_{H_2} P_B P_{H_2}}{(1 + K_B P_B)\left(1 + \sqrt{K_{H_2} K_{H_2}}\right)^2}$$

This equation also gives good fit to the data.

With $K_B = 0.020$, $K_{H2} = 0.016$, and $k = 0.01$, the average error in the predicted rate is 2%.

Assuming the slow step is the reaction of a single hydrogen with benzene fits the data with about 2% average error,

$$r = k\theta_B \theta_H = \frac{(3.37 \times 10^{-6}) P_B P_{H_2}^{1/2}}{\left(1 + 0.0069 P_B + 0.0082 \sqrt{P_{H_2}}\right)^2}$$

If the reactants are on different types of sites, the rate equation is

$$r = k\theta_B \theta_H = \frac{(1.02 \times 10^{-5}) P_B P_{H_2}^{1/2}}{(1 + 0.020 P_B)\left(1 + 0.0051 \sqrt{P_{H_2}}\right)}$$

All four models give a good fit to the data, though with quite different values for the adsorption constants and rate constant. An Eley–Rideal model involving hydrogen from the gas phase gives a poor fit because it predicts first-order behavior for hydrogen.

The data at 77.5°C and 90.5°C lead to similar rate expressions, which can predict the results to within 2 or 3%. The models based on competitive adsorption might be preferred because the values of K_B and K_{H_2} decrease with increasing temperature, as expected, whereas the values of K_B are nearly constant for the noncompetitive adsorption models. Kinetic tests over a wider range of partial pressures and temperatures plus supplemental adsorption measurements would be needed to determine the best model.

Reversible Reactions

When a reaction is reversible, the overall rate expression must be thermodynamically consistent. For a reversible bimolecular reaction where all species are competitively adsorbed, the net rate of reactant consumption, r, is

$$A + B \underset{2}{\overset{1}{\rightleftharpoons}} C + D$$

$$r = k_1 \theta_A \theta_B - k_2 \theta_C \theta_D \tag{2.30}$$

$$r = \frac{k_1 K_A P_A K_B P_B - k_2 K_C P_C K_D P_D}{(1 + K_A P_A + K_B P_B + K_C P_C + K_D P_D)^2}$$

The number of terms in the denominator and the power to which it is raised do not affect the equilibrium conversion.

At equilibrium, $r = 0$, so

$$k_1 K_A P_A K_B P_B = k_2 K_C P_C K_D P_D \quad \text{or} \quad \frac{P_C P_D}{P_A P_B} = \frac{k_1 K_A K_B}{k_2 K_C K_D} = K_{eq} \tag{2.31}$$

When a reversible reaction has a change in the number of moles, the rate expression may have to be adjusted to satisfy the thermodynamic requirement. For example, consider a bimolecular reaction with one product:

$$A + B \underset{2}{\overset{1}{\rightleftharpoons}} C$$

The rate of the forward step might be proportional to $\theta_A \theta_B$, but if the reverse step includes only the term $k_2 \theta_C$, it leads to an incorrect equation:

$$r = k_1 \theta_A \theta_B - k_2 \theta_C$$

$$r = \frac{k_1 K_A P_A K_B P_B}{(1 + K_A P_A + K_B P_B + K_C P_C)^2} - \frac{k_2 K_C P_C}{1 + K_A P_A + K_B P_B + K_C P_C} \tag{2.32}$$

Equation (2.32) does not satisfy the equilibrium requirement because of the different exponents in the denominator terms. A consistent equation is obtained if the reverse reaction is assumed to require a vacant site.

$$A_s + B_s \underset{2}{\overset{1}{\rightleftharpoons}} C_s + s$$

The concentration of vacant sites is proportional to $1 - \theta$, which is found from the equations for θ_A, θ_B, etc.:

$$1 - \theta = \frac{1}{1 + K_A P_A + K_B P_B + K_C P_C} \tag{2.33}$$

The reverse reaction rate and overall rate are

$$r_2 = k_2 \theta_C (1 - \theta)$$

$$r = \frac{k_1 K_A P_A K_B P_B}{(1 + K_A P_A + K_B P_B + K_C P_C)^2} \tag{2.34}$$

The overall rate equation now has the same denominator for both forward and reverse terms, and the equilibrium requirement is met. Often the kinetics for the forward reaction are established by extensive tests, but there is not much data for the reverse reaction. The overall kinetic equation is then adjusted to give zero rate at the equilibrium conversion.

In some kinetic studies, adsorption of reactants or products is so strong that the rate appears inversely proportional to partial pressure rather than having the Langmuir isotherm form. Data for a commercial SO_2 oxidation catalyst were correlated with this equation [8]:

$$r = k \left(\frac{P_{SO_2}}{P_{SO_3}} \right)^{1/2} \left(P_{O_2} - \left(\frac{P_{SO_3}}{K P_{SO_2}} \right)^2 \right) \tag{2.35}$$

It is apparent that SO_3 inhibits the forward reaction, but the rate certainly does not become infinite at $P_{SO_3} = 0$. An equation with a denominator term of $(1 + K_{SO_3} P_{SO_3})$ would avoid this problem. Equation (2.35) is thermodynamically consistent, since the rate goes to zero when

$$P_{O_2} = \left(\frac{P_{SO_3}}{K_{eq} P_{SO_2}} \right)^2 \quad \text{or} \quad \frac{P_{SO_3}}{P_{SO_2} P_{O_2}^{1/2}} = K_{eq} \tag{2.36}$$

RATE OF ADSORPTION CONTROLLING

If adsorption of reactant A is the rate-limiting step of a bimolecular reaction, molecules of A will react almost immediately after being adsorbed, and there will be hardly any A on the surface. The other reactant, B, and the product, C, may be at equilibrium coverage. The reaction rate is then proportional to the partial pressure of A and the fraction of vacant sites:

$$A + s \xrightarrow{\ 1\ } A_s \qquad\qquad \text{slow}$$

$$A_s + B_s \xrightarrow{\ 2\ } C_s \to C_g \qquad \text{fast} \qquad\qquad\qquad (2.37)$$

$$r = k_1 P_A (1 - \theta) = \frac{k_1 P_A}{1 + K_B P_B + K_C P_C}$$

The term $K_A P_A$ is not included in Eq. (2.37) because A is not at adsorption–desorption equilibrium. Equation (2.37) shows that the reaction is first order to A and is inhibited by B and C. However, if the surface reaction between strongly adsorbed B and weakly adsorbed A is the controlling step, with all the species at equilibrium, the reaction might also appear first order to A and negative order to B, as shown by Eq. (2.24). Tests over a wider range of partial pressures of both reactants would be needed to discriminate between these models.

ALLOWING FOR TWO SLOW STEPS

When there are several steps in a reaction sequence, the overall rate may be influenced by two or more slow steps, and there may be no single controlling step. Consider the case where the rate of adsorption is not fast enough to keep θ_A at the equilibrium value and the surface reaction is not fast enough to make θ_A nearly 0. Assume a reaction between an adsorbed molecule of A and a molecule of B from the gas phase. (To simplify this example, only A is adsorbed.)

$$A + s \leftrightarrow A_s$$

$$A_s + B \xrightarrow{\ 3\ } C_s$$

$$C_s \xrightarrow{\ 4\ } C + s \qquad \text{fast}$$

$$r_{\text{ads}} = k_1 P_A (1 - \theta_A)$$

$$r_{\text{des}} = k_2 \theta_A$$

$$r = k_3 \theta_A P_B$$

At steady state, the rate of adsorption is equal to the rate of desorption plus the rate of reaction:

$$k_1 P_A(1 - \theta_A) = k_2 \theta_A + k_3 P_B \theta_A$$
$$k_1 P_A = \theta_A(k_2 + k_3 P_B + k_1 P_A)$$

Dividing all terms by k_2 and letting $K_A = k_1/k_2$, as usual, gives

$$\theta_A = \frac{K_A P_A}{1 + K_A P_A + \left(\dfrac{k_3}{k_2}\right) P_B} \tag{2.38}$$

The reaction rate is

$$r = \frac{k_3 K_A P_A P_B}{1 + K_A P_A + \left(\dfrac{k_3}{k_2}\right) P_B} \tag{2.39}$$

The reaction would appear fractional order to A and to B, though only A was assumed to adsorb. The coefficient for P_B in the denominator is not an adsorption constant but is the ratio of the rate constant for the reaction to the desorption constant for A. The relative importance of the adsorption and reaction steps changes with P_A and with P_B. If P_B is very low so that the denominator term is essentially $(1 + K_A P_A)$, the surface coverage for A is almost the equilibrium value, and the overall rate is controlled by the reaction step. When P_B is very large, there is little A on the surface, and the rate of adsorption of A is the controlling step. For a wider range of intermediate pressures, both the rate of adsorption and the rate of reaction affect the overall rate. Figure 2.8 shows how θ_A and r change with reaction conditions. For a relatively slow reaction, Case I, the desorption rate is faster than the reaction rate for a given θ_A, and the steady-state value of θ_A is only slightly less than θ_A,eq. The surface reaction is the controlling step, and there is little error in assuming $\theta_A = \theta_A$,eq. When the surface reaction is 10 times faster than the rate of desorption, as shown by the steep reaction line for Case II, θ_A is much less than θ_A,eq, and both the rate of adsorption and the rate of the surface reaction affect the overall rate. If the surface reaction had been made 100 times faster than for Case I, θ_A would be about 0.07, and the rate of adsorption of A would be the controlling step.

When reaction occurs between molecules of A and B adsorbed on the same type of site, and when one or both reactants are not at adsorption–desorption equilibrium with the gas phase, the kinetic data may show some unusual effects. The following treatment is for a bimolecular surface reaction with weak adsorption of the product:

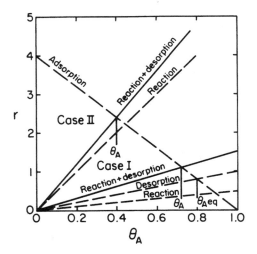

Figure 2.8 Graphical solution for θ_A with nonequilibrium adsorption of A for $A_s + B \rightarrow C$.

$$A + s \overset{1}{\underset{2}{\rightleftharpoons}} A_s$$

$$B + s \overset{3}{\underset{4}{\rightleftharpoons}} B_s$$

$$A_s + B_s \overset{5}{\longrightarrow} C_s \overset{\text{fast}}{\longrightarrow} C_g$$

For both reactants, the surface reaction rate is the difference between the rate of adsorption and the rate of desorption:

$$r = k_1 P_A (1 - \theta_A - \theta_B) - k_2 \theta_A \tag{2.40}$$

$$r = k_3 P_B (1 - \theta_A - \theta_B) - k_4 \theta_B \tag{2.41}$$

$$r = k_5 \theta_A \theta_B \tag{2.42}$$

Equations (2.40)–(2.42) were solved for some arbitrary cases assuming equilibrium adsorption for B and nonequilibrium adsorption for A[9]. Figure 2.9 shows how θ_A is decreased considerably by decreasing P_A, though θ_A,eq for all the pressures shown would be greater than 0.9. Since the surface reaction is a maximum at $\theta_A = 0.5$, a decrease in P_A increases the reaction rate when $\theta_A > 0.50$. However, because θ_A is determined by the intersection of the adsorption line and the steeply curved line for reaction plus desorption, the effect of changing P_A is greater than for equilibrium coverage.

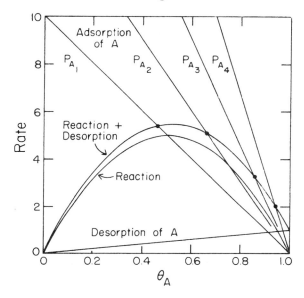

FIGURE 2.9 Graphical solution for θ_A and reaction rate for $A_s + B_s \rightarrow C$. (From Ref. 9 with the permission of Academic Press, NY.)

Figure 2.10 shows the reaction rates for equilibrium and nonequilibrium adsorption. At high values of P_A, the apparent reaction order can be in the range -2 to -3, whereas a limiting order of -1 would be obtained for equilibrium adsorption of both reactants.

DESORPTION CONTROL

If the reaction of A and B on the surface produces a product C that is very strongly chemisorbed, the overall rate may be limited by the rate of product desorption:

$$A_s + B_s \xrightarrow{1} C_s + s \qquad \text{fast} \tag{2.43}$$

$$C_s \xrightarrow{2} C + s \qquad \text{slow} \tag{2.44}$$

The simplest case of desorption control is when the chemical reaction is very fast and irreversible. The reaction is rapid when the catalyst is first exposed to the reactants, but the rate decreases as C accumulates on the surface. A steady state is reached when the surface is nearly completely covered with C, and the slow rate of desorption of C is matched by the

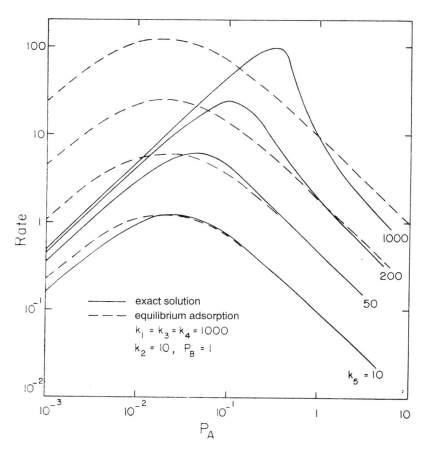

FIGURE 2.10 Comparison of reaction rate with rate based on equilibrium adsorption. (From Ref. 9 with the permission of Academic Press, NY.)

reaction of A and B on the small fraction of remaining sites. The reaction rate increases with temperature but does not depend on the partial pressures of reactants or products:

$$r = k_2 \theta_C \cong k_2 = k_0 e^{-E_{\mathrm{des}}/RT} \tag{2.45}$$

There are not many examples of desorption control in the literature. The dehydrogenation of sec-butyl alcohol to methyl ethyl ketone on brass catalyst was said to be controlled by the desorption of hydrogen for temperatures over 600°F [10], since the reaction rate was independent of total pressure. At lower pressures, the rate went through a maximum with increasing pressure, suggesting a dual-site mechanism with surface reaction as the controlling step. Some of these results are shown in Figure 2.11.

If the surface reaction is reversible and desorption of product controls, the overall rate still depends on the fraction of surface covered by the product, but this may be much less than unity. If the surface reaction is at equilibrium and reactants A and B are also at adsorption–desorption equilibrium but product C is not, the value of θ_C is obtained from modified equations for θ_A and θ_B:

$$\theta_A = \frac{K_A P_A (1 - \theta_C)}{1 + K_A P_A + K_B P_B} \tag{2.46}$$

$$\theta_B = \frac{K_B P_B (1 - \theta_C)}{1 + K_A P_A + K_B P_B} \tag{2.47}$$

Since C is not at equilibrium on the surface because of slow desorption, A and B compete for sites on $(1 - \theta_C)$ fraction of the surface. The surface reaction is assumed to be reversible, with a vacant site needed for the backward step:

$$A_s + B_s \underset{2}{\overset{1}{\rightleftharpoons}} C_s + s \tag{2.48}$$

$$k_1 \theta_A \theta_B = k_2 \theta_C \theta_V$$

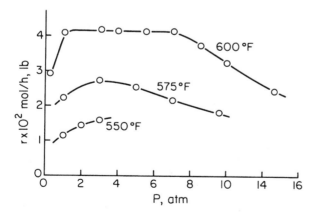

FIGURE 2.11 Reaction rate for dehydrogenation of sec-butyl alcohol. (From Ref. 10.)

$$\theta_C = \frac{K\theta_A\theta_B}{\theta_V} \tag{2.49}$$

where $K = k_1/k_2$. The fraction of vacant sites is

$$\theta_V = \frac{1 - \theta_C}{1 + K_A P_A + K_B P_B} \tag{2.50}$$

Combining Eqs. (2.46), (2.47), and (2.50) gives

$$\theta_C = \frac{KK_A P_A K_B P_B}{1 + K_A P_A + K_B P_B + KK_A P_A K_B P_B} \tag{2.51}$$

When K is large, θ_C is near 1.0; but when K is less than 1.0, θ_C is small and increases with increasing value of P_A and P_B. With a small equilibrium constant (K), increasing K also increases θ_C and the overall rate, even though the desorption of C is the rate-limiting step. However, increasing the catalyst activity to give higher values for both k_1 and k_2 would have no effect on θ_C or the overall rate, so the rate of the surface reaction rate is not limiting.

CHANGES IN CATALYST STRUCTURE

In the kinetic models discussed so far, the catalyst surface was assumed to be uniform and to have constant activity. In practice, catalysts often show a decline in activity with time, because of poisoning or fouling; these changes are discussed later. Catalysts may also change activity because of reactions that alter the chemical composition of the surface. One example of this is the supported silver catalyst used for the partial oxidation of ethylene. Kinetic data suggest that the active catalyst is a partial layer of silver oxide and not metallic silver. But a modified form of the Langmuir–Hinshelwood model can still be used to correlate the data, as shown in the following example.

Example 2.2

The formation of ethylene oxide by partial oxidation of ethylene over a Ag/Al$_2$O$_3$ catalyst was studied by Klugherz and Harriott using a differential reactor [11]. Helium was used as the carrier gas to minimize temperature gradients in the bed. The product gases strongly inhibit the reaction, and the rate data were corrected to the same product partial pressure, $P_p = 0.01$ atm, in order to focus on the effects of ethylene and oxygen concentrations. Because of drifting catalyst activity, the reaction rates were compared with the rate at "standard conditions," which were 220°C, 1.32 atm, 20% C$_2$H$_4$, 20% O$_2$, and 60% He. Some of the data are shown in Figure 2.12, where $R^0_{C_2H_4O}$ is the relative rate of ethylene oxide formation.

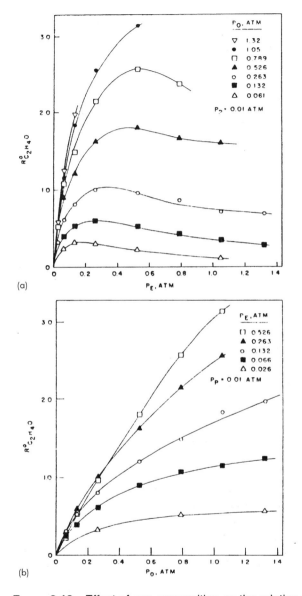

FIGURE 2.12 Effect of gas composition on the relative rate of C_2H_4O formation at 220°C: (a) changing ethylene pressures; (b) changing oxygen pressure. (From Ref. 11.) Reproduced with permission of the American Institute of Chemical Engineers. Copyright 1971 AIChE. All rights reserved.

a. Can the data in Figure 2.12 be correlated with a standard Langmuir–Hinshelwood or Eley–Rideal model for reaction on the silver catalyst?

b. Since oxygen chemisorbs reversibly on silver, test an alternate model based on a silver oxide layer as the catalyst.

Solution:

a. The simplest rate equation for a surface reaction between adsorbed ethylene and adsorbed molecular oxygen is

$$r_{EO} = k_1 \theta_O \theta_E = \frac{k P_E P_O}{(1 + K_E P_E + K_O P_O + K_P P_P)^2} \qquad (2.52)$$

Here O stands for O_2, and K_E and K_O are included in the rate constant k. Rearrangement of Eq. (2.52) yields

$$\left(\frac{P_E P_O}{r_{EO}}\right)^{1/2} = \frac{1 + K_E P_E + K_O P_O + K_P P_P}{\sqrt{k}} \qquad (2.53)$$

Plots of $(P_E P_O/r_{EO})^{1/2}$ versus P_E are shown in Figure 2.13(a). For each value of P_O, a straight line is obtained, but the lines are not parallel, as they should be if Eq. (2.53) holds. The plots of $(P_E P_O/r_{EO})^{1/2}$ versus P_O are curved, and some have negative slopes, which is additional evidence that the model is incorrect.

If the controlling step is the reaction of adsorbed ethylene with an oxygen atom, the rate equation would be

$$r_{EO} = \frac{k P_E P_O^{1/2}}{\left(1 + K_E P_E + (K_O P_O)^{1/2} + K_P P_P\right)^2} \qquad (2.54)$$

Plots of $(P_E P_O^{1/2}/r_{EO})^{1/2}$ versus P_E or $P_O^{1/2}$ (not shown) give poor fits to the data, which is not surprising, since Eq. (2.54) shows a maximum order of 0.5 for oxygen, whereas Figure 2.12(b) shows orders of about 1.0 at low oxygen pressures.

The data for r_{CO_2}, the rate of complete oxidation, are similar to those for r_{EO}, with maxima in the plots against P_E and fractional-order behavior to O_2, and none of the simple models fit the data.

b. If the reactions take place on a silver oxide layer, the fraction of surface covered with oxide, θ_s', can be related to the oxygen pressure. Oxygen is assumed to require two sites for dissociative adsorption:

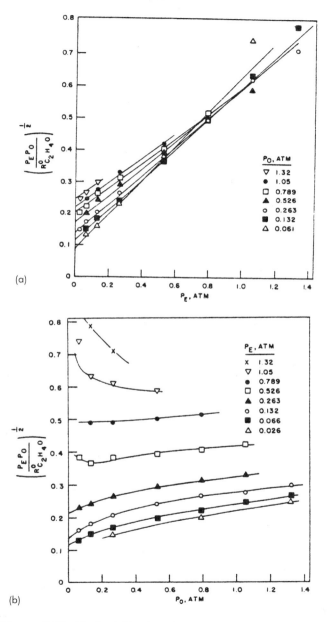

FIGURE 2.13 Test of Eq. (2.53) for ethylene oxide formation. (From Ref. 11.) Reproduced with permission of the American Institute of Chemical Engineer. Copyright 1971 AIChE. All rights reserved.

$$O_2 + 2s \overset{1}{\underset{2}{\rightleftharpoons}} 2O_s$$

$$k_1 P_O (1 - \theta_s')^2 = k_2 \theta_s'^2$$

$$\theta_s' = \frac{(K_S P_O)^{0.5}}{1 + (K_S P_O)^{0.5}}$$

where

$$K_s = \frac{k_1}{k_2}$$

The equation for θ_s' is now added to Eq. (2.52) or Eq. (2.54) to allow for the change in the number of active sites with oxygen pressure. If the formation of ethylene oxide involves an atom of oxygen, Eq. (2.54) is modified to

$$r_{EO} = \frac{k P_E P_O^{1/2}}{\left(1 + K_E P_E + (K_O P_O)^{1/2} + K_P P_P\right)^2} \left(\frac{(K_S P_O)^{1/2}}{1 + (K_S P_O)^{1/2}}\right)^2$$

$$(2.55)$$

Rearranging Eq. (2.55) for a test plot gives

$$\left(\frac{P_E P_O^{3/2}}{r_{EO}}\right)^{1/2}$$
$$= \left(\frac{1}{\sqrt{k}} + \frac{K_E P_E}{\sqrt{k}} + \frac{(K_O P_O)^{1/2}}{\sqrt{k}} + \frac{K_P P_P}{\sqrt{k}}\right)(1 + (K_S P_O)^{1/2})$$

$$(2.56)$$

A plot of $(P_E P_O^{3/2}/r_{EO})^{1/2}$ versus P_E at constant P_O is shown in Figure 2.14. The lines are straight, with slope and intercept increasing with P_O, as expected. The final rate equation for 220°C and $P_P = 0.01$ atm is

$$r_{EO} = \frac{P_E P_O^{3/2}}{\left(0.013 + 0.236 P_E + 0.121 \sqrt{P_O}\right)^2 \left(1 + 0.661 \sqrt{P_O}\right)^2}$$

$$(2.57)$$

Equation (2.57) fits the data with an average error of 5%. The model based on molecular oxygen gives an equation with an average error of 10%. However, the difference in fit is not great, and the exact mechanism of ethylene oxide formation is still uncertain.

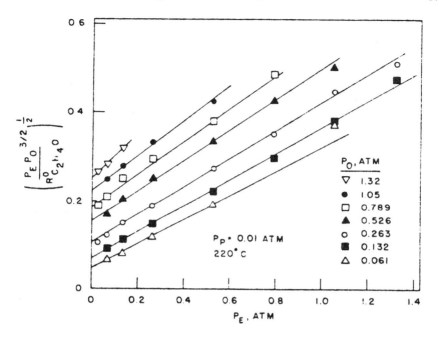

Another example of catalysts that undergo structural changes during reaction are catalysts that act by a redox mechanism, such as vanadium pentoxide used for partial oxidations. As shown by Mars and Van Krevelen [12], the catalyst provides the oxygen needed for the reactions, and it becomes reduced to a lower-valence oxide in the process:

$$A + n\text{Cat—Ox} \xrightarrow{1} B + n\text{Cat—Red}$$

The catalyst is continuously regenerated by oxygen from the gas phase, completing the two-step process:

$$n\text{Cat—Red} + O_2 \xrightarrow{2} n\text{Cat—Ox}$$

If the rate of the first reaction is proportional to θ, the fraction in the fully oxidized state, and the second to $(1 - \theta)$, and if the reactions are first order to A and to O_2, then the equation for the steady-state performance is obtained as follows:

$$r_1 = k_1 P_A \theta$$
$$r_2 = k_2 P_{O_2}(1 - \theta)$$
$$r_1 = nr_2$$
$$k_1 P_A \theta = nk_2 P_{O_2} - nk_2 P_{O_2}\theta \qquad\qquad (2.58)$$
$$\theta = \frac{nk_2 P_{O_2}}{k_1 P_A + nk_2 P_{O_2}}$$
$$r = \frac{nk_1 k_2 P_A P_{O_2}}{k_1 P_A + nk_2 P_{O_2}}$$

At low values of P_A (when $k_1 P_A < < nk_2 P_{O_2}$), the overall reaction appears first order to A and independent of P_{O_2}, and nearly all of the catalyst is in the fully oxidized state. At high values of P_A, the overall reaction is first order to oxygen and nearly zero order to A, and then most of the catalyst is in the reduced state. The change in reaction order with concentration is somewhat similar to that for Langmuir–Hinshelwood kinetics, but the redox mechanism does not give first order to both reactants at low coverage and does not have the possibility of negative orders.

When a reaction takes place by a redox mechanism, other reactions may take place at the same time by Langmuir–Hinshelwood or Eley–Rideal mechanisms or by homogeneous reactions. These other reactions could be eliminated or minimized by using two reactors, one fed with air or oxygen and the other fed with reactant A. The catalyst would be circulated between the two units using moving-bed or fluidized-bed technology.

CATALYST DECAY

Most solid catalysts lose activity under operating conditions, and the decrease should be allowed for in reactor design. If the decline in activity is very slow, the catalyst may be used for several years before being replaced. The initial charge is made large enough to compensate for catalyst decay, or the reaction conditions are gradually adjusted to keep the conversion nearly constant. When rapid fouling occurs, as in the catalytic cracking of petroleum fractions, much of the activity is lost in a few seconds, and continuous regeneration of the catalyst is necessary. When the catalyst retains appreciable activity for several days or weeks, intermittent regeneration can be used; or if the bed has a gradient in activity, the least active portion can be replaced at intervals with fresh catalyst.

The three most common types of processes that cause a loss in catalytic activity are poisoning, sintering, and fouling. *Poisoning* refers to coverage of active sites by trace materials in the feed. A great many types of catalyst poisons have been reported [13], including organic compounds containing sulfur or nitrogen, heavy metals, and even simple molecules such as carbon monoxide, oxygen, and water. In some cases, the poisons are reversibly absorbed and can be removed by treatment with a clean gas, but often the poisoning is essentially irreversible.

When a poison is reversibly adsorbed and Langmuir–Hinshelwood kinetics apply, the rate equation can be modified by adding a term $K_P P_P$ to the denominator:

$$A_s + B_s \rightarrow C_s$$

$$r = \frac{kK_A P_A K_B P_B}{(1 + K_A P_A + K_B P_B + K_P P_P)^2} \tag{2.59}$$

Strong poisoning can change the apparent kinetics of the reaction. If the term $K_P P_P$ in Eq. (2.59) is very large, the reaction could appear first order to A and to B even though fractional or zero orders might be observed in the absence of poison.

Poisoning can affect the selectivity as well as the rate of conversion, and mild poisoning may be beneficial. The oxidation of ethylene is carried out using silver catalysts that are deliberately poisoned with chlorine compounds, and the selectivity is improved, because the total oxidation reaction is suppressed more than the rate of ethylene oxide formation [14]. The presence of sulfur compounds changes the selectivity for competitive hydrogenation, such as the hydrogenation of acetylenes or diolefins in the olefins [15].

A poison that is strongly and irreversibly adsorbed will tend to be localized in the first portion of the catalyst bed at the start of operations and gradually spread through the bed with time. After a while, the bed might have a zone of inactive or dead catalyst near the entrance, a zone of partially deactivated catalyst (similar to the mass transfer zone in an adsorber), and a zone of active catalyst. To further complicate matters, the poison may deposit near the outside of the catalyst pellets and perhaps plug some of the pore mouths. Diffusion and reaction on a partially poisoned pellet is discussed in Chapter 4. Sometimes an extra bed of a special catalyst might be placed upstream of the reactor to act as a guard bed. This catalyst would be chosen to have a high capacity and a high rate of adsorption of the poison.

Sintering is the coalescence or growth of catalyst particles on a support or the fusion of grains of the support itself. The catalyst activity is thereby decreased because larger particles have less surface area per unit mass. When the catalyst has very small metal crystallites ($d < 10$ nm) on a support, the crystallites diffuse randomly on the surface of the support and coalesce when they collide. For larger particles, the surface diffusion is too slow to be significant, and large particles grow at the expense of smaller ones by diffusion of atoms along the surface or by diffusion of gaseous species. The process is similar to the ripening of crystals in a suspension.

Sintering of supported metal catalysts can occur at temperatures of one-third to one-half the melting point [16]. The reduction in surface area is often quite rapid at first and then much slower. The loss in area is sometimes fitted to an empirical equation, where n is like a reaction order:

$$-\frac{dS}{dt} = KS^n \tag{2.60}$$

Tests show values of n from 6 to 15, depending on the type of metal, type of support, and method of catalyst preparation. A theory based on diffusion of tiny crystallites gives $n = 8$ [17]. Sintering can sometimes be retarded by catalyst promoters, but the mechanism of promotion is not clear.

Fouling is the formation of a layer of inert material on part of the catalyst surface. Usually, fouling refers to the deposition of amorphous carbon, which is formed by cracking of hydrocarbons in the feed or product or by polymerization of olefinic or acetylenic compounds adsorbed on the surface. The term *coking* is often used for fouling by carbon deposition, though the deposits are quite different from coke made from coal or heavy oils. The carbon deposits on catalysts can be characterized as $(CH_x)_n$, where x is usually 0.2–1.0. The deposits are very reactive to O_2, H_2, or CO_2, and the amorphous carbon has a greater free energy than graphite, judging from the equilibrium conversion in gasification reactions [18].

In a fixed-bed reactor, if coke is formed primarily from the reactants, there will be a gradient in catalyst activity, with the highest coke content and the lowest activity in the first part of the bed. If a reaction product is the main coke precursor, the coke level will be highest and the activity low toward the end of the bed. Additional complexity arises if the coke-deposition reaction is diffusion limited, which leads to a gradient of coke content within the catalyst pellet. Many theoretical studies have been made showing the shape of profiles of coke content and catalytic activity in catalyst beds subject to fouling, but there are few comparisons of theory and experiment and no way to predict the fouling rate for a new system.

In an early study of coke deposition for natural and synthetic cracking catalyst, the data were correlated with the empirical equation [19]

$$C_C = At^n \tag{2.61}$$

where C_C is the wt% coke, t is the process time, and A and n are constants. Other studies presented similar correlations, but the values of n ranged from

0.4 to 1.0, depending on the feedstock, the flow rate, and the type of catalyst [20]. A value of $n = 0.5$ in Eq. (2.61) would result if the carbon deposition rate, dC_C/dt, varied inversely with the coke content. Although some tests do show $n = 0.5$, there is no basic reason for this dependence. As active sites are covered by carbon, the rate of carbon deposition should decrease, unless the coke is a catalyst for further deposition.

In a fundamental study of coke formation on a chromia-alumina dehydrogenation catalyst, the catalyst activity and coke formation rate were measured in a differential reactor [20]. The equation for the rate of coking allowed for the decrease in rate with increase in coke level and the effects of reactants and products:

$$r_C = \frac{dC_C}{dt} = \frac{k_1 P_B^{0.743} + k_2 P_D^{0.813}}{\left(1 + K_H P_H^{1/2}\right)^2} \exp^{-\alpha C_C} \tag{2.62}$$

Here, P_B, P_D, and P_H are the partial pressures of butane, butadiene, and hydrogen, and C_C is the kg carbon/kg catalyst. The negative effect of hydrogen on the coking rate could mean that hydrogen adsorption on active sites prevents carbon formation on those sites, or it could mean that carbon is being gasified by hydrogen to decrease the net rate of carbon formation.

The regeneration of fouled catalysts is usually done by oxidation, but a gas with only a few percent oxygen is used to avoid overheating the catalysts. Because the heat of reaction is carried through the bed-warming portions of the bed prior to oxidation, temperature changes much above the adiabatic reaction temperature rise may occur.

NOMENCLATURE

Symbols

A_o	constant in Temkin isotherm
a	constant in rate equation
b	constant in rate equation
C_c	coke concentration, wt%
E	activation energy
E_{ad}	activation energy for adsorption
E_{des}	activation energy for desorption
K	equilibrium constant
$K_A, K_B, K_P,$	equilibrium constant for adsorption of A, B, product, respectively
k	reaction rate constant; k_1, k_2, k_3, reaction rate constant for steps 1, 2, 3, respectively

k_{1A}	adsorption rate constant for A
k_{2A}	desorption rate constant for A
n	reaction order
P_A	partial pressure of A
P_{As},	partial pressure of A at surface
R	gas constant
r	reaction rate
r_{ads}	adsorption rate; r_{des}, desorption rate
S	surface area
s	site on catalyst
T	absolute temperature, K
t	time
V	volume adsorbed
V_m	monolayer amount

Greek Letters

α	constant in Eq. (2.62)
θ	fraction of sites occupied
θ_v	fraction of vacant sites

PROBLEMS

2.1 The hydrogenation of propylene was carried out in a semibatch recirculation reactor with a Pt/SiO₂ catalyst (Vorhis, F.H., M.S. thesis, Cornell University, 1968). The reaction was first order to hydrogen and showed a complex dependence on propylene pressure. Data for a typical run are given in Table 2.2.

Test at least two kinetic models that might explain the general feature of these results. Determine the rate constants for each model, and use them to compare actual and predicted rates.

2.2 For Case II of the example illustrated in Figure 2.8, what would be the change in θ_A, θ_{Aeq}, and the reaction rate for the following changes in conditions?

    a. Double P_A
    b. Double P_B
    c. Increase the reaction rate constant 10-fold

2.3 Cumene (A) was cracked to benzene (B) and propylene (C) in a fixed-bed recycle reactor. At constant total pressure, runs were made with three different diluents added to the cumene, and all runs had less than 1% conversion.

TABLE 2.2 Data for Problem 2.1

t, min	r, millimole/min	$P_{propylene}$, atm
0	0.086	0.563
40	0.104	0.520
80	0.138	0.463
120	0.165	0.394
160	0.190	0.312
200	0.214	0.218
240	0.252	0.112
250	0.282	0.083
260	0.320	0.047
264	0.318	0.032
268	0.267	0.017
272	0.129	0.006

i. Cyclohexane dilution for $P_A = 1$–0.5 atm, $r = kP_A^0$
ii. Xylene dilution for $P_A = 1$–0.5 atm, $r = kP_A^{1.0}$
iii. Cumene hydroperoxide dilution, see Table 2.3

a. Can these results be explained using a simple kinetic model?
b. Predict the effect of total pressure on the rate of cracking of cumene diluted with xylene.

2.4   If carbon deposition on a catalyst occurs at a rate proportional to the amount of uncovered surface and full coverage with a monolayer corresponds to 2.0 wt% carbon, how would the carbon content vary with time if 1.2 wt% carbon is obtained in 1 minute? Plot the carbon content versus time

TABLE 2.3 Data for Problem 2.3

Mole fraction hydroperoxide	Relative rate
0	1
0.0004	0.8
0.0014	0.52
0.0030	0.33
0.006	0.18
0.0150	0.12

on logarithmic coordinates and compare with the empirical equation of Voorhies [19].

$$\%C = At^{0.44}$$

2.5   When carbon monoxide and hydrogen react over nickel catalyst to form methane, the rate is almost independent of P_{CO} over a wide range of pressures, and the rate then decreases slightly at higher P_{CO}. Typical results for two temperatures are shown in Figure 2.15 from the Ph.D. thesis of R. W. Fontaine, Cornell University, 1973.

a.   Assuming competitive adsorption of CO and H_2 and surface reaction as the controlling step, compare the shapes of the predicted and experimental curves of r versus P_{CO}.

b.   If the surface is not uniform, could allowing for two types of sites give a better fit for this data in the region of nearly zero-order behavior?

2.6   A differential reactor was used to measure the rate of phosgene formation over 6–8 mesh carbon catalyst. The data were correlated assuming a surface reaction between adsorbed chlorine and weakly adsorbed carbon monoxide with significant adsorption of phosgene:

$$r = \frac{e^{-(2500/T-6.34)} P_{CO} P_{Cl_2}}{\left(1 + e^{(2500/T-7.02)} P_{Cl_2} + e^{(8040/T-26.05)} P_{COCl_2}\right)^2}$$

$$r = \text{moles/hr, gcat}$$

a.   Derive a rate equation for an Eley–Rideal mechanism, and determine the rate constants from the data for 42.7°C and 64°C (Table 2.4). Compare the adsorption constants and the fit of the data with values from the author's equation.

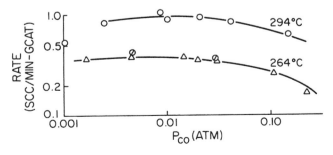

FIGURE 2.15   Effect of carbon monoxide pressure on the rate of methanation at $P_{H2} = 0.43$ atm.

Kinetic Models for Heterogeneous Reactions

77

TABLE 2.4 Data for Problem 2.6

Average Temp. °C	Rate of reaction × 10^3 g moles/ (hr) (g of catalyst)	Partial pressure, atm CO	Partial pressure, atm Cl_2	Partial pressure, atm, $COCl_2$
42.7	5.07	0.206	0.578	0.219
42.7	11.20	0.569	0.194	0.226
42.7	1.61	0.128	0.128	0.845
42.7	9.34	0.397	0.370	0.209
42.7	8.76	0.394	0.373	0.213
64.0	26.40	0.412	0.372	0.216
64.0	26.40	0.392	0.374	0.234
64.0	16.10	0.185	0.697	0.118
64.0	9.40	0.264	0.131	0.605

b. For a given gas composition, show whether the dependence of overall rate versus temperature follows the Arrhenius equation.

2.7 The data in Table 2.5 were obtained in a differential reactor at 150°C for the reaction A + B → C. Obtain an empirical rate equation and discuss possible controlling mechanisms.

2.8 The vapor-phase dehydration of ethanol to diethyl ether was studied in a packed bed with Dowex 50 X-8 resin as catalyst [21]. Initial reaction rates were obtained by extrapolating integral reaction rate data to

TABLE 2.5 Data for Problem 2.7

P, atm P_A	P, atm P_B	P, atm P_C	Rate (mole/hr, lb catalyst)
0.1	0.1	0.1	0.121
0.2	0.2	0.1	0.24
0.2	0.4	0.1	0.263
0.1	0.2	0.4	0.139
0.5	0.2	0.1	0.61
0.5	0.1	0.1	0.46
0.2	0.1	0.3	0.203
0.2	0.5	0.1	0.247

The superficial velocity was 0.5ft/sec for all runs, and the reaction is almost irreversible.

TABLE **2.6** Data for Problem 2.8

P_{A0}, atm	P_{E0}, atm	P_{W0}, atm	$r_0 \times 10^4$ g-moles/min, g cat.
0.381	0	0.619	0.0866
1.0	0	0	1.347
0.947	0.053	0	1.335
0.877	0.123	0	1.288
0.781	0.219	0	1.36
0.471	0.529	0	0.868
0.572	0.428	0	1.003
0.704	0.296	0	1.035
0.641	0.351	0	1.068
1.00	0	0	1.220
0.755	0	0.245	0.571
0.552	0	0.448	0.241
0.622	0.175	0.203	0.535
0.689	0.0	0	1.162

All runs at 120°C; $K_{eq} = 25.2$ at 120°C.

zero conversion. Try to fit the data in Table 2.6 to various rate expressions based on a unimolecular or a bimolecular surface reaction. What mechanisms can be rejected and which of those remaining seem the most likely to be correct? What other tests would you suggest to distinguish between possible mechanisms?

REFERENCES

1. I Langmuir. J Am Chem Soc 40:1301, 1918.
2. CN Hinshelwood. The Kinetics of Chemical Change. Oxford, UK: Clarendon Press, 1940.
3. OA Hougen, KM Watson. Chemical Process Principles. Vol III. New York: Wiley, 1947.
4. EK Rideal. J Soc Chem Ind London 62:335, 1943.
5. DD Eley. Advance Catalysis 1:157, 1948.
6. GC Bond. Catalysis by Metals. San Diego, CA: Academic Press, 1962, p 128.
7. JK Marangozis, BG Mantzournis, AN Sophos. Ind Eng Chem Prod Res Dev 18:61, 1979.
8. R Eklund. The Rate of Oxidation of Sulfur Dioxide with Commercial Vanadium Catalyst. Stockholm: Alumquist and Wiksell, 1966.
9. P Harriott. J Catalysis 22:266, 1971.
10. LH Thaler, G Thodos. AIChE J 6:369, 1960.

11. PD Klugherz, P Harriott. AIChE J 17:856, 1971.
12. P Mars, DW Van Krevelen. Chem Eng Sci 3:sp suppl 41, 1954.
13. R Hughes. Deactivation of Catalysts. San Diego, CA: Academic Press, 1984.
14. GC Bond. Catalysis by Metals. San Diego, CA: Academic Press, 1962, p 452.
15. H Wise, J McCarthy, J Oudar. In: J Oudar, H. Wise, eds. Deactivation and Poisoning of Catalysis. New York: Marcel Dekker, 1985, p 139.
16. JT Richardson. Principles of Catalyst Development. New York: Plenum Press, 1989, p 144.
17. JR Anderson. Structure of Metallic Catalysts. San Diego, CA: Academic Press, 1975.
18. ER Gilliland, P Harriott. Ind Eng Chem 46:2195, 1954.
19 A Voorhies. Ind Eng Chem 37:318, 1945.
20. FJ Dumez, GF Froment. Ind Eng Chem Proc Des Dev 15:291, 1976.
21. RL Kabel, LN Johanson. AIChE J 8:621, 1962.

3

Ideal Reactors

There are two basic types of ideal reactors, stirred tanks, for reactions in liquids, and tubular or packed-bed reactors, for gas or liquid reactions. Stirred-tank reactors include batch reactors, semibatch reactors, and continuous stirred-tank reactors, or CSTRs. The criterion for ideality in tank reactors is that the liquid be perfectly mixed, which means no gradients in temperature or concentration in the vessel.

Tubular reactors, which may be open or packed with catalyst, are considered ideal if there is plug flow of fluid and there are no radial gradients of temperature, concentration, or velocity. In plug-flow reactors, or PFRs, there are axial gradients of concentration and perhaps also axial gradients of temperature and pressure, but in the ideal PFR there is no axial diffusion or conduction.

Most large reactors do not fit the foregoing criteria, but in many cases the deviations from ideal reactors are small, and the equations for ideal reactors can be used for approximate design calculations and sometimes for determining optimum reaction conditions. In this chapter, ideal stirred-tank reactors are considered first and then plug-flow reactors are discussed. The effects of heat transfer, mass transfer, and partial mixing in real reactors are treated in later chapters.

BATCH REACTOR DESIGN

Batch reactors are often designed by direct scaleup from laboratory tests. If the feed concentration and reactor temperature are kept the same as in the lab tests, the time to reach a given conversion in a large ideal reactor should be the same as in lab tests, since the reaction rate ($kmol/hr$-m^3 or lb-mol/hr-ft^3) does not depend on reactor volume. The minimum size of the reactor needed can be calculated from the desired yearly production rate and the time for a complete batch cycle, including times for charging the reactants, heating, reaction, product discharge, and cleaning. A somewhat larger reactor of standard size might be selected to allow for changes in the run schedule or increases in production rate.

In addition to the size of the reactor, there are several other design decisions to be made, including the material of construction, the agitation system, and the method of supplying or removing heat. If the reaction is quite exothermic, heat transfer may be the limiting factor on scaleup. For a jacketed reactor, the surface-to-volume ratio varies inversely with the tank diameter. A reaction that can be carried out isothermally in a 2-liter vessel immersed in a water bath may be difficult to control in a 10-m^3 jacketed reactor, where the surface-to-volume ratio is 17-fold lower. Options to consider when heat removal is a problem include adding a cooling coil, using an external heat exchanger and pump, selecting multiple smaller reactors, and decreasing the reaction rate by using less catalyst or a lower temperature. Examples of these options are given in Chapter 5.

When the reaction kinetics are known, the conversion and yield expected for an ideal batch reactor can be calculated by integrating the rate equation. Then the effects of changing reactant ratio, catalyst concentration, temperature, and mode of operation can be explored to improve the design instead of just duplicating the condition of the laboratory tests. In the following section, equations are derived for a few systems with simple kinetics. The reactions are assumed to be irreversible and to take place at constant temperature. The volume change is neglected, even though there may be a change in the number of moles, since we are dealing with liquid reactants and products. The rates are therefore expressed in terms of concentration changes.

First-Order Reactions

For an irreversible first-order reaction of the type

$$A \rightarrow B + C$$

$$r = kC_A = -\frac{dC_A}{dt} \tag{3.1}$$

$$-\int_{C_{A_0}}^{C_A} \frac{dC_A}{C_A} = k \int_0^t dt$$

$$-\ln\left(\frac{C_A}{C_{A_0}}\right) = kt \quad \text{or} \quad C_A = C_{A_0}e^{-kt} \tag{3.2}$$

The result is often given in terms of the fraction converted:

$$C_A = C_{A_0}(1 - x)$$

$$-\ln(1 - x) = \ln\left(\frac{1}{1 - x}\right) = kt \tag{3.3}$$

Remember the important characteristics of a first-order reaction, which is evident from Eqs. (3.2) and (3.3). If a certain fraction of the reactant is converted in a given time, doubling the time will convert the same fraction of the remaining reactant. Thus for 90% conversion in t seconds, 99% would be converted in $2t$ seconds.

Second-Order Reactions

For a unimolecular second-order reaction,

$$r = -\frac{dC_A}{dt} = kC_A^2 \tag{3.4}$$

$$-\int_{C_{A_0}}^{C_A} \frac{dC_A}{C_A^2} = \frac{1}{C_A} - \frac{1}{C_{A_0}} = kt \tag{3.5}$$

or

$$\frac{x}{1 - x} = kC_{A_0}t \tag{3.6}$$

True second-order reactions of this type are rare, but many bimolecular reactions are first order to both reactants and are sometimes classified as second order.

Consider the reaction $A + B \rightarrow C$, with A the limiting reactant. Let

$$R = C_{B_0}/C_{A_0} \quad \text{where } R > 1$$

$$C_A = C_{A_0}(1 - x), \quad \Delta C_A = C_{A_0}x$$

$$C_B = C_{B_0} - C_{A_0}x$$

$$r = kC_A C_B = kC_{A_0}(1 - x)(C_{B_0} - C_{A_0}x) \tag{3.7}$$

Since $C_B = C_{A_0}(R - x)$ and $-dC_A = C_{A_0}\,dx$,

$$r = -\frac{dC_A}{dt} = C_{A_0}\frac{dx}{dt} = kC_{A_0}(1-x)C_{A_0}(R-x) \qquad (3.8)$$

$$\int_0^x \frac{dx}{(1-x)(R-x)} = \int_0^t kC_{A_0}\,dt \qquad (3.9)$$

Integration between limits gives

$$\ln\left(\frac{R-x}{R(1-x)}\right) = (R-1)kC_{A_0}t \qquad (3.10)$$

If $R = 1.0$, Eq. (3.10) is indeterminate, but x can be calculated from Eq. (3.6), since $C_B = C_A$ and $r = kC_A^2$. Equation (3.10) can be used to show the effect of different feed ratios. Often a very high conversion of one reactant—say A—is desired, and excess B is used to decrease the required reaction time. Sometimes the excess B can be separated and recycled. If 5% excess B is used, then $R = 1.05$ and the initial rate is 5% higher than for $R = 1.0$ and the same value of C_{A_0}. However, at $x = 0.99$, the final concentration of B is $1.05C_{A_0} - 0.99C_{A_0}$, or $0.06C_{A_0}$, and the final reaction rate is six times higher than for $R = 1.0$. Based on Eq. (3.10), the overall effect of the 5% excess is a 65% reduction in the time for 99% conversion.

The conversion curves for several values of R are shown in a semilog plot, Figure 3.1, along with the straight line for a first-order reaction. The abscissa is $k_2 C_{B_0} t$, and kC_{B_0} can be taken as the pseudo-first-order rate constant for A when there is a large excess of B. For $R = 1.5$ or 2.0, the plots are almost linear for $(1 - x) < 0.1$, since C_B doesn't change much in this region and the change in rate with conversion is similar to that for a first-order reaction. Note that the ratio t_{99}/t_{90} is 2.54 and 2.3 for $R = 1.5$ and 2.0, respectively, compared to 2.0 for a true first-order reaction. The plot for $R = 1.0$ shows gradually decreasing curvature, and the very low reaction rates at high conversion leads to $t_{99}/t_{90} = 11$ from Eq. (3.6).

The best value for R cannot be determined just from Figure 3-1 or Eq. (3.10), since increasing R always increases the rate for a given C_{A_0}. In practice, an increase in R increases the raw material cost or the cost of separating unreacted B from the product. Furthermore, if A and B are dissolved in a solvent, increasing C_B may require decreasing C_A, and this would mean less product produced per cycle, even if the batch time was decreased.

Example 3.1

A bimolecular batch reaction that is first order to both reactants is carried out to 99.5% conversion of A in 6.4 hours using 2% excess of reactant B. If 5% excess B is used, what time is needed to reach the same conversion?

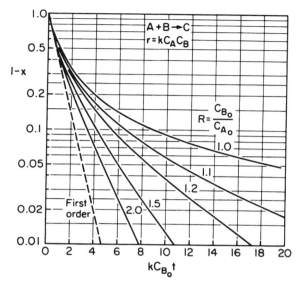

FIGURE 3.1 Conversion for a second-order batch reaction with different reactant ratios.

Solution. Use Eq. (3.10) to get $C_{A_0}k$ for $R = 1.02$:

$$\ln\left(\frac{1.02 - 0.995}{1.02(0.005)}\right) = (1.02 - 1)C_{A_0}kt$$

$$1.5896 = 0.02C_{A_0}k(6.5)$$

$$C_{A_0}k = 12.2$$

With $R = 1.05$,

$$\ln\left(\frac{1.05 - 0.995}{1.05(0.005)}\right) = 2.349 = (1.05 - 1)(12.2)t$$

$$t = 3.85 \text{ hr}$$

Consecutive Reactions

In many addition reactions, one, two, or more molecules of a gas or low-molecular-weight solute react to form a series of products. Examples are the chlorination of benzene or other aromatics, the hydrogenation of organics that have multiple double bounds, the hydrolysis of ethylene oxide, and partial oxidation of hydrocarbons. The general equations for the type of reaction are:

$$A + B \xrightarrow{1} C$$

$$C + B \xrightarrow{2} D \text{ etc}$$

When a reactor is charged with liquid A and B is a gas that is added continuously, it becomes a semibatch reactor. The rates of reaction depend on the concentration of B in the liquid phase, which is a function of gas solubility, pressure, and agitation conditions. However, we are often concerned with the relative reaction rates and the selectivity, which do not depend on C_B if the reaction orders are the same for both reactions. The reactions are treated as pseudo-first-order, and equations are developed for an ideal batch reactor with irreversible first-order kinetics

$$r_1 = k_1' C_A C_B = k_1 C_A \tag{3.11}$$

$$r_2 = k_2' C_C C_B = k_2 C_C \tag{3.12}$$

The concentration of A falls exponentially, as was shown earlier in Eq. (3.2):

$$C_A = C_{A_0} e^{-k_1 t} \tag{3.13}$$

The material balance for product C is

$$\frac{dC_C}{dt} = k_1 C_A - k_2 C_C \tag{3.14}$$

Combining these equations gives

$$\frac{dC_C}{dt} + k_2 C_C = k_1 C_{A_0} e^{-k_1 t} \tag{3.15}$$

If no C is present at the start, integration of Eq. (3.15) gives

$$C_C = \frac{k_1 C_{A_0}}{k_2 - k_1} \left(e^{-k_1 t} - e^{-k_2 t} \right) \tag{3.16}$$

The concentration of C goes through a maximum with time, and $t_{\max}$ can be found by differentiating Eq. (3.16) and setting the derivative to zero:

$$t_{\max} = \frac{\ln(k_2/k_1)}{k_2 - k_1} \tag{3.17}$$

$$C_{C,\max} = C_{A_0} \left(\frac{k_1}{k_2} \right)^{k_2/(k_2 - k_1)} \tag{3.18}$$

The concentration of D is obtained by a material balance:

$$C_{A_0} = C_A + C_C + C_D \tag{3.19}$$

$$C_D = C_{A_0}\left(1 + \frac{k_2 e^{-k_1 t}}{k_1 - k_2} + \frac{k_1 e^{-k_2 t}}{k_2 - k_1}\right) \tag{3.20}$$

Typical concentration curves for consecutive first-order reactions are shown in Figure 3.2. For this example, $k_2/k_1 = 0.2$, and the yield of C reaches a maximum of 67% at $k_1 t = 2.0$, where the conversion of A is 86.5% and the selectivity is 77%. If C is the desired product, the reaction could be stopped at this point: after separation of the products, the unreacted A could be recycled. However, if byproduct D is of little value, the reaction might be stopped at a lower conversion to get a higher selectivity. For this example, at $k_1 t = 1.0$, the conversion is 63% and the yield of C is 56% for a selectivity of 89%. More A must be separated and recycled, but the reaction time is half as great, and byproduct formation is reduced by a factor 2.8 (19.5/7). The optimum conversion would depend on the cost of separating the products and the value of the byproduct relative to the major product.

A classic example of the product distribution in a consecutive reaction system is the study of benzene chlorination by MacMullin [1]. He measured the concentration of mono-, di-, and trichlorobenzene produced in a batch chlorination and determined the relative rate constants. He also showed that batch chlorination gave a higher yield of monochlorobenzene than a single-stage continuous-flow reactor.

Parallel Reactions

Selectivity effects can also be important with parallel reactions having different reaction orders. Consider the case where the main reaction is first

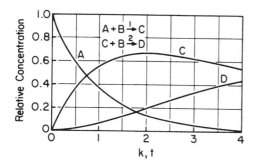

FIGURE 3.2 Relative reactant concentrations for consecutive pseudo-first-order reactions with $k_2/k_1 = 0.2$.

order to both reactants and the byproduct reaction is second order to one of the reactants:

main reaction $\quad\quad\quad\quad$ $A + B \xrightarrow{1} C$

byproduct formation $\quad$ $A + A \xrightarrow{2} D$

$$r_1 = k_1 C_A C_B$$

$$r_2 = k_2 C_A^2$$

The local or instantaneous selectivity is the ratio of r_1 to the total rate of consumption of A:

$$S = \frac{r_1}{r_1 + r_2} = \frac{r_1/r_2}{1 + r_1/r_2} \tag{3.21}$$

$$\frac{r_1}{r_2} = \frac{k_1 C_A C_B}{k_2 C_A^2} = \left(\frac{k_1}{k_2}\right)\left(\frac{C_B}{C_A}\right) \tag{3.22}$$

To minimize byproduct formation in the batch reactor, considerable excess B can be charged, so that the ratio C_B/C_A is high at the start and increases as the reaction proceeds. Calculated concentration curves for $C_{B_0}/C_{A_0} = 2.0$ and $k_2/k_1 = 0.6$ are shown in Figure 3.3. The initial selectivity is 77%, since $r_1/r_2 = 2/0.6 = 3.33$. Since C_A decreases more rapidly than C_B, the local selectivity gradually increases and is nearly 99% at 8 hours, when the conversion of A is 97%. The average selectivity gradually increases to 85%. Higher selectivity could be achieved by decreasing the concentration of A in the initial charge, but this would give a lower concentration of C in the final product and increase separation costs.

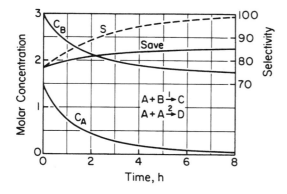

FIGURE 3.3 Selectivity with parallel reactions in a batch rector.

Semibatch Reactions

An alternate approach to improved selectivity for this system is to add A slowly over the course of the reaction and to carry out a semibatch reaction. The reactor might be two-thirds to three-quarters full at the start, and the fluid volume increases as A is added and no product withdrawn. Figure 3.4 shows the calculated concentration curves for the same kinetics and C_{B_0} as for Figure 3.3, with the feed of A at a slow, constant rate for 14 hours.

The value of C_A increases for a few hours until the reaction rate nearly equals the feed rate. Then C_A increases very slowly as C_B decreases and the rate of reaction of A decreases. After the feed of A is stopped, C_A decreases in exponential fashion. With the semibatch operation, the initial selectivity is very high but decreases as C_B decreases and C_A increases slightly. After the feed is stopped, the selectivity again increases. The average selectivity is 95%, a significant improvement over the 85% with batch operation, but the reaction time is increased. If some of the A had been charged at the start to make $C_{A_0} \cong 0.2$, the reaction time would have been decreased with little effect on the average selectivity.

CONTINUOUS-FLOW REACTORS

Operating a stirred-tank reactor with continuous-flow of reactants and products (a CSTR) has some advantages over batch operation. The reactor can make products 24 hours a day for weeks at a time, whereas for a typical cycle, the batch reactor is producing only about half the time. In the CSTR, temperature control is easier because the reaction rate is constant, and the rate of heat release does not change with time, as it does in a batch reactor.

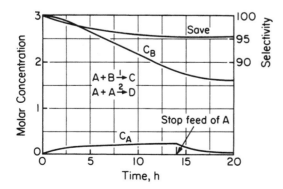

FIGURE 3.4 Selectivity with parallel reactions in a semibatch reactor.

Finally, conversion and selectivity may vary from day to day with a batch reactor, and they are more likely to be constant with a CSTR and a good control system.

The main disadvantage of continuous operation is that the reaction rate is nearly always lower than the average rate for a batch reaction. In most cases, the batch reaction rate decreases as the conversion increases, and in the CSTR the reaction rate is the same as the final reaction rate in the batch reactor. For high conversions, the final rate may be several-fold lower than the average rate, and the average residence time in the CSTR must then be several-fold greater than the reaction time in a batch reactor.

The average residence time in the CSTR is the volume V divided by the volumetric flow rate F, or $t = V/F$. The ratio of CSTR residence time to batch residence time is readily derived for simple kinetic models. For a first-order reaction in a CSTR, the steady-state material balance is

$$\text{in} - \text{out} = \text{amount reacting}$$

$$FC_{A_0} - FC_A = rV = kC_A V \tag{3.23}$$

$$C_{A_0} - C_A = kC_A \frac{V}{F} = kC_A t \tag{3.24}$$

$$C_A(1 + kt) = C_{A_0} \tag{3.25}$$

$$\frac{C_A}{C_{A_0}} = \frac{1}{1 + kt} \tag{3.26}$$

In terms of fraction converted,

$$\text{in} - \text{out} = FC_{A_0} x = kC_{A_0}(1 - x)V$$

$$\frac{x}{1 - x} = kt \tag{3.27}$$

or

$$x = \frac{kt}{1 + kt} \tag{3.28}$$

Equations (3.26), (3.27), and (3.28) are equivalent, and they are used when solving for C_A, t, or x. To compare the batch and CSTR times, the equation for a batch first-order reaction, Eq. (3.3) or Eq. (3.29), is used with Eq. (3.27):

$$\ln\left(\frac{1}{1 - x}\right) = kt \tag{3.29}$$

$$\frac{t_{\text{CSTR}}}{t_{\text{batch}}} = \frac{x/(1 - x)}{\ln(1/1 - x)} \tag{3.30}$$

A few values for this ratio are given in Table 3.1. The ratio of actual reactor volumes for the CSTR and the batch reactor is smaller than the ratio of reaction times, because the total time for a batch reactor is considerably greater than the reaction time. If the CSTR is compared with a continuous pipeline reactor, or PFR, the ratio of times in Table 3.1 is the same as the ratio of reactor volumes.

Similar comparisons could be presented for second-order reactions, but several tables or plots would be needed, because most such reactions involve two reactants, with one fed in excess. The ratio of reaction times is higher than for a first-order reaction, but not much higher when there is a large excess of one reactant. For half-order reactions, there is less change in rate with conversion, and the ratio of reaction times is less than for first-order kinetics.

When batch data are presented as a plot of conversion versus time, the comparison of CSTR and batch reaction times can be made graphically without knowing the reaction order. As shown in Figure 3.5, a tangent to the conversion curve is extended to intersect the time axis. The distance from this intersection to the batch time represents the residence time for the CSTR, as shown here:

$$r = C_{A_0}\frac{dx}{dt} = C_{A_0}(\text{slope}) \tag{3.31}$$

For the CSTR,

$$FC_{A_0}x = rV = C_{A_0}(\text{slope})V \tag{3.32}$$

$$\frac{V}{F} = t_{\text{CSTR}} = \frac{x}{\text{slope}} \tag{3.33}$$

From Figure 3.5, slope $= x/t'$:
or

$$\frac{V}{F} = t_{\text{CSTR}} = \frac{x}{x/t'} = t' \tag{3.34}$$

TABLE 3.1 Relative Reaction Times for First-Order Reactions

X	kt_{CSTR}	kt_{batch}	$t_{\text{CSTR}}/t_{\text{batch}}$
0.5	1	0.693	1.44
0.8	4	1.609	2.49
0.9	9	2.303	3.91
0.95	19	3.00	6.34

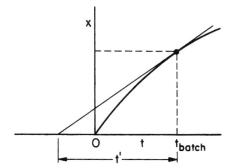

FIGURE 3.5 Determination of residence time for a CSTR from a batch test.

If the rate in the batch reaction at first increases with time and then decreases, as in Figure 3.6, the desired conversion may be reached with a smaller residence time for the CSTR. Operating at x_1 would give $t_{CSTR} < t_{batch}$, but at x_2, $t_{CSTR} > t_{batch}$. Reaction curves of this type could result from an autocatalytic reaction or a reaction with an induction period caused by an inhibitor. Similar plots would result from an exothermic reaction in an adiabatic reactor.

Reactors in Series

When a much larger reactor volume would be needed to go from batch to continuous operation, the required reactor volume can be considerably reduced by using two or more stirred-tank reactors in series. Much of the

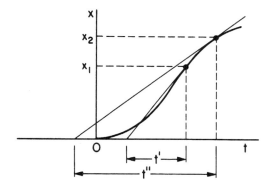

FIGURE 3.6 Conversion plot for an autocatalytic reaction.

reaction would be carried out in the first tank at moderate rate and conversion, and the reaction would be finished in the second tank, where the rate is much lower. The residence times for the tanks are shown as t_1' and t_2' in Figure 3.7. Note that $t_1' + t_2'$ is much lower than the time for a single CSTR operating at x_2. In the example t_2' is slightly greater than t_1', but in practice the tanks would probably be the same size, and x_1 would be slightly larger than shown in Figure 3.7.

Another way of comparing batch and flow reactors is to plot $1/r$ versus x, as shown in Figure 3.8. For a batch or plug-flow reactor, the time is proportional to the area under the curve, the shaded area in Figure 3.8:

$$-\int \frac{dC_A}{r} = \int dt = t \tag{3.35}$$

$$C_{A_0} \int_0^{x_2} \frac{dx}{r} = t \tag{3.36}$$

For a CSTR operating at the same conversion, the rate is constant at the final value, and from Eq. (3.38), and the residence time is proportional to the area of rectangle $abcd$:

$$FC_{A_0}x_2 = rV \tag{3.37}$$

$$C_{A_0}\left(\frac{x_2}{r}\right) = \frac{V}{F} = t \tag{3.38}$$

When the two reactors are used in series, the total volume is proportional to the sum of the rectangular area $ebgh$ and $cdhg$ in Figure 3.8. With

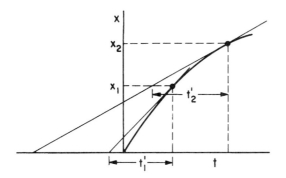

FIGURE 3.7 Conversion plot for two tanks in series.

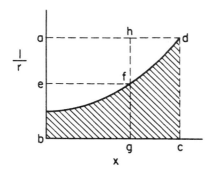

F<small>IGURE</small> **3.8**   Comparison of reactor volumes for one CSTR, two CSTRs, or a PFR.

several reactors in series, the total volume would approach that for a plug-flow reactor.

For a first-order reaction in a series of tanks, the conversion is obtained using Eq. (3.26), since the equation for the concentration ratio is the same for each reactor:

$$\frac{C_{A_1}}{C_{A_0}} = \frac{1}{1 + k_1 t_1}$$

$$\frac{C_{A_2}}{C_{A_1}} = \frac{1}{1 + k_2 t_2} \tag{3.39}$$

$$\frac{C_{A_n}}{C_{A_0}} = \frac{1}{1 + k_1 t_1} \times \frac{1}{1 + k_2 t_2} \times \cdots \times \frac{1}{1 + k_n t_n}$$

When the reactors are equal in size and operate at the same temperature, the equation is

$$\frac{C_{A_n}}{C_{A_0}} = \left(\frac{1}{1 + k_1 t_1}\right)^n \tag{3.40}$$

Figure 3.9 shows the fraction unconverted as a function of total time for one or several tanks in series. For a very large number of tanks, the conversion approaches that for a plug-flow reactor or a batch reactor. With three tanks in series, the total time is 50% more than for plug flow if the desired conversion is 90% and for five tanks the time is only 25% greater.

When a series of stirred tanks is proposed for a reaction, the question of optimum tank size comes up. Would it decrease the total reactor volume to make the first tank larger, since that is where the reaction rate is greatest, or should the last reactor be larger to compensate for the low reaction rate?

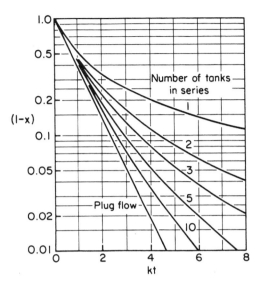

FIGURE 3.9 Conversion for a first-order reaction in a series of stirred tanks.

It is easy to show that for a first-order reaction and two tanks, the volume should be equal:

$$\frac{C_2}{C_0} = \frac{1}{(1 + kt_1)(1 + kt_2)}$$

Taking, for example, $kt_1 = kt_2 = 3$:

$$\frac{C_2}{C_0} = \frac{1}{(1 + 3)^2} = \frac{1}{16} = 0.0625$$

Any other combination with the same total time gives higher C_2/C_0. For example, if $kt_1 = 4$ and $kt_2 = 2$,

$$\frac{C_2}{C_0} = \frac{1}{(1 + 2)(1 + 4)} = \frac{1}{15} = 0.0667$$

For any number of tanks in series and a first-order reaction, the same result is found. The highest conversion is obtained with tanks of equal size. If the reaction is not first order, the maximum total volume is obtained with tanks of different size, but the decrease in total required volume is slight, and the practical solution is to specify tanks of equal size.

Example 3.2

A solution polymerization is to be carried out to 95% conversion in a series of stirred-tank reactors, all operating at the same temperature. Batch tests show that the reaction is first order to monomer, and 95% conversion is reached in 6 hours.

    a.   If four reactors of equal size are used, what total residence time is needed?

    b.   What fraction of the total heat released is generated in each vessel?

Solution

    a.   From batch tests

$$\ln\left(\frac{1}{1-x}\right) = kt$$

$$kt = \ln\frac{1}{0.05} = 3.00$$

$$k = \frac{3.0}{6.0} = 0.5 \text{ hr}^{-1}$$

With four reactors in series, $t_1 = V/F$ for each tank.

$$\frac{C_A}{C_{A_0}} = \frac{1}{(1 + k_1 t_1)^4} = 0.05$$

$$1 + k_1 t_1 = 20^{1/4}$$

$$k_1 t_1 = 1.115$$

$$t_1 = \frac{1.115}{0.5} = 2.23 \text{ hr}$$

$$4t_1 = t, \text{ total residence time} = \underline{8.92\,\text{hr}}$$

$$kt = 4.46 \quad \text{(checks interpolated value, Fig. 3.9)}$$

For tank 1,

$$\frac{C_A}{C_{A_0}} = \frac{1}{1 + 1.115} = 0.473$$

$$x_1 = 1 - 0.473 = 0.527$$

$$\text{Fraction of } Q = \frac{0.527}{0.95} = 0.555$$

Similar calculations give:

$$x_2 = 0.776 \qquad \frac{\Delta Q}{Q} = 0.262$$

$$x_2 = 0.894 \qquad \frac{\Delta Q}{Q} = 0.125$$

$$x_2 = 0.95 \qquad \frac{\Delta Q}{Q} = 0.059$$

Since over half of the heat is generated in the first reactor, care must be taken to ensure adequate capacity for heat removal.

Temperature Optimization

When a sequence of reactions produces a mixture of products, the selectivity for the main product is a major factor in choosing reaction conditions. We have shown that the ratio of reactant concentrations and the conversion can affect the selectivity, particularly when the main and byproduct reactions have different reaction orders. When the reactions have different activation energies, the selectivity will also depend on the temperature. If the main reaction has the higher activation energy, raising the temperature will increase the selectivity and also decrease the time needed to reach the desired conversion. The best operating temperature cannot be chosen from just the kinetics but depends on other factors, such as the cost of supplying or removing heat, vaporization losses, corrosion rate, and safety considerations.

When the byproduct reaction has a higher activation energy than the main reaction, the selectivity is improved by reducing the temperature. However, this means a greater reaction time for a batch reactor or a larger reactor for a flow system. The temperature chosen is again a compromise based on the reactor size, raw material costs, and the cost of product separation. However, for an existing CSTR and a fixed feed rate, an optimum temperature can be defined as the temperature that gives the greatest yield of the main product. Increasing the temperature increases the conversion but decreases the selectivity, so the yield goes through a maximum, as shown in the following example.

Example 3.3

An organic synthesis is carried out continuously in a large stirred reactor at 350 K. The main reaction is first order to A and to B and has an activation energy of 15 kcal. A large excess of B is used to favor the main reaction, and $C_{B_0} = 5\,M$ and $C_{A_0} = 1\,M$.

$$A + B \xrightarrow{1} C \qquad r_1 = k_1 C_A C_B$$

The side reaction is also first order to A, with $E_2 = 20$ kcal.

$$A \xrightarrow{2} D \qquad r_2 = k_2 C_A$$

At 350 K, the conversion of A is 88% and the yield is 81%. Could the yield be improved by changing the temperature?

Solution. At 350 K, $x = 0.88$, $C_A = 0.12$, $C_C = 0.81$, $C_B = 5 - 0.81 = 4.19$. To simplify the analysis, assume C_B is constant:

$$r_1 = \left(k_1' C_B \right) C_A = k_1 C_A$$
$$F C_{A_0} x = (k_1 C_A + k_2 C_A) V$$
$$F C_{A_0} x = C_{A_0}(1 - x)(k_1 + k_2) V$$
$$\frac{x}{1 - x} = (k_1 + k_2)\frac{V}{F} = (k_1 + k_2)t$$
$$(k_1 + k_2)t = \frac{0.88}{0.12} = 7.33$$

$$S = \frac{Y}{x} = \frac{0.81}{0.88} = 0.925 = \frac{r_1/r_2}{1 + r_1/r_2}$$

$$r_1/r_2 = 11.57 = k_1/k_2$$
$$k_1 + k_2 = 11.57 k_2 + k_2 = 12.57 k_2$$
$$k_2 t = \frac{7.33}{12.57} = 0.58$$
$$k_1 t = 6.75$$

Since V/F is constant, the changes in $k_1 V/F$ and $k_1 V/F$ with temperature are calculated from the Arrhenius equation:

$$\frac{k_1 t}{6.75} = \frac{E_1}{R}\left[\frac{1}{350} - \frac{1}{T} \right]; \frac{E_1}{R} = \frac{15000}{1.987}$$
$$\frac{k_2 t}{0.58} = \frac{E_2}{R}\left[\frac{1}{350} - \frac{1}{T} \right]; \frac{E_2}{R} = \frac{20000}{1.987}$$

At 360 K:

$$k_1 t = 12.29$$
$$k_2 t = 1.29$$

$$k_1 t + k_2 t = 13.58$$

TABLE 3.2 Solution for
Example 3.3

T, K	x	S	Y
350	0.88	0.92	0.81
360	0.931	0.905	0.843
365	0.948	0.897	0.850
370	0.961	0.887	0.853
375	0.970	0.878	0.852
380	0.977	0.868	0.848

$$x = \frac{13.58}{14.58} = 0.931$$

$$S = \frac{12.29}{13.58} = 0.905$$

$$Y = Sx = 0.843$$

The results for other temperatures are given in Table 3.2. The maximum yield is obtained at 370 K, but it is a fairly broad maximum, with almost as high a yield at 365 or 375 K.

PLUG-FLOW REACTORS

In a plug-flow reactor (PFR), elements of the fluid are assumed to pass through the reactor with no mixing, and all elements spend the same time in the reactor. For a pipeline reactor, the elements can be pictured as slugs of fluid that move through the pipe at constant velocity, somewhat like bullets moving through a gun barrel. If the flow is fully turbulent and the entrance region (where the velocity profile is developing) is a small fraction of the total length, plug flow is a reasonable assumption. There is a radial velocity profile in turbulent pipe flow, but the maximum velocity is only slightly above the average, and rapid mixing of fluid from the center with fluid from the wall region leads to a narrow distribution of residence times, which justifies the plug-flow assumption. Reactors with laminar flow or with appreciable axial mixing are discussed in Chapter 6.

With a packed-bed reactor, the velocity profile is complex and changing with distance, as the fluid flows around and between the particles. However, when the bed depth is many times the particle diameter ($L/d_p > 40$), the residence time distribution of the fluid is quite narrow, and plug flow can be assumed.

In this chapter, discussion of ideal plug-flow reactors is limited to those that operate either isothermally or adiabatically. It is not easy to get isothermal operation in a tubular reactor unless the heat of reaction is nearly zero. However, if the maximum temperature change is only a few degrees, the ideal-reactor equations are sometimes used, with rate constants evaluated at the average temperature. Adiabatic operation can more readily be achieved by using a well-insulated reactor, and the design techniques are relatively simple even when there is a large temperature change. When the reactor is neither isothermal nor adiabatic, the design must consider heat transfer rates and possible instabilities, which are discussed in Chapter 5.

Homogeneous Reactions

For a homogeneous reaction in an ideal PFR, the material balance is written for a differential volume dV or a differential length dL. The amount reacted in this element is the difference between the input and output flows of the key reactant, A, which is the molar feed rate of A times the incremental conversion:

$$F_A \, dx = r \, dV = r\left(\pi \frac{D^2}{4}\right) dl \tag{3.41}$$

The required reactor volume is found by integration:

$$\int \frac{dx}{r} = \int \frac{dV}{F_A} = \frac{V}{F_A} \tag{3.42}$$

If r is a simple fraction of x, Eq. (3.42) can be integrated directly, as was done for batch reactions [Eqs. (3.3), (3.6), (3.10)]. If r is a complex function of x, the integral can be evaluated numerically or graphically.

To compare the equations for the PFR with a batch reactor, the molar feed rate is expressed as the volume feed rate F times the concentration of A:

$$F_A = FC_{A_0} \tag{3.43}$$

Equation (3.42) can then be presented using the *space velocity* SV or the *space time* τ, which is the reciprocal of the space velocity:

$$\int \frac{dx}{r} = \frac{V}{FC_{A_0}} = \frac{1}{\text{SV}} \times \frac{1}{C_{A_0}} = \frac{\tau}{C_{A_0}} \tag{3.44}$$

where

$$SV \equiv \frac{F}{V} = \frac{\text{volumetric feed rate}}{\text{reactor volume}}$$

$$\tau \equiv \frac{1}{SV} = \frac{V}{F}$$

Other definitions of space velocity have been used, and some of these will be presented later when discussing heterogeneous reactions. For reactions in gases, SV is sometimes defined using the volumetric flow at standard temperature and pressure (STP), which is proportional to the total molar flow, but here it is based on the actual volume flow at the reactor inlet. If there is no change in flow rate as the reaction proceeds, the space time is equal to the average residence time, and this is generally the case for liquids. For gaseous reactions, there are often changes in temperature, pressure, or the number of moles, which change the volumetric flow rate, and then τ is not equal to the average residence time. However, the actual residence time is not important if the total reactor volume for a given conversion has been determined by integration of Eq. (3.42).

In Eq (3.44), the term C_{A_0} in the denominator might seem to indicate that a greater space time or reactor volume would be needed if C_{A_0} is increased. However, for a first-order reaction with no volume change, $r = kC_{A_0}(1 - x)$, and C_{A_0} cancels, giving the familiar equation

$$\int \frac{dx}{k(1-x)} = \frac{V}{F} = t$$

or, if k is constant,

$$\ln\left(\frac{1}{1-x}\right) = kt$$

Example 3.4

The gas-phase reaction $A \rightarrow B + C$ is carried out in a pilot plant tubular reactor at about 2 atm and 300°C. The rate constant is 0.45 sec^{-1}, and the feed rate is 120 cm^3/sec. The feed is 80% A and 20% inert gas (N$_2$). Neglecting the change in pressure and assuming isothermal operation, what reactor volume is needed for 95% conversion?

Solution. Per mole of feed:

$$A = 0.8(1 - x)$$
$$B = C = 0.8x$$
$$N_2 = 0.2$$
$$\text{total moles} = 1.0 + 0.8x$$

$$C_A = \frac{0.8(1-x)}{1+0.8x} C_{\text{total}} = \frac{C_{A_0}(1-x)}{1+0.8x}$$

$$r = kC_A = \frac{kC_{A_0}(1-x)}{1+0.8x}$$

$$\int \frac{dx}{r} = \int \frac{dx(1+0.8x)}{kC_{A_0}(1-x)} = \frac{V}{FC_{A_0}}$$

$$k\frac{V}{F} = \int_0^{0.95} \frac{(1+0.8x)}{1-x} = -0.8x + 1.8\ln\left(\frac{1}{1-x}\right)$$

$$k\frac{V}{F} = -0.76 + 1.8(2.996) = 4.63$$

$$V = \frac{4.63}{0.45} 120 = 1235 \text{ cm}^3, \text{ or } 1.24 \text{ liters}$$

Heterogeneous Reactions

For a heterogeneous catalytic reaction in an ideal packed-bed reactor, the material balance is written for a differential mass of catalyst, dW. The basic equation for the conversion of the key reactant A is the same as for any type of reaction, or combination of reactions, including reversible reactions. For $A \rightarrow B + C$ or $A \rightleftharpoons B + C$ or $A + B \rightarrow C + D$,

$$F_A dx = r\, dW \tag{3.45}$$

where

$$F_A = \text{moles A fed/hr}$$
$$r = \text{total moles A consumed/hr, kg}$$
$$W = \text{mass of catalyst}$$

$$\int \frac{dx}{r} = \int \frac{dW}{F_A} = \frac{W}{F_A} \tag{3.46}$$

Integration of the rate equation gives the mass of catalyst needed per unit feed rate of A for a specified conversion.

The reactor volume is determined from the mass of catalyst and the bed density, ρ_b:

$$V = \frac{W}{\rho_b} \tag{3.47}$$

The dimensions of the reactor are not fixed by these equations, and the same amount of catalyst could be held in a short, wide reactor or a tall, narrow reactor. The reactor dimensions are selected to give reasonable pro-portions and a tolerable pressure drop. Often the mass velocity is chosen first, which gives the cross-sectional area, and the bed length is determined from the required volume. If the calculated pressure drop is too high, a lower mass velocity is chosen, which gives a larger-diameter bed with shorter length and lower pressure drop per unit length.

To relate the conversion to the space velocity, the feed concentration and the bed density are introduced into Eq. (3.46):

$$\int \frac{dx}{r} = \frac{W}{F_A} = \frac{V\rho_B}{FC_{A_0}} = \frac{1}{\text{SV}}\left(\frac{\rho_b}{C_{A_0}}\right) \tag{3.48}$$

where $\text{SV} = F/V, \text{hr}^{-1}$.

It is not really necessary to use the concept of space velocity in design-ing a reactor, since the mass of catalyst needed and the bed volume are determined directly from Eqs. 3.46 and 3.47. However, some patents and technical reports give the conversion as a function of space velocity and temperature rather than presenting fundamental kinetic data. To use such data, the space velocity must be carefully defined and interpreted.

In Eq. (3.48), the space velocity is defined using the volumetric flow rate at the entrance to the reactor, but it could be based on the volume of gas at standard conditions:

$$\text{SV}' = \frac{F(\text{STP})}{V} \tag{3.49}$$

Another definition is based on the void volume of the reactor [2], which corresponds to

$$\text{SV}'' = \frac{F}{V\epsilon} = \frac{1}{\tau'} \tag{3.50}$$

Although $\frac{V\epsilon}{F}$ is closer to the gas residence time than is $\frac{V}{F}$, there is no advantage in using SV'' for calculations, and Eq. (3.50) incorrectly implies that raising ϵ would increase the conversion.

Other terms that are used when feeding liquids to a reactor are the weight hourly space velocity (WHSV) and the liquid hourly space velocity (LHSV) [3]. Both have units of hr^{-1} but are defined differently:

$$\text{WHSV} = \frac{\text{pounds of feed/hr}}{\text{pounds of catalyst}} = \frac{\rho F}{W} \tag{3.51}$$

$$\text{LHSV} = \frac{\text{volume of liquid/hr}}{\text{volume of reactor}} = \frac{F}{V} \tag{3.52}$$

The LHSV is sometimes used when feeding liquids that are vaporized in a preheater before entering the reactor, and of course the LHSV is much lower than the SV based on the actual vapor flow to the reactor. Sometimes WHSV is based on the feed rate of one reactant rather than the total feed rate, as in upcoming Example 3.5.

Even when the space velocity is clearly defined, there may be problems in scaleup or design. It might be thought that if temperature, pressure, and space velocity are kept constant on scaleup, the conversion will be constant. However, as Eq. (3.48) shows, a change in ρ_B or C_{A_0} may affect the conversion. A small-diameter laboratory reactor may have a lower bed density than a large reactor, in which case the large reactor might have a higher conversion for the same SV. Doubling C_{A_0} would double r if the reaction is first order, and the conversion would not change; but for other orders the effects of C_{A_0} would not cancel, and the conversion could change.

Example 3.5

The kinetics of benzene (B) alkylation with propylene (A) to form cumene (C) were studied in a small fixed-bed reactor using a zeolite catalyst [4]. Some of the initial rate data for 220°C and 3.5 Mpa with different feed ratios are given in Table 3.3. Integral reactor data are also given in Table 3.3 for a feed ratio of 10 and several space velocities. Note that the weight hourly space velocity is based on the propylene feed rate, not the total feed rate. The reason for the high feed ratio is to decrease formation of di-isopropyl benzene and to limit the temperature change.

1. Derive a model and a rate equation to fit the initial rate data.
2. Integrate the rate equation to predict the conversion as a function of reciprocal space velocity, and compare the predicted and experimental values.

TABLE 3.3 Data for Example 3.5: Benzene Alkylation at 220°C over MCM-22 Catalyst

a. Initial rate data		b. Integral reactor data, B/A=10	
B/A mole ratio	r_0 mol/hr,g	x	10^3/WHSV, C_3H_6
5	75	0.16	4
7	65	0.31	8.2
10	50	0.40	17
20	33	0.75	39
37	18		

Solution.

1. The initial rate data show a 4.17-fold increase in rate (75/18) as the B/A ratio goes from 37 to 5 and the mole fraction of propylene changes from 1/38 to 1/6, a 6.33-fold increase. Therefore a rate equation based on propylene adsorption is needed to fit the fractional-order behavior. The effect of benzene concentration cannot be determined from the data, since the mole fraction of benzene changes by only 14%. The kinetic data are analyzed using mole fractions instead of actually concentrations, since the reaction takes place in the liquid phase, and the small volume change can be neglected.

    Model I : $A_s + B \rightarrow C$ Eley-Rideal mechanism

    $$r_0 = \frac{kC_B K_A C_A}{1 + K_A C_A}$$

    $$\frac{C_B C_A}{r_0} = \frac{1}{kK_A} + \frac{K_A C_A}{kK_A}$$

    A plot of $C_A C_B / r_0$ vs. C_A is linear with a positive slope and intercept (Fig. 3.10):

    $$\text{intercept} = 0.0013 = \frac{1}{kK_A}$$

    $$\text{slope} = 0.0034 = \frac{1}{k} \quad k = 294$$

    $$K_A = \frac{\text{slope}}{\text{intercept}} = 2.62$$ (i)

    $$r = \frac{769 C_A C_B}{1 + 2.6 C_A}$$

    This equation fits the data with an average error of 3%. The data can also be fitted with a Langmuir–Hinshelwood equation based on reaction between adsorbed benzene and adsorbed propylene but with no term for $K_B C_B$ in the denominator:

    $$r = \frac{772 C_A C_B}{(1 + 1.18 C_A)^2}$$ (ii)

    Both equations fit the data equally well, but Eq. (i) is preferred because it is simpler.

2. For a reaction with a feed ratio of 10, the mole fractions and rate equation are

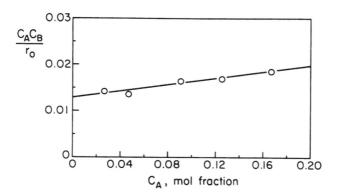

FIGURE 3.10 Test of Eley–Rideal model for benzene alkylation using initial rate data from Example 3.5.

$$C_A = \frac{1 - x}{11}$$

$$C_B = \frac{10 - x}{11}$$

$$FA \, dx = r \, dW = \frac{769(1 - x)(10 - x)dW}{(11)^2(1 + 0.236(1 - x))}$$

To simplify the integration, 9.5 is used as an average value of $(10 - x)$:

$$\int \frac{dx}{1 - x}(1.236 - 0.236x) = \int 769 \frac{(9.5)}{121} \frac{dW}{F_A}$$

$$\ln\left(\frac{1}{1 - x}\right) + 0.236x = 60.4 \frac{W}{F_A} = \frac{60.4}{\text{LHSV}}$$

The predicted values of x are shown along with the experimental values in Figure 3.11. The conversion is less than predicted at high values of 1/WHSV, which might be due to inhibition by the product or to catalyst aging. In tests at other feed ratios, the same trend was noticed—the reaction rate decreased more rapidly than predicted at high conversions.

Adiabatic Reactors

Reactions on solid catalysts are often carried out in adiabatic reactors if there is little change in selectivity or the rate of catalyst aging with tempera-

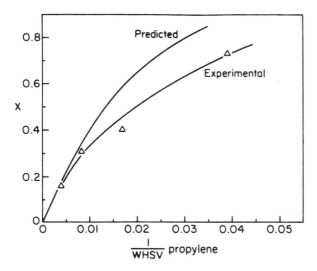

FIGURE 3.11 Integral reactor data for benzene alkylation: Example 3.5.

ture. Examples include the oxidation of sulfur dioxide, the synthesis of ammonia, and the water–gas shift reaction, which are all exothermic, and the formation of styrene from ethylbenzene, an endothermic reaction. The reactor is generally a large-diameter cylindrical vessel containing one or more beds of catalytic particles supported on grids or heavy screens, as shown in Figure 3.12a. Usually the gas or liquid is passed downward through the bed to prevent fluidization. Multiple beds with cooling or heating between stages are used when it is not possible to get high conversion in a single bed.

Another type of reactor has one or more annular beds of catalyst with radial flow of gas either inward or outward [5], as shown in Figure 3.12b. This type may be preferred when the diameter of an axial-flow reactor would be much greater than the required bed depth. By putting the same amount of catalyst in a narrower but longer reactor, the wall thickness can be reduced and the reactor cost decreased. This is particularly important for high-pressure reactions, such as the synthesis of ammonia.

The first step in reactor design is to calculate the equilibrium conversion as a function of temperature for a given pressure and feed ratio. For a bimolecular reversible reaction such as

$$A + B \rightleftharpoons C + D$$
$$\Delta G^\circ = \Delta H^\circ - T \, \Delta S^\circ = -RT \ln K_{eq} \qquad (3.53)$$

(a) **Feed**

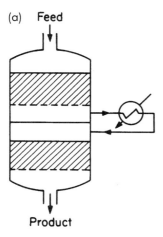

Product

(b) **Feed**

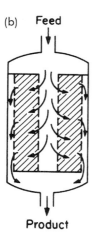

Product

FIGURE 3.12   (a) Two-stage adiabatic reactor with intercooler. (b) Radial-flow adiabatic reactor.

$$K_{eq} = \frac{P_C P_D}{P_A P_B} \tag{3.54}$$

For a specified feed composition, the partial pressures of reactants and products are calculated for several conversions of the limiting reactant, and the corresponding values of K_{eq} are computed from Eq. (3.54). Then from a plot of $\ln K_{eq}$ versus $1/T$ or the corresponding equation, the equilibrium temperature for each conversion is obtained. For an exothermic reaction, an arithmetic plot of T_{eq} versus x will be an S-shaped curve, as shown in Figure 3.13, with lower temperatures needed for high conversions. This type of plot is very useful in determining the maximum conversion possible in a single stage and the number of stages needed for high conversions.

The temperature reached in an adiabatic reactor is determined by a simple heat balance. For an exothermic reaction, the heat generation rate corresponding to conversion x is

$$Q_G = F_A x(-\Delta H) \tag{3.55}$$

At steady state, the energy released is equal to the increase in sensible heat of the feed stream, since there is no heat loss to the surroundings in an adiabatic reactor.

$$Q_s = F\rho_M \bar{c}_p(T - T_F) = Q_G \tag{3.56}$$

The heat capacity of the catalyst and the reactor wall are not included in the heat balance, since once the steady-state temperature profile is established, the solids cannot store any more energy, and all the heat released must be absorbed by the flowing gas. From Eqs. (3.55) and (3.36),

$$T - T_F = \frac{F_A x(-\Delta H)}{F\rho_M \bar{c}_p} \tag{3.57}$$

Since $F_A = FC_{A_0}$, another form of Eq. (3.57) is

$$T - T_F = \frac{C_{A_0} x(-\Delta H)}{\rho_M \bar{c}_p} \tag{3.58}$$

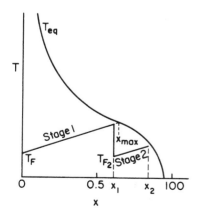

FIGURE 3.13 Equilibrium and operating temperatures for an exothermic reaction in an adiabatic reactor with intercooling.

The change in temperature with conversion is shown as a straight line in Figure 3.13, and the slope is $C_{A_0}(-\Delta H)/\rho_M \bar{c}_p$. Extending this line from T_F to the equilibrium curve gives the maximum possible conversion for a single-stage adiabatic reactor. Since the net reaction rate goes to zero at equilibrium, the actual conversion at the end of the bed is usually somewhat less than $x_{\max}$. If a higher conversion is needed, the gases are cooled in an external exchanger and sent to a second bed. The temperature change in the second bed is proportional to the increase in conversion, and the slope is generally taken to be the same as for the first bed:

$$T_2 - T_{F_2} = \frac{C_{A_0}(-\Delta H)(x_2 - x_1)}{\rho_M \bar{c}_p} \tag{3.59}$$

In some cases, three or four beds in series are used to get nearly complete conversion.

What is the justification for assuming a constant slope for the temperature-conversion lines in Figure 3.13? The heat of reaction, the heat capacities, the molar density change with temperature, and a rigorous derivation would require numerical integration of the equation for dT/dx. However, over a moderate temperature range, the change in dT/dx is small and average values can be used. The recommended approach is to evaluate ΔH at the feed temperature and calculate $\bar{c}_p$ for the product gas mixture at the average temperature in the bed. This corresponds to carrying out the reaction at T_F and using the energy released to heat the moles of product gas to the final temperature, which is thermodynamically equivalent to the actual process. Taking 100 moles of feed as a basis, with y_{A_0} mole fraction A, the heat balance can be written as

$$T - T_F = \frac{100 y_{A_0} x(-\Delta H)_{T_F}}{\sum n_i \bar{c}_{p_i}} \tag{3.60}$$

where

n_i = moles of each gas in the product
$\bar{c}_{p_i}$ = average heat capacity for $(T_F - T)$

An alternative method of operating a multistage adiabatic reactor is to send only part of the feed to the first stage and to use the rest of the feed to mix with and cool the hot gases between stages. The quenching could be done in an external loop or by using multiple mixing nozzles at the bottom of the bed or between the beds in the reactor vessel. This method avoids the cost of heat exchangers, but each quench with fresh cold feed lowers the average conversion. The temperature pattern is then of the type shown in

Figure 3.14 for a three-stage converter with two quenches. This type of converter is used for ICI's low-pressure methanol synthesis [6].

With an *endothermic* reaction, the heat balance equations still apply, but ΔH is positive, and the temperature decreases with increasing conversion. The equilibrium temperature increases as conversion increases, so the plots of bed temperature and equilibrium temperature have opposite slopes, as shown in Figure 3.15. Just as for an exothermic reaction, there is a maximum value of x for a given feed temperature and gas composition. A higher value of T_F would increase x_{max}, but too high a feed temperature can lead to rapid catalyst fouling or greater formation of byproducts.

Styrene production from ethylbenzene is an example of an endothermic reaction carried out in adiabatic reactors. The initial feed temperature is limited to 620°C for current catalysts, but the temperature may be increased with time to compensate for catalyst aging. The reactor feed has several moles of steam per mole of ethylbenzene to lower C_{A_0} and decrease the temperature change for a given conversion [see Eq. (3.58)]. The gases from the first stage are reheated and sent to a second stage to get higher conversion [7].

Optimum Reaction Temperature

For a reversible exothermic reaction, the rate increases with temperature when the mixture is far from equilibrium, but at T_{eq}, the net rate is zero, since the forward and reverse rates are equal. Therefore the reaction rate must go through a maximum at some temperature below T_{eq}. This is illustrated in Figure 3.16. The reverse reaction increases more rapidly with temperature than the forward reaction because of the differences in activation energies. The temperature for maximum rate is the optimum reaction temperature, which may be a few degrees below or quite far from T_{eq}.

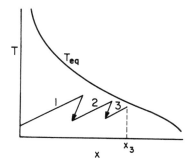

FIGURE 3.14 Temperatures in a three-stage adiabatic reactor with quench cooling.

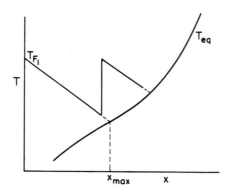

FIGURE 3.15 Temperatures for an endothermic reaction in a two-stage adiabatic reactor.

When both reactions have simple kinetics and follow the Arrhenius relationship, an equation for T_{opt} can be obtained. Consider the reaction $A + B \rightleftharpoons C + D$:

$$r_1 = k_1 P_A P_B = ae^{-E_1/RT} P_A P_B$$
$$r_2 = k_2 P_C P_D = be^{-E_2/RT} P_C P_D$$
$$r = r_1 - r_2$$

At T_{opt}

$$\frac{dr}{dT} = 0 \qquad \text{or} \qquad \frac{dr_1}{dT} = \frac{dr_2}{dT}$$

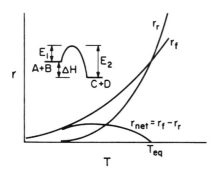

FIGURE 3.16 Net reaction rate for a reversible exothermic reaction.

At constant composition,

$$P_A P_B \frac{dk_1}{dT} = P_C P_D \frac{dk_2}{dT}$$

$$P_A P_B a e^{-E_1/RT} \left(-\frac{E_1}{R}\right)\left(-\frac{1}{T^2}\right) = P_C P_D b e^{-E_2/RT} \left(-\frac{E_2}{R}\right)\left(-\frac{1}{T^2}\right)$$

Canceling terms and replacing the exponential terms with k_1 and k_2 gives

$$P_A P_B k_1 E_1 = P_C P_D k_2 E_2$$

$$\left(\frac{k_1}{k_2}\right)_{\text{at} T_{opt}} = \left(\frac{P_C P_D}{P_A P_B}\right)\left(\frac{E_2}{E_1}\right) \tag{3.61}$$

At equilibrium,

$$k_1 P_A P_B = k_2 P_C P_D \quad \text{and} \quad \frac{P_C P_D}{P_A P_B} = K_{eq} = \frac{k_1}{k_2}$$

Therefore

$$(K_{eq})_{\text{at } T_{opt}} = \left(\frac{P_C P_D}{P_A P_B}\right)\left(\frac{E_2}{E_1}\right) \tag{3.62}$$

The optimum temperature for any conversion is found by evaluating the right-hand side of Eq. (3.62) and finding the temperature at which K_{eq} has that value. To solve directly for T_{opt}, expand both sides of Eq (3.62):

LHS : $$(K_{eq})_{T_{opt}} = \frac{k_1}{k_2} = \frac{a}{b} e^{-\frac{(E_1+E_2)}{RT_{opt}}} = \frac{a}{b} e^{-\frac{\Delta H}{RT_{opt}}}$$

RHS : $$\frac{P_C P_D}{P_A P_B}\left(\frac{E_2}{E_1}\right) = (K_{eq})_{T_{opt}}\left(\frac{E_2}{E_1}\right) = \frac{a}{b} e^{-\frac{\Delta H}{RT_{eq}}}\left(\frac{E_2}{E_1}\right)$$

Equating terms:

$$\frac{a}{b} e^{-\Delta H/RT_{opt}} = \frac{a}{b} e^{-\Delta H/RT_{eq}}\left(\frac{E_2}{E_1}\right)$$

Taking the natural logarithm:

$$-\frac{\Delta H}{RT_{opt}} = -\frac{\Delta H}{RT_{eq}} + \ln\left(\frac{E_2}{E_1}\right)$$

$$\ln\left(\frac{E_2}{E_1}\right) = -\frac{\Delta H}{R}\left[\frac{1}{T_{opt}} - \frac{1}{T_{eq}}\right] \tag{3.63}$$

For a known T_{eq}, Eq. (3.63) can be solved for T_{opt}:

Example 3.6

For a reversible reaction with $E_2 = 2E_1$, $\Delta H = -20$ kcal, and simple kinetics, what is the value of T_{opt} when $T_{eq} = 500$ K and when $T_{eq} = 700$ K?

Solution. From Eq. (3.63) with $R = 1.987$ cal/mole, K:

$$\ln(2) = \frac{20,000}{1.987}\left[\frac{1}{T_{opt}} - \frac{1}{T_{eq}}\right]$$

for $T_{eq} = 500$ K, $T_{opt} = 493$ K, $T_{eq} - T_{opt} = 17$ K
for $T_{eq} = 700$ K, $T_{opt} = 688$ K, $T_{eq} - T_{opt} = 32$ K

The difference between T_{eq} and T_{opt} varies with about the square of the absolute temperature, as can be seen by rearranging Eq. (3.63):

$$T_{eq} - T_{opt} = \ln\left(\frac{E_2}{E_1}\right)\left(\frac{R}{-\Delta H}\right)T_{opt}T_{eq} \tag{3.64}$$

Optimum Feed Temperature

For an exothermic reaction, a plot of T_{opt} versus x will lie beneath the curve for T_{eq}, and the two curves will be close together at low temperature, as shown in Figure 3.17. Much of the catalyst is at a lower-than-optimum temperature, and the rest is above the optimum temperature.

The amount of catalyst needed depends on the feed temperature and the conversion: if x_1 is fixed, it can be shown that there is an optimum feed temperature. If the feed temperature is so high that x_1 is reached only when $T \cong T_{eq}$ [line (a) in Fig. 3.17], the final rate is almost zero, and a very large mass of catalyst would be needed. Changing the feed temperature to T_{F_2} also requires a large amount of catalyst, since every part of the catalyst is

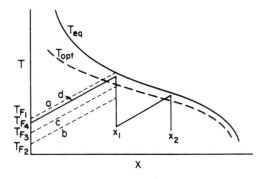

FIGURE 3.17 Optimum reaction temperature for a reversible exothermic reaction.

below the corresponding optimum temperature. Going from T_{F_2} to T_{F_3} would decrease the amount of catalyst needed, since each part of the catalyst is operating at a higher rate, and there must be an optimum feed temperature between T_{F_1} to T_{F_3}. A more detailed analysis and simple calculations show that for $T_F = T_{opt}$, the final temperature is quite close to the equilibrium value, as indicated by the solid line for case (d) and T_{F_4}. Much of the conversion takes place below the optimum reaction temperature, and for a small part of the conversion the temperature is above the optimum.

For a multistage reactor, there is an optimum feed temperature for each stage, but the intermediate conversions are also variables to be optimized. For a three-stage converter with a fixed total conversion, there are three feed temperatures and two intermediate conversions to be optimized to get the minimum amount of catalyst.

When a reactor has been built with a deeper bed of catalyst than the design amount to allow for possible increase in feed rate or for catalyst decay, the proper feed temperature may vary from the calculated value of $T_{F,opt}$. Operating with $T_F = T_{Fopt}$ may give a temperature profile like that of line (a) in Figure 3.18, where there is no temperature change in the last part of the bed because the gas is at equilibrium. In that case, a higher conversion could be reached by lowering the feed temperature, as shown by line (b), to increase the value of T_{eq}. The goal should be to maximize the temperature change across the bed, which corresponds to maximizing the conversion.

Example 3.7

A sulfuric acid plant will use the double-contact, double-adsorption (DC/DA) process [8] to get an overall SO_2 conversion of 99.5% for a feed with 11% SO_2 and 10% O_2. The first converter will operate at about 1.5 atmo-

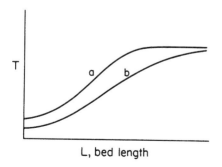

FIGURE 3.18 Temperature profiles for an adiabatic reactor. (a) Feed temperature too high, $T_{final} = T_{eq}$. (b) Lower feed temperature gives higher conversion.

spheres with three adiabatic beds and two external coolers to give a conversion of about 97%.

    a.   Plot the equilibrium temperature and the optimum reaction temperature for the first converter as a function of SO_2 conversion using the kinetic data of Eklund [9] for commercial catalyst pellets (7 mm × 30 mm).

    b.   If the conversion in the first stage is set at 0.68, what is the optimum feed temperature? How close does the exit temperature come to T_{eq}?

Thermodynamic and kinetic data are given in Tables 3.4 and 3.5.

$$K_{eq} = \frac{P_{SO_3}}{P_{SO_2} P_{O_2}^{1/2}}$$

$$\ln K_{eq} = \frac{11{,}412}{T} - 10.771$$

Solution:

$$SO_2 + \frac{1}{2}O_2 \rightleftharpoons SO_3$$

    a.   Basis: 100 moles of feed

$$SO_2 = 11(1-x) \qquad\qquad P_{SO_2} = \frac{11(1-x)P}{100 - 5.5x}$$

$$SO_3 = 11x \qquad\qquad\quad P_{SO_3} = \frac{11xP}{100 - 5.5x}$$

$$O_2 = 10(10 - 5.5x) \qquad P_{O_2} = \frac{10(10 - 5.5x)P}{100 - 5.5x}$$

$$N_2 = 79$$

$$\sum = \text{total moles} = 100 - 5.5x$$

TABLE 3.4 Thermodynamic Data for Example 3.7

Temperature	c_p, cal/mol, °C				$-\Delta H$, kcal/mol
	SO_2	SO_3	O_2	N_2	
700 K	12.17	17.86	7.89	7.35	23.27
800 K	12.53	18.61	8.06	7.51	23.08

TABLE 3.5 Kinetic Data for Example 3.7

$T°C$	420	440	460	480	500	520	540	560	580	600
$10^6 k$	2	5.1	10.3	18	27	37.5	48	59	69	77

$$r = k\left(\frac{P_{SO_2}}{P_{SO_3}}\right)^{1/2}\left(P_{O_2} - \left(\frac{P_{SO_3}}{P_{SO_2} K_{eq}}\right)^2\right); \; k = gmol/g\,cat,\,sec,\,atm$$

Let

$$\Phi = \frac{P_{SO_3}}{P_{SO_2} P_{O_2}^{1/2}} = \left(\frac{x}{1-x}\right)\left(\frac{100-5.5x}{10-5.5x}\right)^{1/2}\frac{1}{P^{1/2}}$$

For $P = 1.5$, $x = 0.6$,

$$\Phi = \frac{0.6}{0.4}\left(\frac{96.7}{6.7}\right)^{1/2}\frac{1}{1.5^{1/2}} = 4.65$$

$$\ln 4.65 = \frac{11,412}{T_{eq}} - 10.771$$

$$T_{eq} = 927\,K = 654°C$$

Repeating the calculations for other values of x gives the results shown in Table 3.6.

The optimum temperature for each conversion is found by trial, since the kinetics are complex and k does not follow the Arrhenius equation.

For $x = 0.6$,

$$P_{O_2} = \frac{6.7}{96.7}1.5 = 0.1039\,atm$$

TABLE 3.6 Equilibrium and Optimum Temperatures for Example 3.7

x	Φ	T_{eq},K	$T_{eq},°C$	$T_{opt},°C$
0.5	2.99	962	689	612
0.6	4.65	927	654	592
0.7	7.53	892	619	562
0.8	13.49	853	580	536
0.9	31.88	802	529	498
0.95	69.11	760	487	466

$$r = k\left(\frac{0.4}{0.6}\right)\left(0.1039 - \left(\frac{0.6}{0.4K_{eq}}\right)^2\right)$$

Trials show r is a maximum at $T = 592°C$. The optimum and equilibrium temperatures are shown in Figure 3.19. The difference between T_{opt} and T_{eq} is about $60°C$ at $x = 0.5$, and it gradually decreases to about $20°C$ at $x = 0.95$. The change in T_{eq}-T_{opt} is greater than for Example 3.6, because the kinetic data show a pronounced decrease in apparent activation energy as the temperature increases.

b. A feed temperature of 700 K is chosen for the first trial and ΔH at 700 K is used to calculate the heat generated. The heat capacities are evaluated at 800 K, since the temperature rise is about 200°C. For 100 moles of feed, the product gas contains

$$SO_2 = 0.32(11) = 3.52 \text{ mol}$$
$$SO_3 = 0.68(11) = 7.48 \text{ mol}$$
$$O_2 = 10 - 5.5(0.68) = 6.28 \text{ mol}$$
$$N_2 = 79 \text{ mol}$$

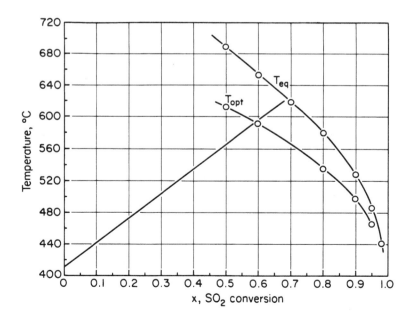

FIGURE 3.19   Temperature in Stage 1 of an SO_2 converter: Example 3.7.

$$\sum n_i c_{pi} = 3.52(12.53) + 7.48(18.61) + 6.28(8.06) + 79(7.51)$$

$$= 827 \text{ cal/}^\circ\text{C}$$

From Eq. (3.60) for $x = 0.68$,

$$T - T_F = \frac{100(0.11)(0.68)(+23,270)}{827} = 210^\circ\text{C}$$

For $T_F = 700 \text{ K} = 427^\circ\text{C}$, $T = 427 + 210 = 637^\circ\text{C}$ at $x = 0.68$, but $T_{eq} = 626^\circ\text{C}$, so a feed temperature of 427°C is too high. Try $T_F = 683 \text{ K} = 410^\circ\text{C}$. Neglecting the slight change in ΔH and using the same heat capacities, $(T - T_F)$ is still 210°C. $T = 410 + 210 = 620^\circ\text{C}$, 6°C less than T_{eq}. The maximum conversion for $T_F = 410^\circ\text{C}$ is 69%, the intersection of the T-versus-x line with the equilibrium curve.

To calculate the amount of catalyst needed for $T_F = 410^\circ\text{C}$, the rate equation is integrated graphically after using the heat balance line to get T for each x:

$$F_A \, dx = r \, dW$$

$$\int \frac{dx}{r} = \frac{W}{F_A}$$

For $x = 0$, the rate equation is indeterminate, since $P_{SO_3} = 0$, so the calculation is started at $x = 0.1$. For $x = 0.1$, $T = 441^\circ\text{C}$, and $K_{eq} = 1.83$,

$$P_{O_2} = \frac{10 - 0.55}{100 - 0.55} 1.5 = 0.143 \text{ atm}$$

By interpolation, $k \cong 5.25$:

$$10^6 r = 5.25 \left(\frac{0.9}{0.1}\right)^{1/2} \left[0.143 - \left(\frac{0.1}{0.9 \times 183}\right)^2\right] = 2.25$$

Further calculations led to the plot of $1/r$ vs. x in Figure 3.20a. The value of W/F_A, the area under the curve, is 0.196. Calculations for other feed temperatures were used to prepare a plot of W/F_A versus T_F, shown in Figure 3.20b. The optimum feed temperature is 412°C, which makes the final temperature 622°C, only 4°C below T_{eq}.

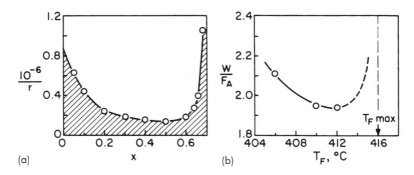

Figure **3.20** Determining optimum feed temperature for an SO_2 converter: (a) Getting W/F_A for $T_F = 410°C$. (b) W/F_A for different feed temperatures.

PRESSURE DROP IN PACKED BEDS

Integration of the rate equations gives the total mass of catalyst needed, and from the bulk density the reactor volume can then be calculated. The actual dimensions of the catalyst bed depend on the mass velocity, which must be chosen considering the pressure drop in the bed. For an axial-flow design, a high mass velocity means a small reactor diameter but a long bed, which might have quite a high pressure drop. A lower mass velocity means a larger-diameter reactor with a shorter bed and lower pressure drop, but the reactor diameter may be impractically large. A radial-flow design can be considered in place of a very large-diameter axial-flow unit.

The pressure drop in a fixed bed can be calculated from the Ergun equation [10]:

$$\frac{\Delta P}{L} = \frac{150 u_0 \mu}{\left(\phi_s d_p\right)^2} \frac{(1-\epsilon)^2}{\epsilon^3} + \frac{1.75 \rho u_o^2}{\phi_s d_p} \left(\frac{1-\epsilon}{\epsilon^3}\right) \tag{3.65}$$

The pressure drop per unit length should be predicted for inlet conditions and for the estimated outlet conditions; if these are not very different, an average value can be used to get the overall pressure drop for an approximate design. If there is a large pressure drop across the catalyst bed, Eq. (3.65) can be incorporated in the stepwise solution to determine the amount of catalyst needed and the total pressure drop.

Possible errors in the predicted pressure drop come from the strong dependence on external void fraction ϵ and the shape factor ϕ_s. The void fraction for beds of spheres or short cylinders is about 0.35 when

$d_p/D \leq 0.1$, but it increases with d_p/D. Void fractions for rings and special particle shapes may be much higher. The shape factor is 1.0 for spheres and short cylinders but is only 0.3–0.6 for rings and saddle-shaped particles [11].

NOMENCLATURE

Symbols

C	molar concentration
CSTR	continuous stirred-tank reactor
c_p	heat capacity,
$\bar{c}_p$	average heat capacity for gas mixture
D	reactor diameter
d_p	particle diameter
E	activation energy
F	volumetric feed rate
F_A	molar feed rate of reactant A
K_A	adsorption constant
K_{eq}	Equilibrium constant
k	rate constant
L, l	Reactor length
LHSV	liquid hourly space velocity
n	number of moles
P_A, P_B	partial pressure of A, B
PFR	plug-flow reactor
Q, Q_G	heat generation rate
Q_S	Sensible heat
R	gas constant, reactant ratio
r	reaction rate
S	selectivity
SV	space velocity
SV', SV''	other space velocities, Eqs. (3.49), (3.50)
T	temperature, T_F of feed, T_{eq} equilibrium value, T_{opt}, optimum temperature
t	time
u_o	superficial velocity
V	volume of reactor
W	mass of catalyst
x	fraction converted
Y	yield

Greek Letters

ΔG	free-energy change
ΔH	enthalpy change
ΔS	entropy change
ϵ	void fraction in bed
μ	viscosity
π	pi
ρ	fluid density
ρ_b	bed density
ρ_m	molar density
τ	space time $= 1/SV$
ϕ_s	shape factor

PROBLEMS

3.1   A reaction that is first order to both reactants is carried out isothermally in a well-mixed batch reactor with initial concentrations $C_{A_0} = 1.5\ M, C_{B_0} = 4.0\ M$.

$$A + 2B \rightarrow C$$
$$k = 0.024\ \text{L/mol, min}$$

    a.   What reaction time is needed for 95% conversion of A?
    b.   What average residence time is required if the same conversion is obtained in a CSTR?

3.2 Chlorine gas is bubbled through an organic liquid to produce monochloro and dichloro derivatives. The pseudo-first-order rate constants for the two reactions are 6.2 hr^{-1} and 1.6 hr^{-1}:

$$A + Cl_2 \rightarrow B + HCl$$
$$B + Cl_2 \rightarrow C + HCl$$

    a.   In a batch reaction, what is the maximum yield of B and the value of t_{max}?
    b.   What is the conversion of A and the selectivity at t_{max}?
    c.   If the reaction is stopped when the net rate of B formation is 10% of the initial rate of B formation, what would be the yield, conversion, and selectivity?

3.3   A polymerization is carried out continuously in a series of six identical stirred reactors.

a. For an overall conversion of 98%, what fraction of the monomer is converted in each stage?

b. How does total residence time compare with the reaction time for a batch reactor operating at the same conversion?

3.4 The cracking of organic compound A to give B and C is carried out continuously in two 1000-gallon reactors using a soluble catalyst. The reactors are operated in series, and both are kept at 80°C. At normal conditions, 92% of the feed forms B, 3% forms tar, and 5% is unreacted. The main reaction is first order with an activation energy of 14 kcal/mol. Lab test with pure A showed that the tar formation reaction has an activation energy of 21 kcal/mol.

Could the conversion to B be increased by changing one or both reaction temperatures? What conditions are optimum?

3.5 A free-radical polymerization is carried out in a CSTR with an average residence time of 240 min. The reaction is first order to monomer M and half order to initiator I, and the pseudo-first-order rate constant for the monomer reaction is 2.3×10^{-2} min^{-1} when the initiator concentration is 0.01 M. The decomposition rate constant for initiator is 4.1×10^{-3} min^{-1}.

a. With feed concentrations $M_0 = 4$ M and $I = 0.01$ M, what conversion is predicted for a residence time of 300 minutes?

b. About how much time would be needed in a batch reactor to reach the same conversion?

3.6 A gas-phase hydrogenation was studied in a $\frac{1}{2}$-inch lab reactor using $\frac{1}{8}$-inch catalyst pellets, and a conversion of 60% was found at a space velocity of 1250 hr^{-1}, based on the actual gas feed rate and the reactor volume. The pellet density is 1.42 g/cm^3, and the void fraction in the bed was 0.46. The superficial velocity in the lab reactor was 0.5 ft/sec.

A large reactor with many two-inch tubes will operate at the same space velocity but with a superficial velocity of 3.0 ft/sec. The estimated void fraction in the bed is 0.40. What conversion is expected if the average temperature is the same as in the lab reactor?

3.7 The H$_2$S-promoted oxidative dehydrogenation of butene was studied by Vodekar and Pasternak [12]. Typical run conditions were 970°F, 1 atm, Al$_2$O$_3$ catalyst with 20 m^2/g, a gas feed with 1H$_2$S, 0.75O$_2$, and 2.8N$_2$ per mole C$_4$H$_8$. At a butene hourly space velocity of 300 hr^{-1}, the butene conversion was 83% and the butadiene yield 44%.

a. At the given conditions, what reactor volume would be needed to produce 500 lb/hr of butadiene? (Assume plug flow and neglect temperature gradients.)

b. Estimate the average residence time of gases in the reactor, and compare with the authors' "actual contact time" of 0.8 sec.

3.8 A sulfuric acid plant will have a capacity of 2000 tons/day when the SO_2 converter feed is 11% SO_2, 10% O_2, and 79% N_2.

a. For a mass flow rate of 500 lb/hr, ft², what is the converter diameter?
b. The catalyst is 0.8-cm × 0.8-cm cylinders that pack with a void fraction of 0.40. What is the pressure drop in psi/ft in the converter? Assume $P = 1.5$ atm and $T = 500°C$ as average conditions.
c. If a flow rate of 600 lb/hr, ft² is used to reduce the vessel diameter, how much increase would there be in $\Delta P/L$ and in ΔP for the entire converter, assuming the same space velocity?
d. For a ring-type catalyst with a 4-mm hole in the 8-mm × 8-mm pellets, estimate the pressure drop per foot of bed for $G = 500$ lb/hr, ft².

3.9 Styrene is produced by the catalytic dehydrogenation of ethyl benzene (EB) in a two-stage fixed-bed adiabatic reactor. Equilibrium and heat capacity data are as given shortly.

a. Plot the equilibrium temperature as a function of conversion for a feed with 2 kg H_2O/kg EB at a total pressure of 1.5 atm.
b. If the feed enters at 620°C, what is the maximum conversion in a single-stage reactor? (Neglect the effects of byproduct formation and assume constant total heat capacity.)
c. If the reactor is operated under vacuum, and the average pressure is 0.6 atm, what is the maximum conversion in a single-stage reactor?

$$\text{EB} \rightleftharpoons \text{S} + \text{H}_2 \qquad \Delta H = 125\,\text{kJ/mol}$$
$$c_{p,\text{H}_2\text{O}} = 38.75\,\text{J/mol},\,°C \text{ at } 550°C$$
$$c_{p,\text{EB}} \cong 270\,\text{J/mol},\,°C$$
$$\ln K_{\text{eq}} = 16.12 - 15,380/T$$

3.10 In the second stage of the SO_2 converter described in Example 3.7, the conversion increased from 68% to 92%.

a. What is the maximum inlet temperature?
b. What is the recommended inlet temperature?

REFERENCES

1. RB MacMullin. Chem Eng Progr 44:183, 1948.
2. O Levenspiel. Chemical Reaction Engineering. New York: Wiley, 1962, p 101.
3. CG Hill Jr. An Introduction to Chemical Engineering Kinetics and Reactor Design. New York: Wiley, 1978, p 256.
4. A Corma, V Martinez-Soria, E Schnoeveld. J Catalysis 192:163, 2000.
5. JR Jennings, SA Ward. In: MV Twigg, ed. Catalyst Handbook. 2nd ed. London: Wolfe, 1989, p 431.
6. GW Bridger, MS Spencer. In: MV Twigg, ed. Catalyst Handbook. 2nd ed. London: Wolfe, 1989, pp 441–468.
7. DH James, WM Castor. In: Ullman's Encyclopedia of Industrial Chemistry. 5th ed. Vol A25. pp 329–344.
8. TJ Browder. Chem Eng Progr 67(5):45, 1971.
9. R-B Eklund. The Rate of Oxidation of Sulfur Dioxide with a Commerical Vanadium Catalyst. PhD dissertation, Stockholm, 1956.
10. S Ergun. Chem Eng Prog 48:89, 1992.
11. WL McCabe, JC Smith, P Harriott. Unit Operations of Chemical Engineering. 6th ed. New York: McGraw-Hill, 2000, p 161.
12. M Vodekar, IS Pasternak. Can J Chem Eng 48:664, 1970.

4

Diffusion and Reaction in Porous Catalysts

CATALYST STRUCTURE AND PROPERTIES

Solid catalysts are generally used as small particles of porous solid with most of the active sites on the internal surfaces. Metal catalysts can be prepared by impregnating porous supports such as Al_2O_3, SiO_2, and TiO_2 with an aqueous solution of a metal salt, followed by drying and decomposition or reduction to give tiny crystallites of metal deposited on the support. Examples of this type include Ni/Al_2O_3, Ag/Al_2O_3, and Pt/SiO_2. The total surface area of the support is often 100–500 m^2/g, and almost all of this surface is internal, since the external surface is less than 1 m^2/g. For a spherical particle, the external area per unit mass is

$$a_{ext} = \frac{6}{d_p \rho_p} \tag{4.1}$$

For $d_p = 0.1$ cm and $\rho_p = 2 g/cm^3$,

$$a_{ext} = \frac{6}{0.1(2)} = 30 \, cm^2/g = 0.003 m^2/g$$

Thus, if impregnation leaves metal deposited over the same fraction of external and internal surfaces of the support, a 0.1-cm particle with $S_g = 300 m^2/g$ would have 99.999% of the metal on internal surfaces.

Other catalysts are prepared as nearly pure porous metals (Raney nickel, for example) or as precipitated mixtures of oxides, sulfides, or other salts. These porous materials also have large internal surface areas and much greater catalytic activity than if they were dense solids with reaction limited to the external surface. However, active sites inside the particle are less accessible than those on the outside, and the reaction rate often depends on the rates of diffusion of reactants and products in the pores of the solid.

The rate of internal diffusion depends on the porosity of the catalyst and the size, shape, and orientation of the pores, as well as on the properties of the diffusing species. The pores are nearly always assumed to be randomly oriented cylindrical capillaries, since diffusion of gases or liquids can then be treated with simple models. However, this is not an accurate representation of pores in real catalysts. Many supports and catalysts are made of tiny crystals or nearly spherical grains fused together at the points of contact, and the pores are similar to the irregular void spaces in a packed bed. If the catalyst is made by burning or leaching out part of the solid, the pores will be irregular in cross section, and some of the pores will be dead-ended. Bottlenecks may exist between large pores, and the pore walls may be rough or smooth. Typical pore structures are shown in Figure 4.1.

A regular pore structure is found in crystalline zeolites or molecular sieves; but when these materials are used as catalysts, tiny zeolite crystals (1–2 μm) are combined with a binder to make practical-size pellets (1–5 mm). Spaces between the cemented crystals are macropores of irregular shape and size, and diffusion in these macropores has to be considered as well as diffusion in the micropores of the zeolite crystals. The cylindrical capillary model is used to describe diffusion in zeolite catalyst and other catalysts and porous solids because of its simplicity and because most of the literature values for average pore size are based on this model. However, the

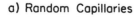

a) Random Capillaries   b) Fused Spheres   c) Irregular Pores with
                                                    Dead-end Pockets

Figure 4.1 Possible pore structures.

differences between the model and pore structures of real catalysts, such as those shown in Figure 4.1, should be kept in mind when correlating data or predicting catalyst performance.

RANDOM CAPILLARY MODEL

Although most porous solids have a distribution of pore sizes, calculations are usually based on an average pore size. The surface mean pore radius can be calculated from the surface area and the pore volume. The surface area per gram, S_g, is generally obtained from a nitrogen adsorption isotherm at 77K, using the BET method [1] to determine the amount corresponding to a monolayer of adsorbed nitrogen. Continuing the adsorption test to values of P/P_0 close to 1.0, where the pores fill by capillary condensation, gives the pore volume per gram, V_g. The pore volume can also be measured by weighing a dry pellet, soaking in water or other wetting liquid, blotting to remove excess liquid from the surface, and weighing again to get the amount of liquid taken up. The pore volume and surface area are then expressed in terms of the number of pores, the average pore length, and the average pore radius:

$$V_g = n\pi\bar{r}^2 L \tag{4.2}$$

The surface area of n cylindrical pores open at the ends would be $n\pi$ $(2\bar{r})L$ if the pore walls were smooth and there were no pore intersections. However, with a moderately high internal void fraction ($\epsilon = 0.3$–0.6) and randomly oriented pores, intersections are frequent, and the wall area of a pore is reduced where two pores intersect. Since any slice through a solid with a random pore structure will have ϵ open fraction, the surface of the cylindrical pores is assumed to be reduced by the same fraction, and the surface is proportional to $(1 - \epsilon)$. Allowing for an irregular surface of the pore walls, which increases the area by a roughness factor, r.f., the total surface per gram is

$$S_g = n\pi(2\bar{r})L(1 - \epsilon)(\text{r.f.}) \tag{4.3}$$

Taking the ratio S_g/V_g and solving for $\bar{r}$ gives

$$\bar{r} = \frac{2V_g}{S_g}(1 - \epsilon)(\text{r.f}) \tag{4.4}$$

The roughness factor has not been measured for typical porous catalysts, but it might range from 1 to 2, and since ϵ is often 0.3–0.5, the product $(1 - \epsilon)(\text{r.f.})$ could be about 1. Most authors ignore both terms and use a simpler equation for the average pore size, which would be correct for smooth-walled, nonintersecting, cylindrical pores:

$$\bar{r} = \frac{2V_g}{S_g} \qquad (4.5)$$

Equation (4.5) is used for examples in this text to be consistent with other authors, but the assumptions underlying this equation limit the accuracy of the calculations concerning diffusion and reaction in porous solids.

DIFFUSION OF GASES IN SMALL PORES

Diffusion of gases in the pores of a catalyst pellet takes place because of concentration gradients created by the chemical reactions. Consumption of reactants leads to lower reactant concentrations and higher product concentrations inside the pellet than outside the pellet. The effective diffusion coefficients for reactants and products are predicted using models of the pore structure, and the first step is estimating the diffusion coefficient for a gas in a straight cylindrical pore. If the pore diameter is much smaller than the mean free path in the gas phase (about 1000 Å for air at 200°C, 1 atm), diffusion is affected mainly by collisions of molecules with the pore wall rather than by collisions between molecules. In the limit of negligible molecule–molecule collisions, the transport process is called *Knudsen flow*. The ratio of flux to concentration gradient is the *Knudsen diffusivity*, D_k, which is proportional to the pore size and the average molecular velocity. For a cylindrical pore,

$$D_K = \frac{2}{3}r\bar{v} \qquad (4.6)$$

A convenient form of the equation is

$$D_K = 9700r\left(\frac{T}{M}\right)^{1/2} \qquad (4.7)$$

where

D_K = Knudsen diffusivity, cm^2/sec

r = pore radius, cm

T = absolute temperature, K

M = molecular weight

$\bar{v}$ = average molecular velocity, cm/sec

In the derivation for Knudsen flow, molecules striking the wall are assumed to be temporarily adsorbed and then to leave the surface at random angles (diffuse reflection). Molecules move down a pore in a series of random jumps, some shorter than the pore diameter and some much longer, and some jumps will be in the reverse direction, as illustrated in Figure (4.2).

Diffusion in the bulk gas is also a random walk process, and the similarity is brought out by the following approximate equation for binary gas diffusion:

$$D_{AB} = \frac{1}{3}\lambda\bar{v}$$ (4.8)

where

D_{AB} = bulk diffusivity for a mixture of A and B

λ = mean free path in gas phase

$\bar{v}$ = mean molecular velocity

Note that if $2r$ is taken as a reasonable estimate for the mean jump length or mean free path in a cylindrical pore, Eqs. (4.8) and (4.6) are equivalent. A table of bulk diffusivities for some pairs of gases is given in the appendix.

For pores smaller than the mean free path but not small enough to have only Knudsen flow, the diffusion flux in the pore is affected by molecule–wall collisions and molecule–molecule collisions. Considering both types of collisions as resistances to diffusion leads to the following equation for the pore diffusion coefficient:

$$\frac{1}{D_{\text{pore}}} = \frac{1}{D_K} + \frac{1}{D_{AB}}$$ (4.9)

For very small pores, D_{pore} approaches D_K, and Eq. (4.7) can be used for D_{pore}. For pores much larger than the mean free path, D_K is much greater than D_{AB}, and D_{pore} approaches D_{AB}. However, the change from

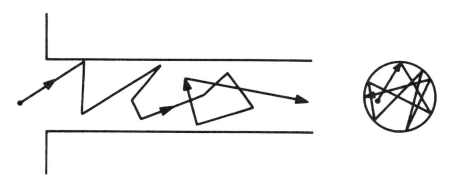

FIGURE 4.2 Knudsen diffusion in a cylindrical pore.

Knudsen flow to bulk diffusion is a gradual transition that covers about two orders of magnitude in pore size, as shown by Figure (4.3). Both terms in Eq. (4.9) are significant in many cases of industrial importance.

The effects of several variables on D_K and D_{AB} are compared in Table 4.1.

Temperature has a greater effect on D_{AB} than on D_K, because the mean free path increases with about $T^{1.0}$ in addition to the $T^{0.5}$ change in average molecular velocity. Raising the pressure decreases D_{AB}, because the increase in density decreases the mean free path. However, pressure has no effect on D_K, because bimolecular collisions are neglected in the derivation for Knudsen flow. The binary diffusivity D_{AB} depends on the velocities and diameters of both species and decreases when the molecular weight of either gas increases, though not according to a simple power function. When several gases are present, D_{AM}, the diffusivity of A in the mixture must be calculated from the diffusivities of each pair of components. However, when Knudsen diffusion dominates, D_A is independent of the properties of other gases in the mixture, which makes calculations much easier. For reactions with a change in the number of moles, any net flow into or out of the catalyst should be considered if bulk diffusion dominates, but the net flow

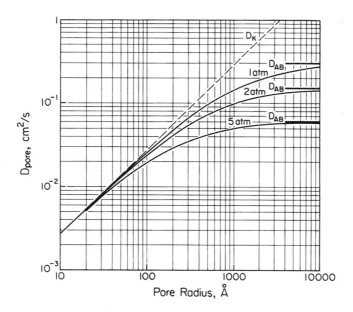

FIGURE 4.3   Transition from Knudsen flow to bulk diffusion.

TABLE 4.1 Comparison of Knudsen and Bulk Diffusion

Variable	Knudsen diffusion	Bulk diffusion
Temperature	$D_K \propto T^{1/2}$	$D_{AB} \propto T^{1.5-2}$
Pressure	No effect	$D_{AB} \propto P^{-1}$
Pore size	$D_K \propto r$	No effect
Other molecules	No effect	$D_{AB} \uparrow$ as $M_B \downarrow$
Net flow	No effect	Reverse flow of B decreases flux of A

has no effect on Knudsen diffusion. The net flow effect is neglected for examples in this chapter.

For pores only slightly larger than the size of the molecules, the diffusivity is much less than that predicted by Eq. (4.7). For example, measured diffusion coefficients of n-alkanes in zeolite crystals (molecular sieves) range from 10^{-9} to 10^{-12} cm^2/sec, compared to about 10^{-4} cm^2/sec from Eq. (4.7) [2]. The shape of the molecule is very important, and branched or cyclic molecules such as isobutane and cyclohexane are excluded from zeolites that permit entry of linear paraffins of the same or higher molecular weight. There is also a large effect of temperature, with activation energies of 3–15 kcal for hydrocarbon diffusion in zeolites.

Theories for diffusion in zeolites are not yet sufficiently developed to give reliable predictions, but experimental data are available for many systems.

EFFECTIVE DIFFUSIVITY

The pore diffusion coefficient, D_{pore}, is the flux per unit cross-sectional area normal to the pore axis divided by the concentration gradient along the axis. The pores are generally at random angles to the external surface, but the effective diffusivity D_e is based on the gradient normal to the surface (radial gradient for a spherical particle) and on the entire cross section of the particle. For a solid with internal void fraction ϵ and a random pure structure, the fraction open area for any slice through the solid is also ϵ. Therefore the effective diffusivity can be no higher than ϵD_{pore}. For a pore that is not perpendicular to the outer surface, the area normal to the pore axis is less than the open area exposed by a cut parallel to the surface. If all the pores are at random angles to the outer surface, the cross-sectional area available for diffusion is less than the open area by a factor that depends on the average angle of the pores. This same angle factor makes the average concentration gradient along the pore axis less than the gradient measured normal to the surface. These two factors are combined into a tortuosity τ, which is included in the equation for D_e:

$$D_e = D_{\text{pore}} \frac{\epsilon}{\tau} \tag{4.10}$$

The tortuosity is sometimes considered to be the average length of the tortuous path through a network of pores relative to the straight-line distance through the particle. However, because of the difference between the open area and the area normal to the pore axis, as explained previously, the tortuosity is approximately the square of the path length ratio, and τ values of 2–3 would be reasonable. Measured values of τ [found from Eq. (4.10)] generally range from 2 to 6, but values as low as 0.6 and higher than 10 have been reported. The tortuosity should be considered an empirical correction factor for random orientation of the pores and for errors in predictions of D_{pore}. Very low or very high values of τ are a warning of possible experimental error or errors in predicting D_{pore}. A very high τ could be caused by a large fraction of dead-end pores or a low-porosity skin on the pellet surface. A value of τ less than 1.0 can arise if the distribution of pore sizes is not properly accounted for.

The effective diffusivity can be measured directly by the Wicke–Kallenbach method [3]. A cylindrical pellet is forced into a short plastic sleeve connecting two gas streams so that opposite faces of the pellet are exposed to different flowing gases at the same pressure, as shown in Figure 4.4. Inexpensive gases such as N_2 and He are often used. The flow rates are adjusted to be much greater than the diffusion flux, and D_e is determined for each gas from the flow rate and exit gas analysis of the other stream and the concentration gradient. The method can also be used for a sphere if the plastic tubing is stretched to give a tight fit over much of the surface.

Another method of obtaining D_e is to measure the dispersion of a pulse of tracer gas introduced into a carrier gas passing through a packed bed of particles. Tests at different gas velocities are used to separate the broadening due to external mass transfer and axial dispersion from that

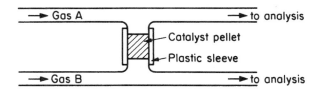

FIGURE 4.4   Wicke–Kallenbach method for measuring effective diffusivity.

caused by internal diffusion [4]. This method can be applied to particles of any shape.

Measurements of D_e are usually made at ambient conditions using simple gases such as N_2, He, H_2, and CO_2. To predict D_e for the same catalyst under reaction conditions, the effects of changes in temperature, pressure, and gas composition must be accounted for. One approach is to predict D_{pore} for the test gas (say, He) from Eqs. (4.5), (4.7), and (4.9) and to calculate τ from Eq. (4.10) using measured values of D_e and r. Then D_{pore} is predicted for the reactants at various reaction temperatures and pressures, and the same values of ϵ and τ are used to get D_e from Eq. (4.10). The relative importance of Knudsen diffusion and bulk diffusion may change with reaction conditions, but ϵ and τ should be constant.

A simpler approach is possible if either bulk diffusion or Knudsen diffusion is dominant at both ambient (reference) conditions and reaction conditions. For Knudsen flow, the effective diffusivity for any component can be calculated just from the ratio of temperatures and molecular weights, as indicated by Eq. (4.7). When He is the test gas,

$$D_{eA} = D_{eHe}\left(\frac{4}{M_A} \times \frac{T}{T_{ref}}\right)^{0.5} \tag{4.11}$$

If bulk diffusion dominates at reference conditions and reaction conditions, the ratio of bulk diffusivities is used to convert measured values to reaction conditions, or ϵ/τ is obtained from the D_e measurement at reference conditions.

Example 4.1

The gas-phase chlorination of methane is carried out using 1/4" by 1/4" cylindrical catalyst pellets. The catalyst properties are $S_g = 235$ m^2/g, $V_g = 0.29$ cm^3/g, and $\rho_p = 1.41$ g/cm^3. The effective diffusivity for He in N_2 at 1 atm at 20°C was 0.0065 cm^2/sec, measured by the Wicke–Kallenbach test.

    a. Predict D_e for Cl_2 at the typical reaction conditions of 300°C, 1.5 atm, 20% Cl_2, 80% CH_4.
    b. Calculate the tortuosity.
    c. Predict D_e for Cl_2 at 300°C, 15 atm, 20% Cl_2, 80% CH_4.

Solution. Calculate the mean pore radius using Eq. (4.5).

a. $\bar{r} = \dfrac{2(0.29)}{235 \times 10^4} = 2.47 \times 10^{-7}$ cm $= 24.7$ Å

$T = 573$ K; assume mostly Knudsen flow. At 300°C,

$$D_{e,Cl_2} = D_{e,He}\left(\frac{M_{He}}{M_{Cl_2}} \times \frac{T}{293}\right)^{0.5} = 0.0065\left(\frac{4}{70.9} \times \frac{573}{293}\right)^{0.5}$$

$$= 2.16 \times 10^{-3} \text{ cm}^2/\text{sec}$$

b. Get D_{pore} at 20°C for He.

$$D_K = 9700(2.47 \times 10^{-7})\left(\frac{293}{4}\right)^{0.5} = 2.05 \times 10^{-2} \text{ cm}^2/\text{sec}$$

$$D_{AB} = 0.73 \text{ cm}^2/\text{sec} \qquad \text{at 1 atm, 298 K [5]}$$

$$D_{AB} = 0.73\left(\frac{293}{298}\right)^{1.7} = 0.71 \text{ cm}^2/^2\text{sec} \qquad \text{at 1.5 atm, 293 K}$$

$$\frac{1}{D_{\text{pore}}} = \frac{1}{2.05 \times 10^{-2}} + \frac{1}{0.71}$$

$$D_{\text{pore}} = 1.99 \times 10^{-2} \text{ cm}^2/\text{sec}$$

$$\epsilon = V_g \rho_p = 0.29 \text{ cm}^3/\text{g} \times 1.41 \text{ g/cm}^3 = 0.409$$

$$\tau = \frac{D_{\text{pore}}\epsilon}{D_e} = \frac{1.99 \times 10^{-2}(0.409)}{0.0065} = 1.25$$

This value of τ is quite low, and it suggests that a significant fraction of the pores may be much larger than 25 Å in radius. The effect of pore size distribution is illustrated later (Examples 4.2, 4.5).

c. At 1 atm, 273 K,

$$D_{Cl_2/CH_4} = 0.15 \text{ cm}^2/\text{sec}$$

At 15 atm, 300°C,

$$D_{Cl_2/CH_4} \cong \frac{0.15}{15}\left(\frac{573}{273}\right)^{1.7} = 0.035 \text{ cm}^2/\text{sec}$$

$$D_{K,Cl_2} = 9700(2.47 \times 10^{-7})\left(\frac{573}{70.9}\right)^{0.5} = 6.81 \times 10^{-3} \text{ cm}^2/\text{sec}$$

$$\frac{1}{D_{pore}} = \frac{1}{0.035} + \frac{1}{6.81 \times 10^{-3}} = 175$$

$$D_{pore} = 5.7 \times 10^{-3} \text{ cm}^2/\text{sec}$$

$$D_e = \frac{5.7 \times 10^{-3}(0.409)}{1.25} = 1.87 \times 10^{-3} \text{ cm}^2/\text{sec}$$

At the higher pressure, bulk diffusion becomes significant and lowers the value of D_e.

PORE SIZE DISTRIBUTION

Although the average pore size is often given as one of the important physical properties of a catalyst, the distribution of pore sizes can have a major effect on the performance and should also be reported. If some of the surface area is in very small pores, large molecules can be excluded, which may or may not be desirable. Even if all the interior surface is accessible to reactants, reactions that are limited by the rate of diffusion may benefit from having a wide distribution of pore sizes even if the total surface area is thereby reduced.

Pore size distributions are routinely measured using the mercury penetration test. Liquid mercury, which does not wet most solids, is forced into the pores under successively higher pressures. The pressure required to fill a cylindrical pore of radius r is [6]

$$P = \frac{2\sigma \cos(\theta)}{r} = \frac{5.95 \times 10^{-4}}{r} \tag{4.12}$$

where

P = pressure, atm

σ = surface tension

r = pore radius, cm

θ = contact angle, 130 to 140°

This method can be used to get the pore size distribution for pores down to about 100-Å radius ($P = 594$ atm). However, the method is based on the cylindrical capillary model and does not allow for bottlenecks or for pores of irregular cross section.

The second method of obtaining the pore size distribution is based on interpretation of the adsorption isotherms for N_2 or other gases. The calculations allow for multilayer adsorption and eventually pore filling by capillary condensation [1]. Sometimes the adsorption and desorption plots show a hysteresis loop at high relative pressures, and it is not clear which branch of the isotherm should be used to calculate pore sizes. The two methods are complementary, since adsorption isotherms give data for small pores and mercury penetration is applicable to large pores. If only adsorption measurements are available, the size distribution of small pores can be calculated and the rest of the pore volume lumped together as macropores. The exact size distribution of the macropores ($r > 1000$ Å) is usually unimportant, as long as the pores are large enough to have bulk diffusion.

Data on pore size distribution can be presented as a plot of cumulative pore volume against pore radius, as shown in Figure 4.5 for pellets made from porous alumina particles. The particles have micropores of 20- to 100-Å radius that contribute about 0.4 cm^3/g to the total pore volume. The remainder of the pore volume is in macropores, the spaces between the small particles, and the macropore volume depends on the pelletizing pressure. Figure 4.5b is an alternate plot of the data, which shows more clearly the pronounced bimodal pore size distribution obtained at low pelletizing pressure. Increasing the pressure reduces the macropore volume, but it does not affect the micropores.

Where large and small pores are distributed throughout a catalyst and the pores are randomly oriented as in Figure 4.1a, a molecule diffusing into the catalyst will follow a tortuous path through many short, interconnected pore segments. A key to modeling the flux in such a solid is the recognition that pore intersections must be very frequent because of the high porosity. Therefore, the average length of the pore segments is close to the pore diameter, and since this is usually several orders of magnitude smaller than the particle size, a molecule diffusing an appreciable distance into a catalyst will pass through a great many pore segments. Although pore segments of small diameter will give higher incremental conversions for a given length than large pores (because of the greater surface area per unit volume and the lower flow into the pore), mixing will occur by molecular diffusion at pore intersections, and the reactant concentration at any distance from the external surface should be essentially the same in pores of all sizes. The radial concentration gradient for a spherical pellet or the gradient normal to the surface for a flat slab is then the driving force for diffusion into the catalyst, and gradients in individual pores do not need to be considered. The diffusion flux is the sum of the fluxes in pores of all sizes, and an average pore diffusivity is obtained by using either the pore volume or the void fraction as a weighting factor:

$$\bar{D}_{\text{pore}} = \frac{\sum D_{\text{pore},i} V_i}{\sum V_i = V_g} \tag{4.13}$$

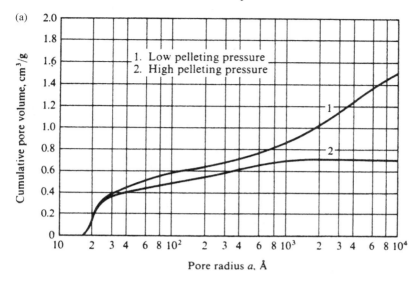

(a)

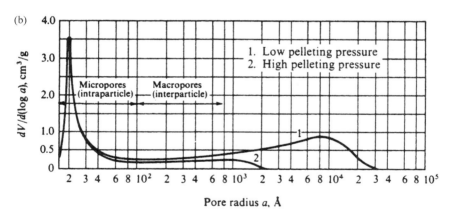

(b)

FIGURE 4.5 Pore volume in alumina pellets: (a) Cumulative pore volume. (b) Pore volume distribution. (From Ref. 7 with permission from author and publisher.)

or

$$\bar{D}_{pore} = \frac{\sum D_{pore,i} \epsilon_i}{\epsilon} \tag{4.14}$$

The effective diffusivity and flux are defined in the same way as for uniform pores:

$$D_e = \frac{\bar{D}_{\text{pore}}\epsilon}{\tau} \tag{4.15}$$

$$J_A = D_{eA}\left(\frac{dC_A}{dr}\right) \tag{4.16}$$

The method of allowing for pore size distribution is called the random intersecting pore model. A similar result for D_e was obtained based on a "parallel-path model" [8,9], but this name is misleading, since if the pores were parallel and nonintersecting, the reactant concentration gradient would be different for each pore size. An alternate model for D_e, the "random pore model" of Wakao and Smith [4,10] is based on the probability that short segments of macropores or micropores (all normal to the surface) will overlap. This model applies only to a bimodal distribution; and although it does not include a tortuosity, it gives reasonable results for moderate values of ϵ_a the macropore void fraction. Their model includes a term proportional to ϵ_a^2 and may not be valid for low or high values of ϵ_a.

To use Eq. (4.13), the pore size distribution is approximated using a small number of size intervals or by using just two or three categories; micropores $(d < 100\text{Å})$, mesopores $(100–1000\text{Å})$, and macropores $(d > 1000\text{Å})$. When the catalyst has a strongly bimodal distribution, the pore volume is generally divided into micropores and macropores. Knudsen diffusion often dominates in the micropores. The average size of the micropores must then be carefully determined, but the size of the macropores is not critical, since the bulk diffusivity does not depend on pore size. Since D_a, the diffusivity in the macropores, may be more than 10 times D_i, the micropore diffusivity, a small fraction of macropores can greatly increase the average pore diffusivity [11]. If the distribution of pore sizes is not allowed for in such a case and D_{pore} is based on $\bar{r}$, then an abnormally low value of τ may be obtained, as shown in Example 4.2.

Example 4.2

a. Predict the effective diffusivity of O_2 in air at 200°C and 1 atm for a catalyst with the following properties:

$S_g = 150 \text{ m}^2/\text{g}$

$V_g = 0.45 \text{ cm}^3/\text{g}$

$V_i = 0.30 \text{ cm}^3/\text{g}$ micropores, $\bar{r}_i = 40 \text{ Å}$

$V_a = 0.15 \text{ cm}^3/\text{g}$ macropores, $\bar{r}_a = 2000 \text{ Å}$

$\rho_p = 1.2 \text{ g/cm}^3$

$\tau = 2.5$

b. Calculate the surface mean pore radius for this catalyst and the corresponding value of D_{pore}. If this value of D_{pore} is used with the value of D_e predicted in part (a), what would be the tortuosity?

Solution.

a. $$\epsilon = V_g \rho_p = 0.45 \times 1.2 = 0.54$$

For N_2–O_2 at 1 atm, 200°C,

$$D_{AB} \cong 0.49 \text{ cm}^2/\text{sec}$$

For micropores,

$$D_K = 9700(40 \times 10^{-8})\sqrt{\frac{493}{32}} = 0.0152 \text{ cm}^2/\text{sec}$$

$$D_{pore} = \frac{1}{1/0.0152 + 1/0.49} = 0.0147 \text{ cm}^2/\text{sec}$$

For macropores,

$$D_K = 9700(2000 \times 10^{-8})\sqrt{\frac{493}{32}} = 0.761 \text{ cm}^2/\text{sec}$$

$$D_{pore} = 0.298 \text{ cm}^2/\text{sec}$$

$$\bar{D}_{pore} = \frac{0.30(0.0147) + 0.15(0.298)}{0.45} = 0.109 \text{ cm}^2/\text{sec}$$

$$D_e = \frac{\epsilon \bar{D}_{pore}}{\tau} = \frac{0.54(0.109)}{2.5} = 2.35 \times 10^{-2} \text{ cm}^2/\text{sec}$$

b. $$\bar{r} = \frac{2V_g}{S_g} = \frac{2(0.45)}{150 \times 10^4} = 6 \times 10^{-7}\text{cm} = 60 \text{ Å}$$

$$D_K = 9700(60 \times 10^{-8})\sqrt{\frac{293}{32}} = 2.28 \times 10^{-2} \text{ cm}^2/\text{sec}$$

$$D_{pore} = 2.18 \times 10^{-2} \text{ cm}^2/\text{sec}$$

$$\tau = \frac{2.18 \times 10^{-2}}{2.35 \times 10^{-2}}(0.54) = 0.50$$

A tortuosity less than 1.0 has no physical significance and indicates what can happen if the pore size distribution is not allowed for.

DIFFUSION OF LIQUIDS IN CATALYSTS

Liquid-phase catalytic reactions can be carried out using a packed bed of catalyst pellets or by suspending fine particles of catalyst in a stirred reactor or bubble column. Diffusion coefficients in liquids are generally 10^{-6}–10^{-4}cm^2/sec, three to five orders of magnitude lower than for gases at standard conditions. This makes pore diffusion effects quite important for liquid-phase reactions and may lead to the choice of powdered catalysts rather than large particles or pellets. However, even for particles as small as 20 microns, there may be significant pore diffusion effects, as in the case for some catalytic hydrogenations [12].

There is no equation comparable to the Knudsen equation for diffusion of liquids in small pores, but the pore walls do limit the movement of molecules and cause a decrease in the diffusivity. Diffusion coefficients can be predicted from the bulk diffusivity and a hindrance factor that depends on the pore size and the solute size. For moderate-molecular-weight solutes, the empirical Wilke–Chang equation is used for the bulk diffusivity [13]:

$$D_{AB} = 7.4 \times 10^{-8} \frac{T\,(xM_B)^{1/2}}{\mu}\,\frac{}{V_A^{0.6}} \tag{4.17}$$

where

$$D_{AB} = \text{diffusivity of solute A in solvent B, cm}^2/\sec$$
$$M_B = \text{molecular weight of solvent}$$
$$T = \text{absolute temperature, K}$$
$$\mu = \text{viscosity, cP}$$
$$V_A = \text{molar volume at boiling point, cm}^3/\text{mol}$$
$$x = \text{association parameter, (2.6 for H}_2\text{0,}$$
$$\text{1.9 for CH}_3\text{OH, 1.5 for C}_2\text{H}_5\text{OH,}$$
$$\text{and 1.0 for unassociated solvents)}$$

The hindrance factor for liquid-filled pores depends on the size of the solute molecule relative to the diameter of the pore. For a spherical molecule in a cylindrical pore [14],

$$\frac{D_{\text{pore}}}{D_{\text{bulk}}} = (1 - \lambda)^2\left(1 - 2.1\lambda + 2.01\lambda^2 - 0.95\lambda^3\right) \tag{4.18}$$

where

$$\lambda = \frac{d_s}{d_p}$$

ds = solute diameter

d_p = pore diameter

A simpler equation that fits most data about as well is

$$\frac{D_{\text{pore}}}{D_{\text{bulk}}} = (1 - \lambda)^4 \tag{4.19}$$

For nonspherical molecules, it is not clear what effective diameter to use in Eq. (4.18). Measurements of D_e for n-paraffins diffusing into silica-alumina beads gave the same hindrance factor for C_6, C_{10}, and C_{16}, and the data were correlated using the same d_s, the diameter of the smallest cylinder that the molecules would fit in without distortion [15]. However, other molecules that adsorbed on the surface gave much lower hindrance factors.

EFFECT OF PORE DIFFUSION ON REACTION RATE

First-Order Reaction

As the reaction rate in a catalyst particle is increased, the concentration gradients needed to bring reactants to the particle and the gradients inside the particle become larger. Usually internal diffusion is more important than external diffusion, because D_e is much less than D_{AB}, and the average diffusion distance inside the particle is greater than the external film thickness. Therefore external mass transfer effects are neglected for a preliminary analysis, and the concentrations at the particle surface are assumed to be the same as in the bulk fluid. Since mass transfer through the external film is a process in series with mass transfer and reaction inside the porous catalyst, the two effects can easily be combined later.

The equations for simultaneous pore diffusion and reaction were solved independently by Thiele and by Zeldovitch [16,17]. They assumed a straight cylindrical pore with a first-order reaction on the surface, and they showed how pore length, diffusivity, and rate constant influenced the overall reaction rate. Their solution cannot be directly adapted to a catalyst pellet, since the number of pores decreases going toward the center and assuming an average pore length would introduce some error. The approach used here is that of Wheeler [18] and Weisz [19], who considered reactions in a porous sphere and related the diffusion flux to the effective diffusivity, D_e. The basic equation is a material balance on a thin shell within the sphere. The difference between the steady-state flux of reactant into and out of the shell is the amount consumed by reaction.

Consider the volume element between r and $r + dr$, as shown in Figure 4.6. Assume a simple first-order irreversible reaction, and assume the temperature is uniform so that D_e and k are constant. The effect of internal and external temperature gradients is discussed in the next chapter.

The reaction rate is expressed *per unit volume of pellet*, which makes the units of k sec^{-1}.

$$A \longrightarrow B + C$$
$$r = kC$$

where

$$r = \text{moles A/sec, cm}^3\text{pellet}$$
$$C = \text{moles A/cm}^3\text{gas}$$
$$k = \text{sec}^{-1}(\text{moles A reacting/sec, cm}^3 \text{ pellet, moleA/cm}^3 \text{ gas})$$

The material balance is

in by diffusion − out by diffusion = amount reacting

$$4\pi(r + dr)^2 D_e\left(\frac{dC}{dr} + \frac{d^2C}{dr^2}\,dr\right) - 4\pi r^2 D_e\left(\frac{dC}{dr}\right) = 4\pi r^2 dr\ kC \qquad (4.20)$$

Neglecting terms with $(dr)^2$ and rearranging gives

$$\frac{2}{r}\left(\frac{dC}{dr}\right) + \frac{d^2C}{dr^2} = \frac{kC}{D_e} \qquad\qquad (4.21)$$

The boundary conditions are

$$C = C_s \qquad \text{at } r = R$$
$$\frac{dC}{dr} = 0 \qquad \text{at } r = 0 \text{ (no diffusion past the center)}$$

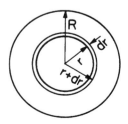

FIGURE 4.6   Model for diffusion and reaction in a sphere.

Integration of Eq. (4.21) shows that the concentration profile depends on the distance ratio (r/R) and a dimensionless parameter ϕ called the Thiele modulus:

$$C = C_s \frac{\sinh(\phi r/R)}{r/R \, \sinh(\phi)} \tag{4.22}$$

$$\phi = R\left(\frac{k}{D_e}\right)^{1/2} \tag{4.23}$$

Concentration profiles are plotted in Figure (4.7) for several values of ϕ. For $\phi = 1$, the concentration drops to $0.85 C_s$ at the center, which means 15% lower reaction rate. However, most of the pellet has a concentration much closer to C_s, and taking $r/R = 0.8$ as an average distance (since this divides the sphere into two nearly equal parts), the average concentration is about $0.94 C_s$, giving a 6% reduction in rate compared to that at the surface. A value of $\phi = 1$ is generally taken to indicate the onset of diffusion limitations.

For $\phi = 3$, the concentration decreases to $0.3 C_s$ at the center, and the profile suggests a moderate effect of diffusion on the average rate. For $\phi \geq 5$, the concentration at the center is almost zero, and the reaction rate will be very low in the central part of the catalyst. The average reaction rate could be found by integrating the concentration profile, but a simpler approach is based on the diffusion flux. The amount of reactant entering

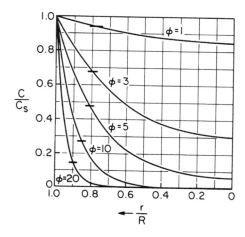

FIGURE 4.7 Reactant concentration gradients in a sphere for a first-order reaction.

the particle must equal that consumed at steady state, so Eq. (4.22) is differentiated to get the flux.

$$\frac{dC}{dr} = \frac{C_s(\phi/R)\cosh(\phi r/R) - (C_s/r)\sinh(\phi r/R)}{(r/R)\sinh(\phi)} \tag{4.24}$$

The derivative is evaluated at the pellet surface:

$$\left(\frac{dC}{dr}\right)_R = \frac{C_s}{R}\left(\frac{\phi\cosh(\phi) - \sinh(\phi)}{\sinh(\phi)}\right) = \phi\frac{C_s}{R}\left(\frac{1}{\tanh(\phi)} - \frac{1}{\phi}\right) \tag{4.25}$$

The overall rate is the diffusion flux into the pellet:

$$\text{rate per pellet} = 4\pi R^2 D_e\left(\frac{dC}{dr}\right)_R \tag{4.26}$$

$$= 4\pi R^2 D_e\phi\frac{C_s}{R}\left(\frac{1}{\tanh(\phi)} - \frac{1}{\phi}\right) \tag{4.27}$$

The overall rate per pellet (or per unit volume or mass of pellet) divided by the rate that would exist if there were no diffusion limitations is defined as the effectiveness factor, η:

$$\eta = \frac{\text{overall reaction rate}}{\text{reaction rate based on external conditions}} \tag{4.28}$$

For the first-order case,

$$\eta = \frac{4\pi R D_e\phi C_s\left(\frac{1}{\tanh(\phi)} - \frac{1}{\phi}\right)}{\frac{4}{3}\pi R^3 k C_s} \tag{4.29}$$

$$\eta = \frac{3D_e\phi}{R^2 k}\left(\frac{1}{\tanh(\phi)} - \frac{1}{\phi}\right) \tag{4.30}$$

since $\phi^2 = R^2 k/D_e$,

$$\eta = \frac{3}{\phi}\left(\frac{1}{\tanh(\phi)} - \frac{1}{\phi}\right) \tag{4.31}$$

Equation (4.31) gives the effectiveness factor for a first order, irreversible, isothermal reaction in a sphere. The actual reaction rate is the effectiveness factor times the ideal rate, the rate for reactant concentration throughout the pellet equal to the concentration at the external surface:

$$r = \eta k C_s \tag{4.32}$$

where

$r = \text{moles/sec, cm}^3$ pellet

$k = \text{moles/sec, cm}^3$ pellet, $\text{mole/cm}^3 = \text{sec}^{-1}$

For gas-phase reactions, it is often more convenient to base the rate on a unit mass of catalyst and to use partial pressure as the driving force for reaction. The rate equation then becomes:

$$r = \eta k P_A \qquad (4.33)$$

where

$r = \text{moles/sec, g catalyst}$

$k = \text{moles/sec, g, atm}$

Primes or subscripts could be used to identify rate constants based on a unit mass of catalyst or on a unit volume of catalyst and those based on partial pressure or on concentration of reactant, but this notation would be cumbersome and perhaps confusing. In most cases, a simple k denotes the kinetic constant, but the units must be carefully checked for consistency. In the Thiele modulus, k must be expressed in sec^{-1}, [Eq. (4.23)] but once ϕ and η are evaluated, other definitions of k can be used, as in Eq. (4.33).

Values of η can be obtained from Eq. (4.31) or from the plot in Figure 4.8, where the solution for first-order kinetics is the middle one of the three solid lines. For $\phi = 1.0, \eta = 0.94$, and this marks the onset of significant diffusion effects. For $\phi < 1$, the diffusion effect can be neglected for design calculations. For the intermediate region, $1 < \phi < 5$, there is a small to moderate effect of diffusion. For $\phi > 5$, η varies almost inversely with ϕ, since $\tanh(\phi)$ approaches 1.0 and η approaches $3/\phi$. In this region, the reaction is sometimes said to be diffusion controlled, but this could be misleading, since the rate also depends on the kinetic rate constant. However, the term *pore diffusion limitation* is generally used for this region.

For $\phi > 10, \eta \cong 3/\phi$,

$$r = \eta k C_s = \frac{3kC_s}{R\sqrt{k/D_e}} = \frac{3C_s}{R}\sqrt{kD_e} \qquad (4.34)$$

Thus the rate varies with the square root of the rate constant and the square root of the diffusivity, and there is no single controlling step.

To illustrate how both D_e and k influence the rate when ϕ is large, consider an example where $\phi = 10$ and the concentration profile is as shown in Figure 4.7. The horizontal bars on the curves mark the effectiveness factor or average internal concentration ratio. If D_e is increased fourfold by changing the pore structure without changing k, ϕ would be 5, and the

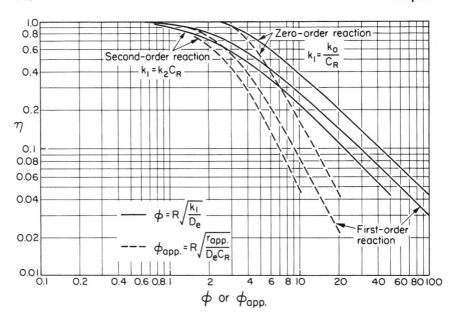

FIGURE 4.8   Isothermal effectiveness factors for spheres.

concentration gradient at the surface would be only about half as steep. Then the actual flux (and reaction rate) would increase only about twofold in spite of the fourfold increase in D_e. The exact factor of increase would be the ratio of η values, or $0.48/0.27 = 1.78$. If k was increased fourfold to make $\phi = 20$, the reaction rate per particle would go up only about twofold, because the faster consumption of reactant requires a steeper concentration profile so that diffusion can keep up. As shown in Figure 4.7, the reaction takes place closer to the surface for $\phi = 20$, and the average concentration decreases to about half the original value. The net change in reaction rate is $4 \times 0.14/0.27 = 2.1$.

In the region of strong pore diffusion effects, the rate varies inversely with particle size, as shown by Eq. (4.34). Experimental tests with different particle sizes are often used to check for pore diffusion limitations. If the rate varies with R^{-1}, the effectiveness factor is low, but the value of η cannot be determined. If the rate increases less than twofold when R is halved, the data can perhaps be fitted to the appropriate curve in Figure 4.8 to determine ϕ and η for both sizes and thus obtain the true kinetic constant. However, more accurate values of k, ϕ, and η are obtained when crushed catalyst is tested and the particle size is reduced until there is no further increase in rate as R is decreased. The system is then free of diffusion effects, and $\eta = 1.0$. For other sizes, η can be obtained directly from the ratio of reaction rates.

Although crushed catalyst or very small pellets can be used in lab tests to get intrinsic kinetic data, much larger pellets may be desired in the commercial reactor. For fixed-bed reactors, the lower pressure drop, better heat transfer, and lower cost per pound of large pellets must be considered as well as the effectiveness factor. If 1/8" pellets give $\phi = 0.5$ and $\eta = 0.98$, increasing d_p to 1/4", with $\phi = 1.0$ and $\eta = 0.94$, might give savings in operating cost that would more than offset the need to use slightly more catalyst.

Effect of Temperature

In the region of strong pore diffusion limitations, the reaction rate will increase less rapidly with increasing temperature than if diffusion effects are absent. If the rate constant follows the Arrhenius relationship and Knudsen diffusion dominates, the apparent activation energy is about half the true value:

$$r \propto \sqrt{kD_e}$$
$$k = Ae^{-E/RT}, \qquad D_e \propto T^{1/2}$$

At high ϕ,

$$r \propto e^{-E/2R}T^{1/4} \tag{4.35}$$

When kinetic data are taken over a very wide range of temperatures, the plot of $\ln(k)$ vs. $1/T$ may show a straight line at low temperatures (slope $= E/R$) followed by a gradual transition to another straight line with a slope half as great. Such a plot would be good evidence for pore diffusion limitations at the higher temperatures. A plot with a slight curvature does not prove that pore diffusion is becoming important, since reactions with complex kinetics may not follow the Arrhenius equation; with several rate constants and adsorption constants all changing with temperature, the controlling step may change with temperature, or no one step may be controlling. Also, a low value of the apparent activation energy does not indicate pore diffusion limitation unless the true activation energy has been clearly established by other tests with fine particles of the same catalyst.

As the reaction rate increases in the region of low effectiveness factors, external mass transfer becomes important and eventually controls the overall rate. Since bulk diffusivity in gases depends on $T^{1.5-2.0}$, the apparent activation energy will show a further decrease, and the plot of rate vs. $1/T$ will be as shown in Figure 4.9. Sometimes the transition from the pore diffusion region to the external mass transfer region occurs over a small

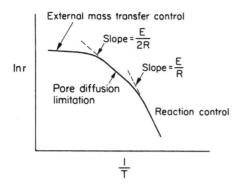

FIGURE 4.9   Effect of temperature on reaction in porous catalysts.

temperature range, and the rate plot is a smooth curve with no straight line portions.

Other Reaction Orders

For a *second-order* reaction, the basic equation is similar to Eq. (4.21) but with the right-hand term replaced with k_2C^2/D_e. The Thiele modulus is based on k_2C_s, which has the units of a first-order rate constant:

$$\phi_2 = R\sqrt{\frac{k_2C_s}{D_e}} \tag{4.36}$$

The concentration profiles are similar to those in Figure 4.7, and the effectiveness factors are plotted in Figure 4.8. For a given ϕ, the effectiveness factor for second-order is lower than for a first-order reaction, since a decrease in average concentration has more effect for the higher-order reactions. The solution for second-order reactions applies only where the rate is proportional to the square of the concentration of a single reactant, which is rarely the case. For a bimolecular reaction with a rate proportional to both reactant concentrations, $r = k_2C_AC_B$, the solution for second-order kinetics can be used if the initial concentrations are equal and if the diffusivities of A and B are nearly the same. If reactant B is in considerable excess, the reaction could be considered pseudo-first-order with $k = k_2C_B$ and Eq. (4.31) used for a reasonable approximation of η.

Some heterogeneous reactions are nearly *zero order* because of strong reactant adsorption. The basic equation for pore diffusion with a zero-order reaction is

$$\frac{2}{r}\frac{dC}{dr} + \frac{d^2 C}{dr^2} = \frac{k_0}{D_e} \qquad (4.37)$$

Concentration profiles develop inside the pellet, but there is no change in reaction rate as long as any reactant is present. The concentration profile when reactant is present throughout the pellet is

$$C = C_s \left[1 - \frac{\phi_0^2}{6}\left(1 - \frac{r}{R} \right)^2 \right] \qquad (4.38)$$

$$\phi_0 = R\sqrt{\frac{k_0}{C_s D_e}} \qquad (4.39)$$

Note that k_0/C_s has the units of a first-order rate constant.

As ϕ_0 increases, the concentration at the center decreases and becomes zero at a critical value of ϕ. The critical modulus is $\phi_0 = \sqrt{6}$, which corresponds to $C = 0$ at $r = 0$. For higher values of ϕ_0, Eq. (4.38) cannot be used, since it gives negative values for C. With a zero-order reaction and $\phi_0 > \sqrt{6}$, C becomes zero at a radius r_0, and there is no reaction in the particle from this point to the center. The value of ϕ_0 for a fixed r_0 is given by the equation

$$\left(\frac{r_0}{R} \right)^3 - \frac{3}{2}\left(\frac{r_0}{R} \right)^2 + \frac{1}{2} = \frac{3}{\phi_0^2} \qquad (4.40)$$

The effectiveness factor is the fraction of the sphere volume where reaction takes place:

$$\eta_0 = 1 - \left(\frac{r_0}{R} \right)^3 \qquad (4.41)$$

The effectiveness factor curve for a zero-order reaction is included in Figure (4.8). The curve lies about 40% above that for a first-order reaction and has a slope of -1 at high values of ϕ_0. Note that for high values of ϕ_0, the reaction rate varies with $C_s^{0.5}$, and the reaction appears to be 0.5 order.

A unimolecular reaction that follows Langmuir–Hinshelwood kinetics with the reactant strongly adsorbed might be nearly zero order at high concentrations and first order at low concentrations. The effectiveness factor would fall between the values predicted for zero- and first-order reactions. Curves for a few simple Langmuir–Hinshelwood models have been published [20, 21], but often an approximate value obtained by interpolation between the limiting cases is satisfactory for estimating the effectiveness factor.

Reversible Reactions

When a reversible reaction takes place in a porous catalyst, the product concentration inside the particle exceeds the external concentration, and this makes the effectiveness factor lower than if the reaction was irreversible and depended only on the reactant concentration. A simple treatment is possible if both forward and reverse steps are first order:

$$A \underset{2}{\overset{1}{\rightleftharpoons}} B \tag{4.42}$$

$$r = k_1 C_A - k_2 C_B \tag{4.43}$$

Since $k_1/k_2 = K_{eq}$,

$$r = k_1 \left(C_A - \frac{C_B}{K_{eq}} \right) \tag{4.44}$$

The total concentration of A + B is a constant, C_T:

$$C_B = C_T - C_A \tag{4.45}$$

$$r = k_1 \left[C_A - \frac{C_T - C_A}{K_{eq}} \right] \tag{4.46}$$

$$r = k_1 \left[\frac{C_A(K_{eq} + 1)}{K_{eq}} - \frac{C_T}{K_{eq}} \right] \tag{4.47}$$

The term C_T is replaced using C_A^*, the final equilibrium value for the mixture:

$$K_{eq} = \frac{C_B^*}{C_A^*} = \frac{C_T - C_A^*}{C_A^*} = \frac{C_T}{C_A^*} - 1 \tag{4.48}$$

$$r = k_1 \left[C_A \left(\frac{K_{eq} + 1}{K_{eq}} \right) - \frac{(K_{eq} + 1)C_A^*}{K_{eq}} \right] \tag{4.49}$$

$$r = k_r(C_A - C_A^*) \tag{4.50}$$

where

$$k_r = k_1 \left(\frac{K_{eq} + 1}{K_{eq}} \right) \tag{4.51}$$

Equation (4.50) shows that the driving force for reaction is $(C_A - C_A^*)$; and since C_A^* is constant, the equation for diffusion and reaction in a sphere becomes

$$D_e \left[\frac{d^2(C_A - C_A^*)}{dr^2} + \frac{2}{r} \frac{d(C_A - C_A^*)}{dr} \right] = k_r(C_A - C_A^*) \qquad (4.52)$$

The concentration profile given by Eq. (4.22) and the effectiveness factor of Eq. (4.31) are applicable, but the rate constant in the Thiele modulus is k_r, not k_1. For $K_{eq} = 1$, $k_r = 2k_1$ and the effect of pore diffusion is significantly greater than if only the forward reaction is considered.

For more complex reversible reactions, such as the oxidation of SO_2 or the synthesis of NH_3, numerical calculations are needed to determine η, because the kinetics are complex and the diffusivities of reactants and products are different.

Determining Effectiveness Factors from Apparent Rate Constants

It is often necessary to determine whether experimental rate data were influenced by pore diffusion effects and, if so, what the effectiveness factor was. If the reaction rate and order are known, it might seem a straightforward procedure to calculate ϕ and η using these data and an estimated value of D_e. However ϕ is defined based on the true rate constant; if the calculated ϕ indicates η is less than 1.0, the measured rate constant is not the true rate constant, and the values of ϕ and the corresponding η are also incorrect. The measured rate constant can then be adjusted by dividing by η to get a new, larger ϕ. The new ϕ leads to a still lower estimate for η, which requires a further correction of the rate constant. Carrying out this procedure several times gives final values of k, ϕ, and η.

Instead of the trial-and-error correction procedure, a simple method permits η to be determined directly from a new set of plots. The rate constant and modulus calculated directly from experimental data are called apparent values, k_{app} and ϕ_{app}, and these parameters are equal to the true values only if $\eta = 1.0$. For any reaction order, k_{app} is the rate per unit external concentration and has the units sec^{-1}.

$$k_{app} = \frac{\text{rate per unit pellet volume}}{\text{external concentration}} = \frac{r}{C_s} \qquad (4.53)$$

The apparent and true values of k are related by η:

$$k_{app} = k\eta \qquad (4.54)$$

$$\phi_{app} = R \left(\frac{k_{app}}{D_e} \right)^{1/2} = R \left(\frac{k\eta}{D_e} \right)^{1/2} \qquad (4.55)$$

$$\phi_{app} = \eta^{1/2} R \left(\frac{k}{D_e} \right)^{1/2} = \eta^{1/2} \phi \tag{4.56}$$

For reaction orders of 0, 1, and 2, values of η were obtained for a range of ϕ values, and values of ϕ_{app} were calculated from Eq. (4.56). The plots of η vs. ϕ_{app} are shown as dashed lines in Figure (4.8). The dashed lines converge to the corresponding solid lines as η approaches 1, since then $\phi \cong \phi_{app}$. At large values of ϕ_{app}, the lines have slopes of -2, and the difference between ϕ and ϕ_{app} can become quite large. In using Figure (4.8) with experimental data, k_{app} is calculated from Eq. (4.53) using the measured rate and the external concentration of the reactant, or the key diffusing species if more than one reactant is involved. The ϕ_{app} is calculated from k_{app}, R, and D_e for the key species. If the reaction is first order, η is obtained directly from the middle dashed line. The true k and ϕ can then be calculated and η obtained from the solid line in Figure 4.8 as a check. If the reaction has an apparent order less than 1 or greater than 1, the dashed line for the zero- and second-order cases can be used along with that for first order to give a range of possible values for η. When pore diffusion limits the reaction rate, the apparent order also changes unless the reaction is first order, and determining the exact value of η is difficult but perhaps not necessary.

Example 4.3

The reaction $A \longrightarrow B + C$ was studied in a differential reactor using 1/4" spherical catalyst pellets. At 150°C and 1.2 atm, the reaction rate was 7.6×10^{-3} mol/hr,g with 10% A and 90% N_2 and 14×10^{-3} mol/hr,g with 20% A and 80% N_2 in the feed. The estimated diffusivity of A is $D_e = 0.0085$ cm^2/sec, and the pellet density is 1.4 g/cm^3. Did pore diffusion influence the rate?

Solution. From gas laws for 10% A:

$$C_A = \frac{0.10}{22,400} \times \frac{273}{423} \times \frac{1}{1.2} = 2.4 \times 10^{-6} \text{ mol/cm}^3$$

$$k_{app} = \frac{7.6 \times 10^{-3}(1.4)}{2.4 \times 10^{-6}} \frac{1}{3600} = 1.23 \text{ sec}^{-1}$$

$$R = 0.125 \times 2.54 = 0.3175 \text{cm}$$

$$\phi_{app} = 0.3175 \sqrt{\frac{1.28}{0.0085}} = 3.82$$

The rate increased 1.84 times when C_A was doubled, so the order is quite close to 1. Use the curve for ϕ_{app} and a first-order reaction in Figure 4.8:

$$\eta = 0.42$$

$$k = \frac{1.23}{0.42} = 2.93 \text{ sec}^{-1}$$

Check:

$$\phi = 0.3175 \sqrt{\frac{2.93}{0.0085}} = 5.89 \quad \eta = 0.42$$

The pore diffusion effect decreased the rate about 60%.

Complex Reactions in Porous Catalysts

When the reaction rate depends on the concentrations of several species, or when more than one reaction is involved, analytical solutions of the pore diffusion equations are impossible or too complicated to be useful. The equations for simultaneous diffusion and reaction of several species can be solved numerically if concentrations at the center are specified, but then many cases must be solved to match given external concentrations. For such cases, a simplified method can be used instead to show the approximate effect of gradients for each species.

The first step is to determine which reactant is most important as far as possible diffusion effects are concerned. The effectiveness factor for that reactant is calculated, usually assuming pseudo-first-order kinetics. The gradients for other reactants and products are then estimated and used to correct the predicted rate and effectiveness factor.

Consider the following general reaction:

$$A + bB \longrightarrow cC + dD \tag{4.57}$$

The flux of B into the catalyst is b times that of A, and the fluxes are proportional to the diffusivities:

$$J_A = D_{eA}\left(\frac{dC_A}{dr}\right) = \frac{1}{b} D_{eB}\left(\frac{dC_B}{dr}\right) \tag{4.58}$$

[No term for net volumetric flow is included in Eq. (4.58), which is permissible if Knudsen diffusion predominates.]

Integration of Eq. (4.58) shows that the ratio of concentration changes for any distance into the pellet is the stoichiometric coefficient times the diffusivity ratio:

$$\frac{\Delta C_B}{\Delta C_A} = b\left(\frac{D_{eA}}{D_{eB}}\right) \tag{4.59}$$

Converting to partial pressures for gas-phase reactions, Eq. (4.59) can be rewritten to show the relative partial pressure differences:

$$\Delta P_B = \Delta P_A b\left(\frac{D_{eA}}{D_{eB}}\right) \tag{4.60}$$

The ratio of fractional pressure changes is

$$\frac{\Delta P_B/P_B}{\Delta P_A/P_A} = b\left(\frac{D_{eA}}{D_{eB}}\right)\left(\frac{P_{A_s}}{P_{B_s}}\right) \tag{4.61}$$

If the reaction is approximately first order to both A and B, the Thiele modulus should be based on the reactant with the higher fractional change in pressure. For example, if $b = 2$, $P_{B_s} = 2.5P_{A_s}$, and $D_{eA} = 1.7D_{eB}$, the ratio in Eq. (4.61) is $2 \times 1.7/2.5 = 1.36$, so the modulus should be based on D_{eB} and k calculated from C_{Bs}. When the reaction rate is fast enough to cause a 10% change in P_A, the change in P_B will be 13.6%. The reaction orders should also be considered, but it is difficult to include this in a simple formula, and, also, the effective order often changes with concentration when Langmuir–Hinshelwood kinetics apply. For the case when the reaction is nearly first order to A and zero order to B, the modulus is usually based on A, since diffusion of B could limit the rate only for values of ϕ large enough to make $C_B = 0$ at some point in the pellet.

If A is the limiting reactant and ϕ has been evaluated using D_{eA} and k based on C_{A_s}, then η_A, the effectiveness factor based on A, is used to get a first estimate of $\bar{C}_A$ and ΔC_A, or ΔP_A:

$$\bar{C}_A = \eta_A C_{A_s} \tag{4.62}$$

$$\Delta C_A = C_{A_s} - \bar{C}_A = (1 - \eta)C_{A_s} \tag{4.63}$$

$$\Delta P_A = (1 - \eta)P_{A_s} \tag{4.64}$$

It may help to think of $\bar{C}_A$ or $\bar{P}_A$ as the concentration or partial pressure at a typical point inside the catalyst pellet, which is $r/R \cong 0.8$ for $\phi \leq 5$, as shown by Figure (4.7). This radius divides the sphere into two nearly equal volumes. The change in partial pressure for B is calculated from Eq. (4.60), and a similar calculation gives ΔP_C and ΔP_D, where product pressure differences are taken as positive:

$$\Delta P_C = \Delta P_A c\left(\frac{D_{eA}}{D_{eC}}\right) \tag{4.65}$$

$$\Delta P_D = \Delta P_A d\left(\frac{D_{eA}}{D_{eD}}\right) \tag{4.66}$$

The average pressures in the particles are:

$$\bar{P}_A = \eta P_{A_s} = P_{A_s} - \Delta P_A \tag{4.67}$$

$$\bar{P}_B = P_{B_s} - \Delta P_B$$
$$\bar{P}_C = P_{C_s} + \Delta P_C \tag{4.68}$$

$$\bar{P}_D = P_{D_s} + \Delta P_D$$

With these average partial pressures, the actual reaction rate can be predicted and compared with the ideal rate based on $P_{A_s}, P_{B_s}, P_{C_s},$ and P_{D_s} to get a new value for η. If this differs significantly from η_A, the calculated rate can be divided by C_{A_s} to get new estimates of ϕ and η_A, and the procedure is repeated until convergence is obtained. Example (4.4) illustrates the procedure for a case where the major effect of pore diffusion comes from the buildup of product in the particle combined with strong product inhibition.

Example 4.4

The solid catalyzed reaction $A + B \longrightarrow C$ is carried out at 4 atm and 150°C. Laboratory tests with crushed catalyst led to the following kinetic equation:

$$r = \frac{2.5 \times 10^{-5} P_A P_B}{(1 + 0.1 P_A + 2 P_C)^2} \text{ gmol/sec, cm}^3$$

The catalyst being considered for the plant reactor comes in 0.6-cm spheres, and the effective diffusivities are estimated to be:

$$DeA = 0.02 \text{ cm}^2/\text{sec}$$
$$D_{eB} = 0.03 \text{ cm}^2/\text{sec}$$
$$D_e C = 0.015 \text{ cm}^2/\text{sec}$$

For a feed with 30% A and 70% B, estimate the effectiveness factor at the start of the reaction and at 50% conversion.

Solution. Calculate the ideal rate and a pseudo-first-order rate constant based on C_A. At $x = 0$,

$$P_A = 0.3(4) = 1.2 \text{ atm}, \qquad T = 150 + 273 = 423 \ K$$

$$PB = 0.7(4) = 2.8 \text{ atm}, \qquad R = 0.3 \text{ cm}$$

$$\text{ideal rate } r^* = \frac{2.5 \times 10^{-5}(1.2)(2.8)}{(1 + 0.12)^2} = 6.70 \times 10^{-5} \text{ mol/sec, cm}^3$$

$$C_A = \frac{P_A}{RT} = \frac{1.2}{(82.056)(423)} = 3.4 \times 10^{-5} \text{ mol/cm}^3$$

$$k = \frac{6.7 \times 10^{-5}}{3.46 \times 10^{-5}} = 1.94 \text{ sec}^{-1}$$

Based on A,

$$\phi = 0.3\left(\frac{1.94}{0.02}\right)^{0.5} = 2.95$$

From Figure 4.8, $\eta_A \cong 0.68$ for a first-order reaction. If there was no adsorption of C, the effectiveness factor would be about 0.68, since the decrease in P_B inside the pellet would tend to offset the slight increase in η due to the fractional-order dependence on P_A. The actual effectiveness factor is lower than 0.68 because of the inhibiting effect of C. To estimate the average pressures inside the pellet, $\eta = 0.68$ is used to get a first estimate of $\bar{P}_A$ and then $\bar{P}_B$ and $\bar{P}_C$:

$$\bar{P}_A \cong 0.68(1.2) = 0.82$$
$$\Delta P_A = 1.2(0.32) = 0.38$$
$$\Delta P_B = \Delta P_A\left(\frac{D_{eA}}{D_{eB}}\right) = 0.38\left(\frac{0.02}{0.03}\right) = 0.25$$
$$\bar{P}_B = 2.8 - 0.25 = 2.55$$
$$\Delta P_C = \Delta P_A\left(\frac{D_{eA}}{D_{ec}}\right) = 0.38\left(\frac{0.02}{0.015}\right) = 0.51$$
$$\bar{P}_C = 0.51$$

Based on these average pressures, the rate is

$$r = \frac{2.5 \times 10^{-5}(0.82)(2.55)}{(1 + 0.082 + 2(0.51))^2} = 1.18 \times 10^{-5}, \text{ mol/sec, cm}^3$$

$$\eta = \frac{r}{r^*} = \frac{1.18 \times 10^{-5}}{6.70 \times 10^{-5}} = 0.176$$

This value of η is much lower than that based on just the gradient for A because of the large change in the denominator term. The true value of η is between 0.176 and 0.68, and an approximate value can be obtained by repeating the calculations with revised estimates for r. Further calculations show that $\eta \cong 0.4$ at the start of the reaction and $\eta \cong 0.86$ at 50% conversion. The effectiveness factor is over twice as large as at the start of the reaction, since the rate decreases by a factor of about 4 for less than a twofold change in P_A, and the effect of the internal gradient in C is not as large when the external gas already has some C.

Multiple Reactions

When a reactant is consumed in more than one reaction, the overall rate must be used in evaluating ϕ, even if the reactions are of different order. For example, consider the partial oxidation of a hydrocarbon A to product B with byproducts formed by a parallel reaction:

$$A + O_2 \longrightarrow B \qquad\qquad r_1 = k_1 P_A P_{O_2}$$

$$A + 3O_2 \longrightarrow 2C + D \qquad r_2 = \frac{k_2 P_A P_{O_2}}{\left(1 + K_A P_A + K_{O_2} P_{O_2}\right)^2}$$

If A is the limiting reactant, the modulus should be calculated using D_{eA} and the total consumption rate of A:

$$k = \frac{r_1 + r_2}{C_{A_s}}$$

$$\phi = R\left(\frac{k}{D_{eA}}\right)^{0.5}$$

As a check, the modulus based on O_2 could be calculated using D_{eO_2} and a different rate constant:

$$k = \frac{r_1 + 3r_2}{C_{O_2 s}}$$

If the modulus based on A is higher than that based on O_2, the calculation would be based on A and perhaps corrected for the O_2 gradient by the method shown in Example (4.4).

When consecutive reactions occur and the desired product is the intermediate, gradients due to pore diffusion lower the selectivity. Consider the following sequence of reactions:

$$A + B \xrightarrow{1} C \qquad \text{desired product} \qquad\qquad (4.69)$$

$$C + B \xrightarrow{2} D + E \qquad \text{byproducts} \qquad\qquad (4.70)$$

The overall selectivity for this example is the net moles of C produced divided by the moles of A consumed. An instantaneous or local selectivity can be defined as the net rate of C formation divided by the rate of consumption of A:

$$S = \frac{r_1 - r_2}{r_1} = 1 - \frac{r_2}{r_1} \tag{4.71}$$

The local selectivity decreases with increasing conversion, since r_1 decreases as A is consumed and r_2 increases as C is formed. The overall selectivity is an integrated average of the local values. Internal concentration gradients lower the selectivity because r_1 is decreased and r_2 is increased. If the reactions have similar simple kinetics, the change in local selectivity can be easily estimated using partial pressure differences. If

$$\frac{r_2}{r_1} = \frac{k_2 P_C P_B}{k_1 P_A P_B} \tag{4.72}$$

then

$$\frac{r_2}{r_1} = \frac{k_2(P_C + \Delta P_C)}{k_1(P_A - \Delta P_A)} \tag{4.73}$$

If the selectivity is high, so that only a few percent of C is converted to D + E, the flux of C out of the catalyst is about equal to the flux of A into the catalyst, and the gradients for A and C are related to the diffusivities:

$$\Delta P_C \cong \Delta P_A \left(\frac{D_{eA}}{D_{eC}} \right) \tag{4.74}$$

When the conversion is low, say, 10%, a moderate effect of pore diffusion on the main reaction has a large effect on the local selectivity. If $\eta = 0.8$ and $x = 0.1$, $\Delta P_A = 0.2 P_A$ and $\Delta P_C \cong 0.2 P_A$; the average partial pressure of C in the catalyst is $(0.1 + 0.2) P_A$, giving a rate of byproduct formation three times that for the ideal case of no internal gradients.

As the conversion increases, the change in P_C becomes smaller relative to the external value, and the change in byproduct formation is not as great. However, the maximum yield of product B is significantly lower than what could be achieved in the ideal case.

Effect of Catalyst Shape

Porous catalysts are used in a variety of shapes, including spheres, cylinders, rings, irregular particles, and thin coatings on tubes or flat surfaces. Diffusion plus reaction in a flat slab is a simple case to analyze, since the area for diffusion does not change with distance from the external surface. The equation is similar to that for diffusion and reaction in a straight cylindrical pore [16], but D_e is used in place of D_{pore}, and k is based on the rate per unit volume of catalyst rather than per unit area of pore surface. For a first-order reaction,

$$D_e\left(\frac{d^2C}{dl^2}\right) = kC \qquad (4.75)$$

The Thiele modulus for a flat slab is based on the thickness of the slab L if only one side is exposed to the gas, or on the half thickness $L/2$ if both sides are exposed:

$$\phi_L = L\left(\frac{k}{D_e}\right)^{1/2} \qquad (4.76)$$

The concentration profiles are similar to those shown in Figure 4.7, and the gradients go to zero at $l = L$. The effectiveness factor is defined in the same way as for a sphere [Eq. (4.28)] and is a simpler function of the modulus:

$$\eta_L = \frac{\tanh(\phi_L)}{\phi_L} \qquad (4.77)$$

For high values of ϕ_L, η_L varies inversely with ϕ_L,

$$\eta_L \cong \frac{1}{\phi_L} \qquad (4.78)$$

The limiting effectiveness factor for a sphere is $\eta \cong \frac{3}{\phi}$ [Eq. (4.31)], and this agrees with the flat-slab solution if the length term in the modulus [Eq. (4.76)] is replaced with $R/3$, the volume/surface ratio for a sphere (L is the volume/surface ratio for the slab). For low effectiveness factors, the concentration gradient is confined to the region very near the surface, and the curvature of the surface is unimportant.

For first-order reactions and moderate values of ϕ, η for a sphere is slightly less than the value of η for a slab with the same volume/surface ratio. As shown in Table 4.2, the maximum difference is about 14%. This small difference means that solutions for complex kinetic models that were obtained for the flat-slab case can be used to get approximate effectiveness factors for spherical catalysts.

If the catalyst is deposited as a thin layer on the inside or outside of a tube, the slab model can be used if the thickness of the catalyst layer is much less than the tube radius. The slab model is also used to analyze the performance of eggshell catalysts, which have a layer of active catalyst near the outer surface of the pellet, and of catalyst monoliths, which have a thin layer of catalyst on the inside of square, triangular, or hexagonal passages. However, when the catalyst layer is very thin, pore diffusion effects are

TABLE 4.2   Effectiveness Factors
for Spheres and Slabs

Sphere		Slab		Ratio
ϕ	η	$\phi_L = \phi/3$	η_L	η/η_L
1	0.939	0.333	0.965	0.973
2	0.806	0.667	0.874	0.922
5	0.480	1.667	0.559	0.859
10	0.270	3.330	0.302	0.893
20	0.143	6.670	0.150	0.950
50	0.059	16.67	0.060	0.980

likely to be accompanied by effects of mass transfer in the external boundary layer, which may be as thick or thicker than the layer of catalyst. The combined effects of external and internal mass transfer will be treated in the next chapter.

Cylindrical catalyst pellets can be prepared by extrusion or by pelletization. For short, stubby cylinders ($L/d_p \cong 1$), the effectiveness factor equation for a sphere is generally used with R equal to the cylinder radius. (The surface-to-volume ratio for the cylinder is $R/3$, the same as for the sphere, though the cylinder has 30% more surface area than a sphere of the same volume.) For long cylinders, the effectiveness factor is less than for a sphere of the same radius, since the surface-to-volume ratio is lower. Numerical solutions are available for different L/d_p ratios, but the solution for spheres can be used for approximate values of η if R in the modulus is replaced with $1.2R$ for $L/d = 2$, $1.33\ R$ for $L/d = 4$, and $1.5R$ for a very long cylinder.

Ring-type catalysts are used for some processes where rapid reaction rates would give very low effectiveness factors for large spheres or cylinders yet large particles are needed for pressure drop or heat transfer considerations. Diffusion through the inner surface of the ring increases the effectiveness factor and may more than offset the decrease in mass of catalyst per unit reactor volume. The effectiveness factor can be estimated using Figure 4.8 and a modulus corrected to allow for the greater surface-to-volume ratio. For a ring with the inner radius $r = 0.5R$ and $L = 2R$, the surface-to-volume ratio is $5/R$, compared to $3/R$ for a sphere or a short cylinder. Therefore, the appropriate modulus for this ring catalyst is

$$\phi_{\text{ring}} = R\left(\frac{3}{5}\right)\left(\frac{k}{D_e}\right)^{1/2} = 0.6\phi_{\text{sphere}} \tag{4.79}$$

To show the advantage of the ring catalyst, consider a case where $\phi = 10$ for a solid cylinder and $\eta = 0.27$. For the ring with $\phi = 6$, η is about 0.42; since the ring has 25% less volume than the cylinder, the ratio of reaction rates per pellet is $0.75(0.42/0.27) = 1.17$. Thus the ring would give 17% higher rate than the solid cylinder. For higher values of ϕ, the ratio of rates approaches 1.24. Thinner-wall rings would give somewhat higher ratios for large values of ϕ. A difference of 10–30% might not seem enough to justify using ring catalysts, but the lower pressure drop is an additional benefit, and the catalyst cost should be less per unit volume of reactor. Catalyst rings are used for the steam reforming of methane at 800–1050 K, and effectiveness factors are estimated to be in the range 0.02–0.05 [22].

For irregular particles produced by crushing or grinding, the external surface is larger than for spheres of the same nominal size, which tends to increase the effectiveness factor but also increases the pressure drop through a bed of particles. Shape factors based on pressure drop are given for some typical granular solids [23], but these shape factors are not used for pore diffusion calculations, and R in the Thiele modulus is taken as 0.5 times the nominal or screen size.

OPTIMUM PORE SIZE DISTRIBUTION

When a reaction is limited by pore diffusion, there are several approaches to increasing the effectiveness factor. Using smaller particles would help, but this may be impractical because of high pressure drop, poor heat transfer, or increased cost of the catalyst. The use of rings or other shapes with a high surface-to-volume ratio should also be considered. A third alternative is to change the pore structure. If the catalyst has most of its surface area in micropores, changing the distribution of pore sizes by choosing a different support or by altering the method of catalyst preparation may lead to a higher effectiveness factor. The basic idea is to provide some macropores to increase the effective diffusivity and make the interior reaction sites, which are mainly in the micropores, more accessible to the reactants.

In the following analysis, the catalytic activity is assumed to be proportional to the surface area per gram, S_g, which varies inversely with the pore radius [Eq. (4.5)]. The surface area is also proportional to the pore volume V_g, but increasing V_g gives a weaker pellet, and so V_g and ϵ are assumed constant. Although a broad distribution of pore sizes is better than a narrow distribution for fast reactions, analysis shows that a bimodal distribution is likely to be the optimum. If there are just two types of pores, macropores and micropores, the total surface area is

$$S_g = \frac{2V_a}{\bar{r}_a} + \frac{2V_b}{\bar{r}_b} \tag{4.80}$$

$$V_a + V_b = V_g \tag{4.81}$$

The effective diffusivity according to the parallel path model is [Eq. (4.13, 4.14)]

$$D_e = \frac{(D_a V_a + D_b V_b)}{V_g} \frac{\epsilon}{\tau} \tag{4.82}$$

A catalyst with all micropores will have a large surface area and high intrinsic activity but a low value of D_e. Replacing some of the micropores with macropores reduces S_g but increases D_e and η. The goal is to maximize the rate per pellet, which is proportional to $k\eta$. The macropores should be large enough so that bulk diffusion predominates and the average pore diffusivity is much larger than that in the micropores. The fraction of pore volume in macropores can then be varied to find the distribution that gives a maximum reaction rate, as shown in the following example.

Example 4.5

A first-order irreversible reaction is to be carried out under conditions where pore diffusion affects the overall reaction rate. Calculate the optimum pore size distribution for a spherical pellet 5 mm in diameter using the following rate data and diffusion coefficients.

$$A \longrightarrow B + C$$

Data for powdered catalyst $r = kC_A$:

$$k = 3.6 \text{ sec}^{-1}$$
$$V_g = 0.60 \text{ cm}^3/\text{g}$$
$$S_g = 300 \text{ m}^2/\text{g}$$
$$d_p = 0.02 \text{ cm}$$
$$\rho_p = 0.8 \text{ g/cm}^3$$
$$\bar{r} = 40\text{Å (narrow distribution)}$$
$$D_{KA} = 0.012 \text{ cm}^2/\text{sec}$$
$$D_{AB} = 0.40 \text{ cm}^2/\text{sec}$$

Solution. Check the effectiveness factor for powdered catalyst: 1 g catalyst has $1/0.8 = 1.25 \text{ cm}^3/\text{g}$.

$$\epsilon = \frac{\text{pore volume}}{\text{total volume}} = \frac{0.60}{1.25} = 0.48$$

$$\bar{r} = \frac{2V_g}{S_g} = \frac{2(0.60)}{300 \times 10^4} = 4 \times 10^{-7}\,\text{cm} = 40\,\text{Å}$$

$$\frac{1}{D_{\text{pore}}} = \frac{1}{D_K} + \frac{1}{D_{AB}} = \frac{1}{0.012} + \frac{1}{0.40}$$

$$D_{\text{pore}} = 1.16 \times 10^{-2}\,\text{cm}^2/\text{sec}$$

Assume $\tau = 3$:

$$D_e = \frac{1.16 \times 10^{-2}(0.48)}{3} = 1.86 \times 10^{-3}$$

$$\phi_{\text{app}} = R\left(\frac{k_{\text{app}}}{D_e}\right)^{1/2} = 0.01\left(\frac{3.6}{1.86 \times 10^{-3}}\right)^{0.5} = 0.44$$

$\eta = 1.0$ for powdered catalyst

$k = 3.6\,\text{sec}^{-1}$ is the true rate constant

Choose $r = 2000$ Å for macropores, or 2×10^{-5} cm :

$$D_{K_b} = 0.012\left(\frac{2000}{40}\right) = 0.60$$

$$\frac{1}{D_{\text{pore,b}}} = \frac{1}{0.4} + \frac{1}{0.6}$$

$$D_{\text{pore,b}} = 0.24\,\text{cm}^2/\text{sec}$$

Keep 40-Å micropores, so

$$D_{\text{pore,a}} = 1.16 \times 10^{-2}\,\text{cm}^2/\text{sec}$$

Vary V_b, keeping $V_a + V_b = 0.60$ cm^3/g. For $V_b = 0.1$ cm^3/g and $V_a = 0.5$ cm^3/g:

$$S_g = 2\left(\frac{0.5}{4 \times 10^{-7}}\right) + 2\left(\frac{0.1}{2 \times 10^{-5}}\right) = 2.51 \times 10^6 \text{ cm}^2/\text{g} = 251 \text{ m}^2/\text{g}$$

$$k = 2.6\left(\frac{251}{300}\right) = 3.01 \text{ sec}^{-1}$$

$$D_e = \frac{[1.16 \times 10^{-2}(0.5) + 0.24(0.1)]}{0.6} \times \frac{0.48}{3} = 7.95 \times 10^{-3} \text{ cm}^2/\text{sec}$$

$$\phi = 0.25\left(\frac{3.01}{7.95 \times 10^{-3}}\right)^{0.5} = 4.86$$

From Figure 4.8, using the curve based on true k, $\eta = 0.49$:

$$\eta k = 0.49(3.01) = 1.47$$

If only micropores were present in the 5-mm pellet,

$$\phi = 0.25\left(\frac{3.6}{1.86 \times 10^{-3}}\right)^{0.5} = 11, \qquad \eta = 0.248$$

$$\eta k = 0.245(3.6) = 0.89$$

Introducing $0.1\text{-cm}^3/\text{g}$ macropores increases the overall rate about 65%. Results for other values of V_b are shown in Table 4.3. The maximum rate is for a pellet with about 1/4 of the pore volume in macropores and 3/4 in micropores, and the rate per pellet is about 70% greater than with all micropores.

If a smaller pellet size were chosen in Example 4.5, the optimum pore size distribution would shift to a lower fraction of macropores, since the effectiveness factors would be closer to 1.0. When the rate constant is much higher or the pellet size much larger, so that η varies inversely with ϕ, an

TABLE 4.3 Solution for Example 4.5

V, cm^3/g		Surface area, m^2/g			k	D_e			
V_a	V_b	S_a	S_b	S_g	s^{-1}	cm^2/sec	ϕ	η	ηk
0.6	0	300	0	300	3.6	1.86×10^{-3}	11	0.25	0.89
0.5	0.1	250	1	251	3.01	7.95×10^{-3}	4.86	0.49	1.47
0.45	0.15	225	1.5	226.5	2.72	1.10×10^{-2}	3.93	0.57	1.55
0.4	0.2	200	2	202	2.42	1.40×10^{-2}	3.29	0.63	1.52
0.35	0.25	175	2.5	177.5	2.13	1.71×10^{-2}	2.79	0.70	1.49

analytical solution is possible for the optimum distribution. If the surface area is almost entirely in micropores and f is the fraction of macropore volume,

$$S = S_0(1 - f) \tag{4.83}$$

$$k = k_0(1 - f) \tag{4.84}$$

If the diffusion flux in the micropores is negligible,

$$D_e \cong \frac{D_b V_b \epsilon}{\tau V_g} = \frac{D_b \epsilon f}{\tau} \tag{4.85}$$

$$\eta \cong \frac{3}{\phi} = \frac{3}{R}\left[\frac{D_b \epsilon f}{k_0(1 - f)\tau}\right]^{1/2} \tag{4.86}$$

$$\eta k = \frac{3}{R}\left[\frac{D_b \epsilon f}{k_0(1 - f)\tau}\right]^{1/2} k_0(1 - f) \tag{4.87}$$

or

$$\eta k = \beta[f(1 - f)]^{1/2} \tag{4.88}$$

The maximum rate is obtained for $f = 0.5$, or with half of the pore volume in macropores. In Example 4.5, the macropore size was arbitrarily chosen as $r_b = 2000$ Å, giving $D_{pore} = 0.24$ cm^2/sec, which is 0.6 D_{AB}. Assuming much larger macropores would increase D_{pore} and D_e, but the maximum value of D_{pore} would be the bulk diffusivity D_{AB}, which is 0.4cm^2/sec for this example. The model should not be used for very large macropores, because the distance between macropores becomes too large. Molecular sieve pellets have a bimodal pore size distribution with a very large difference between the sizes of the pores in the zeolite crystal and the macropores, which are the spaces between the crystals. This case can be treated by applying the solutions for pore diffusion plus reaction first to the diffusion in the macropores using the pellet radius and D_e to calculate ϕ_1. This gives the macropore effectiveness factor η_1, which is a measure of the average concentration in the macropores [24,25]. Then a second modulus ϕ_2 is calculated using the crystal radius and the effective diffusivity in the micropores. The corresponding value of η_2 is then multiplied by η_1 to get an overall effectiveness ratio:

$$\eta = \eta_1 \times \eta_2 \tag{4.89}$$

NOMENCLATURE

Symbols

a	surface area, pore radius
a_{ext}	external surface area
b,c,d	stochiometric coefficients
C	concentration
C_A	concentration of species A
C_A^*	equilibrium concentration
$C_{s,}$	concentration at external surface
D_{AB}	bulk diffusivity in a binary mixture
D_{AM}	diffusivity of A in a mixture
D_e	effective diffusivity in a porous solid
D_K	Knudsen diffusivity
D_{pore}	diffusivity in a pore
D_{pa}	diffusivity in micropores
D_{pb}	diffusivity in macropores
d_p	particle diameter, pore diameter
d_s	solute diameter
E	activation energy
f	fraction of pore volume in macropores
J	molar flux
K_{eq}	equilibrium constant for reaction
k	reaction rate constant
k_{app}	apparent reaction rate constant
k_r	effective rate constant for reversible reaction
L	length of pore, thickness of catalyst slab
l	distance from surface
M	molecular weight
M_A	molecular weight of solute
M_B	molecular weight of solvent
n	number of pores
P	pressure
P_o	vapor pressure
R	radius of sphere, cylinder, or particle, gas constant
r	radius of pore, distance from center of sphere
$\bar{r}$	mean pore radius
r	reaction rate
r^*	ideal reaction rate
r.f.	roughness factor
S	selectivity
S_g	total surface area, usually m^2/g
T	absolute temperature K

T_{ref}	reference temperature
$\bar{v}$	mean molecular velocity
V_A	molar volume of solute
V_g	pore volume per gram
V_a	pore volume in micropores
V_b	pore volume in macropores
x	fractional conversion, association parameter

Greek Letters

β	constant in Eq. (4.88)
ϵ	void fraction, porosity
η	effectiveness factor
η_0	effectiveness factor for zero-order reaction
η_L	effectiveness factor for slab
λ	mean free path, ratio of molecule size to pore size
μ	viscosity
θ	contact angle
π	pi
ρ	density
ρ_ρ	particle density
σ	surface tension
τ	tortuosity
ϕ	Thiele modulus for sphere
ϕ_{app}	apparent modulus
ϕ_0	modulus for zero-order reaction
ϕ_2	modulus for second-order reaction
ϕ_L	Thiele modulus for slab

PROBLEMS

4.1 The catalytic hydrogenation of aldehydes was studied by Oldenburg and Rase using a differential reactor [26]. For n-butyraldehyde they reported

$$r = \frac{kP_U}{P_{H_2}^{1/2}}$$

where

r = g-mole/hr, g cat
P_U = aldehyde pressure, psia (range $10-30$)
P_{H_2} = hydrogen pressure, psia (range $10-30$)
$\ln k = 5.31 - 7800/RT$ for $T = 150-180°C$

The catalyst pellets were $\frac{1}{8} \times \frac{1}{8}$-in cylinders with $S_g \cong 70 \text{ m}^2/\text{g}$, $V_g \cong 0.20 \text{ cm}^3/\text{g}$, and $\rho_p = 2.6 \text{ g/cm}^3$.

a. Estimate the effectiveness factor for the run conditions most likely to have produced a pore diffusion limitation.
b. Predict the reaction rate for $\frac{1}{4} \times \frac{1}{4}$-in cylinders at 150°C with $P_U = 20$ psia, $P_{H_2} = 30$ psia.

4.2   The partial oxidation of ethylene was studied in a differential reactor using 20–28 mesh particles of an 8% Ag/Al$_2$O$_3$ catalyst [27]. At the standard conditions of $P_E = 0.263$ atm, $P_{O_2} = 0.263$ atm, $P_{H_e} = 0.789$ atm, and $T = 220°C$, the reaction rates were:

$r_1 = 6{-}7 \times 10^{-6}$ moles EO/min, g cat $E = 20\text{kcal/mol}$

$r_2 = 4{-}6 \times 10^{-6}$ moles E to CO$_2$/min, g cat $E_2 = 27\text{kcal/mol}$

The catalyst had a surface area of 0.5 m^2/g and a pore volume of 0.175 cm^3/g. The skeletal density of the α-Al$_2$O$_3$ support is 4.0 g/cm^3.

a. Estimate the isothermal effectiveness factor at the standard conditions from the foregoing rates.
b. If a catalyst of the same composition were prepared in the form of $\frac{3}{8}$-inch spheres and used at 240°C with the following gas, what would be the effectiveness factor?

$$P = 15 \text{ atm}, \qquad P_{O_2} = 1 \text{ atm},$$
$$P_E = 0.8 \text{ atm}, \qquad P_{CO_2} = 1.2 \text{ atm}, \qquad P_{N_2} = 12 \text{ atm}$$

4.3   The cracking of cumene was carried out at 950°F over pelleted silica-alumina catalysts [28]. The change in rate with particle size showed the importance of pore diffusion, but it is not known whether the data for the smallest particles are free from diffusion effects.

a. Try to determine the effectiveness factors for the different sizes by plotting the data.
b. Predict the effectiveness factor for the smallest particles using a calculated diffusivity and an assumed tortuosity.

Catalyst type	d_p,cm	W/F at 28% conversion lb/lb mole, hr
Crushed	0.045	1.3
Bead	0.33	5.7
Bead	0.43	7.6
Bead	0.53	10.0

The tests were carried out at 950°F and 1 atm. Assume a first-order reaction. The catalyst properties are:

S	$= 342 \text{ m}^2/\text{g}$
catalyst density	$= 2.3 \text{ g/cc}$
pellet density	$= 1.14 \text{ g/cc}$
bed density	$= 0.68 \text{ g/cc}$

4.4   Dart et al. studied the oxidation of carbon deposited on cracking catalysts at temperatures of 850–1200°F [29]. Data for one run are:

1100°F
0.208 atm, O_2 pressure

Wt. % Carbon	Time (sec)
1.72	0
0.85	40
0.55	120
0.08	700

Catalyst Properties

True density	2.42 g/cm^3
Pellet density	1.20 g/cm^3
Bulk density	0.74 g/cm^3
Ave. diameter	3.5 mm
Area	$82 \text{ m}^2/\text{g}$

a.   Calculate the average pore diameter of the catalyst.
b.   To what extent might diffusion into the pores have limited the reaction rate, assuming that the carbon was deposited evenly over the catalyst surface?
c.   Can any additional evidence about the role of diffusion be obtained from the first-order dependence on oxygen pressure or the apparent activation energy of 26,600 calories?
d.   The authors reported no difference in rate for 2.0-mm and 3.5-mm particles with 0.73% carbon at 1100°F. What conditions might be more likely to produce a diffusion effect? (See also Ref. 30.)

4.5   A reaction carried out with a catalyst in the form of $\frac{3}{8} \times \frac{3}{8}$-in cylinders has an apparent Thiele modulus of 4.5.

    a.   What is the effectiveness factor and the true modulus assuming the reaction is first order?

    b.   If the catalyst is prepared in the forms of $\frac{3}{8}$- in rings with an $\frac{1}{8}$-in central hole, what is the predicted effectiveness factor?

    c.   For the rings, how would the rate per unit mass of catalyst and per unit volume of reactor compare with values for the cylinders?

4.6   A gas-phase hydrogenation reaction was studied using a catalyst fine enough to avoid any effect of pore diffusion. The rate equation is:

$$A + 2H_2 \rightarrow B$$

$$r = \frac{kP_A P_{H_2}}{(1 + 12P_B)^2}$$

    a.   For the following gas compositions, sketch the concentration gradients for A, H_2, and B for the case where the reaction is rapid enough to make $P_{A_{ave}}$ 10% less than P_A. Assume $D_A = D_B$
$$= \frac{1}{4} D_{H_2}$$

Case 1	Case 2
$P_A = 1$ atm	$P_A = 1$ atm
$P_B = 0$	$P_B = 0.5$ atm
$P_{H2} = 2$ atm	$P_{H2} = 1.2$ atm

    b.   Estimate the effectiveness factors for the two cases.

4.7   Some commercial catalysts are prepared by a skin-deep impregnation of a porous support.

    a.   If the thickness of the active layer is 10% of the pellet diameter, how would the effectiveness factors compare with those for a uniformly active pellet? Assume a first-order reaction and spherical pellets, and compare over a wide range of reaction rates.

    b.   How would the reaction rates per particle compare over a wide range of rates, expressed as ϕ sphere?

4.8   The catalytic reduction of nitric oxide with ammonia was studied using 20–32 mesh copper mordenite catalyst [31].

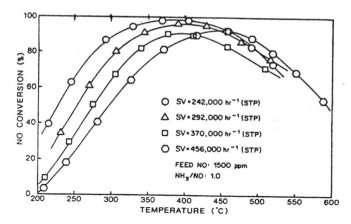

FIGURE 4.10 Effect of space velocity on temperature dependence of NO conversion with CuHM catalyst.

    a.   Determine reaction rate constants for temperatures up to 400°C from the data in Figure 4.10. Plot the rate constants to determine the activation energy.

    b.   Use the average rate constant at 350°C to estimate the apparent Thiele modulus. Was the reaction rate limited by pore diffusion at 350°C? Does the value of the activation energy support your conclusion?

REFERENCES

1. S Brunauer, PH Emmett, E Teller. J Am Chem Soc 60:309, 1938.
2. DM Ruthven. Principles of Adsorption and Adsorption Processes. New York: Wiley, 1984, pp 140–163.
3. E Wicke, R Kallenbach. Kolloid Z 97:135, 1941.
4. N Wakao, JM Smith. Chem Eng Sci 17:825, 1962.
5. RC Reid, TK Sherwood. The Properties of Gases and Liquids. 2nd ed. New York: McGraw-Hill, 1966.
6. HL Ritter, LE Drake. Ind Eng Chem Anal Ed 17:782–787, 1945.
7. JM Smith. Chemical Engineering Kinetics. 3rd ed. New York: McGraw-Hill, 1981, p 346.
8. MFL Johnson, WE Stewart. J Catalysis 4:248, 1965.
9. CN Satterfield. Mass Transfer in Heterogeneous Catalysis. MIT Press, 1970, p 64.
10. N Wakao, JM Smith. Ind Eng Chem Fund 3:123, 1964.
11. RS Cunningham, CJ Geankoplis. Ind Eng Chem Fund 7:535, 1968.

12. W Cordova, P Harriott. Chem Eng Sci 30: 1201, 1975.
13. CR Wilke, P Chang. AIChE J 1:264, 1955.
14. EM Renkin. J Gen Physiol 38:225, 1954.
15. CN Satterfield, CK Colton, WH Pitcher Jr. AIChE J 19:628, 1973.
16. EW Thiele. Ind Eng Chem 31:916, 1939.
17. YB Zeldovitch. Acta Physicochim URSS 10:583, 1939.
18. A Wheeler. Advances in Catalysis III: New York: Academic Press, 1951, p 249.
19. PB Weisz. Chem Eng Progr Symp Ser 55:25-29, 1959
20. GW Roberts, CN Satterfield. Ind Eng Chem Fund 4:288, 1965.
21. GW Roberts, CN Satterfield. Ind Eng Chem Fund 5:317, 1966.
22. JC DeDeken, EF Devos, GF Froment. Chem Reaction Eng 7. ACS Symposium Ser 196, 1982, p 181.
23. WL McCabe, JC Smith, P Harriott. Unit Operations of Chemical Engineering. 6th ed. New York: McGraw-Hill, 2000, p 946.
24. JO Mingle, JM Smith. AIChE J 7:243, 1961.
25. MR Rao, N Wakao, JM Smith. Ind Eng Chem Fund 3:127, 1964.
26. CC Oldenburg, HF Rase. AIChEJ 3:462, 1997.
27. PD Klugherz, P Harriott. AIChEJ 17: 856, 1971.
28. TE Corrigan, JC Garver, HF Rase, RS Kirk. Chem Eng Progr 49:603, 1953.
29. JC Dart, RT Savage, CG Kirkbride, Chem Eng Progr 45:102, 1949.
30. PB Weisz, RD Goodwin. J Catalysis 2:397, 1963.
31. FG Medros, JW Eldridge, JR Kittrell. Ind Eng Chem Res. 28:1171, 1989.

5

Heat and Mass Transfer in Reactors

The discussion of ideal reactors in Chapter 3 was limited to reactors that were either isothermal or adiabatic. Isothermal operation is often desirable, but it is hard to achieve in a large reactor with an exothermic reaction because of the difficulty in removing the heat that is released. For stirred-tank reactors of fixed proportions, the rate of heat generation goes up with the cube of the diameter, but the wall area goes up only as the diameter squared. Similarly for a tubular reactor, the heat generation rate per unit length varies with the square of the diameter, but the wall area is proportional to the diameter. Therefore, for both reactor types, heat transfer becomes more difficult when the diameter is increased. Furthermore, there is more chance of unstable behavior for the larger reactors operating with a higher-temperature driving force for heat transfer. A reactor operating satisfactorily at the desired temperature may be close to the stability limit and may jump to a much higher temperature after just a slight change in feed concentration or jacket temperature.

In this chapter, correlations for heat transfer in reactors are presented, and the requirements for stable operation are discussed. The continuous stirred-tank reactor is treated first, since it is the simplest case, and uniform temperature and concentration are assumed for the fluid in the tank. For a homogeneous reaction in a pipeline, there are axial gradients of temperature and concentration to deal with, which complicates the analysis. A still more difficult problem is the design of a packed-bed tubular reactor, where there are radial as well as axial gradients to consider.

STIRRED-TANK REACTOR

A stirred-tank reactor is generally equipped with an external jacket covering the sides and bottom of the tank, as illustrated in Figure 5.1a. Steam or hot water is sent to the jacket to bring the initial charge to the desired temperature; once reaction starts, cooling water is used to remove the heat of reaction. The tank temperature is maintained at the desired value with a feedback control system.

For heat transfer to or from the jacket of a baffled tank, the following equation applies if a standard six-blade turbine is used; similar equations are available for other types of impellers [1]:

$$\frac{h_j D_t}{k} = 0.76 \left(\frac{D_a^2 n \rho}{\mu}\right)^{2/3} \left(\frac{c_p \mu}{k}\right)^{1/3} \left(\frac{\mu}{\mu_w}\right)^{0.24} \tag{5.1}$$

The coefficient h_j is the film coefficient for the inner surface of the vessel, D_t is the tank diameter, and k is the thermal conductivity of the fluid. The other

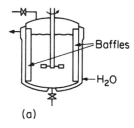

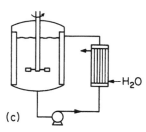

(a)

(c)

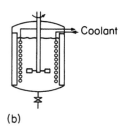

(b)

FIGURE 5.1   Heat removal in stirred-tank reactors: (a) Jacket; (b) cooling coil; (c) external exchanger.

terms are the agitator Reynolds number, the Prandtl number, and a factor correcting for the difference between wall and bulk viscosities.

The overall heat transfer coefficient U depends on h_j, the wall resistance, and h_o, the film coefficient for coolant in the jacket:

$$Q = UA\ \Delta T_{\text{ave}} \tag{5.2}$$

$$\frac{1}{U} = \frac{1}{h_j} + r_{\text{wall}} + \frac{1}{h_o} \tag{5.3}$$

The outside coefficient, h_o, is quite low for water in a simple open jacket. It can be increased by using agitation nozzles, a spiral baffle in the jacket, or a half-pipe coil welded to the tank [1,2]. The wall resistance is usually negligible for a steel tank, but it may be significant for a thick stainless steel wall or for a glass-lined reactor.

As an alternative to jacket cooling or to supplement the jacket, a helical cooling coil can be installed between the impeller and the wall, as shown in Figure 5.1b. The film coefficient for the outside of the coil is given by [1]

$$\frac{h_c D_c}{k} = 0.17 \left(\frac{D_a^2 n\rho}{\mu}\right)^{0.67} \left(\frac{c_p\mu}{k}\right)^{0.37} \left(\frac{D_a}{D_t}\right)^{0.1} \left(\frac{D_c}{D_t}\right)^{0.5} \left(\frac{\mu}{\mu_0}\right)^{0.24} \tag{5.4}$$

The coil inside coefficient is given by the Dittus–Boelter equation with a correction for curvature of the helical coil [3]:

$$\frac{h_i D}{k} = 0.17 \left(\frac{Du\rho}{\mu}\right)^{0.8} \left(\frac{c_p\mu}{k}\right)^{1/3} \left(1 + 3.5\frac{D}{D_{\text{He}}}\right) \tag{5.5}$$

For very exothermic reactions, the best solution may be to recirculate the fluid through an external heat exchanger, usually of the shell-and-tube type (Figure 5.1c). The heat transfer area is not limited by the tank dimensions and can be much greater than the jacket area.

When the reaction is carried out in a low-boiling-point solvent or when one of the reactants is quite volatile, heat can be removed by allowing the solvent or reactant to vaporize. The vapors are condensed in an overhead condenser, and the liquid is returned to the reactor. The limiting factor in this design may be the allowable vapor velocity in the reactor. Too high a velocity will cause foaming or excessive entrainment of liquid, and the reaction rate per unit volume of reactor will decrease because of greater gas holdup.

REACTOR STABILITY

With an exothermic reaction in a CSTR, there may be a stability problem if the reactor operates at low to moderate conversion with a large ΔT for heat removal. The tendency to instability occurs because of the exponential increase in reaction rate with increasing temperature compared to a linear increase in heat transfer rate. This is illustrated by Figure 5.2, where the steady-state rates of heat generation Q_G and heat removal Q_R are plotted against reactor temperature for a given feed rate and reactor volume. The heat generation curve is S-shaped, with the lower part showing increasing slope because the rapid increase in rate constant has more effect than the slight decrease in steady-state reactant concentration. Values of Q_G are obtained from the feed rate, conversion, and heat of reaction:

$$Q_G = F_A x (-\Delta H) \tag{5.6}$$

The conversion is calculated by material balance:

$$F_A x = rV \tag{5.7}$$

For a first-order reaction,

$$x = \frac{kV/F}{1 + kV/F} \tag{5.8}$$

The conversions are not indicated in Figure 5.2, but since Q_G is proportional to x, a plot of x versus T would have the same shape as the plot of Q_G versus T.

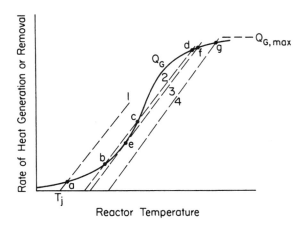

FIGURE 5.2   Steady-state heat balance for a CSTR.

For conversions greater than about 50%, the effect of decreasing the reactant concentration makes Q_G increase less rapidly with temperature, and Q_G approaches a limiting value corresponding to complete or equilibrium conversion. The curve shown in Figure 5.2 is typical of a first-order reaction, but similar shapes would be found for other reactions with positive orders. With a zero-order reaction, the slope of the curve would continue to increase until complete conversion was obtained.

With a jacket for heat removal and a constant jacket temperature, which could be achieved with a boiling fluid in the jacket, the heat transfer rate is

$$Q_j = UA(T - T_j) \tag{5.9}$$

With cooling water in the jacket, the jacket temperature depends on the coolant flow rate and the rate of heat transfer, but with a very high flow of coolant, an average value of T_j can be used in Eq. (5.9).

Some heat removal comes from the increase in sensible heat of the liquid flowing into and out of the tank, so the total rate of heat removal is

$$Q_R = UA(T - T_j) + F\rho c_p(T - T_F) \tag{5.10}$$

A plot of Q_R versus T is a straight line with slope $(UA + F\rho c_p)$. The x-axis intercept is slightly below the value of T_j if $T_F < T_j$, but it is equal to T_j if $T_F = T_j$. The dashed lines in Figure 5.2 are for different values of T_j and the same $(UA + F\rho c_p)$.

At steady state, the rates of heat generation and removal must be equal, so the operating temperature is found from the intersection of the Q_G and Q_R lines. There may be one, two, or three intersections for a given T_j, some of which indicate unstable conditions. When T_j is quite low, as with line 1, there is only one intersection, point a, at a low value for Q_G and a low conversion. The driving force for heat transfer, $(T - T_j)$, is also low, because not much heat is generated. The reactor is inherently stable, since a slight increase in temperature would make Q_R greater than Q_G, and the temperature would decrease toward the normal operating point. A slight decrease in T would make $Q_G > Q_R$, which would tend to increase T and restore normal conditions.

With a higher T_j, there could be three intersections, as shown by line 2 in Figure 5.2. The lower and upper intersections, b and d, are stable operating points according to the previous reasoning, but the middle intersection at c indicates an unstable condition. A slight increase in temperature leads to more heat generated than removed, and the imbalance worsens as the temperature continues to rise toward the upper stable point d. The rapid increase in temperature may lead to boiling of the liquid, with consequent foaming and entrainment, or to a dangerous increase in pressure. This type of excursion, called a *runaway reaction*, has been responsible for many explosions and fires in industrial reactors.

With a somewhat higher T_j, line 3, the Q_R line, becomes tangent to the Q_G curve at e, which is an unstable operating point, and there is a stable operating intersection f at high temperature and high conversion. For quite high values of T_j, there is only one intersection, g, indicating stable operation at high conversion, high temperature, and high driving force for heat removal.

Figure 5.2 illustrates the steady-state criterion for stability, which is that the rate of change of heat removal must be greater than the rate of change of heat generation at the normal operating temperature:

$$\frac{dQ_R}{dT} > \left(\frac{dQ_G}{dT}\right)_{ss} \tag{5.11}$$

Equation (5.11) is a necessary but not a sufficient criterion for stability. At low to moderate conversion, there is a large reservoir of reactant in the tank, and a sudden increase in T could increase Q_G more than indicated by $(dQ_G/dT)_{ss}$. The total derivative or the slope of the Q_G curve includes the effect of changing reactant concentration as well as the effect of changing rate constant. The maximum possible rate of change in Q_G is given by the partial derivative, which for a first-order reaction is related to the total derivative by the fraction unconverted [4]:

$$\left(\frac{\partial Q_G}{\partial T}\right)_x = \left(\frac{dQ_G}{dT}\right)_{ss}\left(\frac{1}{1-x}\right) \tag{5.12}$$

If the normal conversion is about 50%, a sudden small increase in the temperature could increase Q_G by up to twice the amount predicted by steady-state analysis. In practice, there will be some decrease in concentration during a rise in temperature, and the change in Q_G will depend on the order of the reaction and the nature of the upset. A formal analysis of reactor stability for a first-order reaction yields an additional criterion [4–6]:

$$UA + F\rho c_p\left(\frac{2-x}{1-x}\right) > \left(\frac{\partial Q_G}{\partial T}\right)_x \tag{5.13}$$

A conservative criterion for all reaction orders is

$$UA + F\rho c_p > \left(\frac{\partial Q_G}{\partial T}\right)_x \tag{5.14}$$

Equations (5.13) and (5.14) are not written using (dQ_R/dT), since the jacket temperature is assumed constant and the heat capacity of the wall is taken to be negligible. For operation close to the stability limit, the dynamics of the wall and jacket should be accounted for, especially if there is a large temperature change from inlet to outlet in the jacket and appreciable jacket heat capacity.

A special case arises when there is only one temperature where the rate of heat removal equals the steady-state rate of heat generation and the heat removal line is slightly steeper than the heat generation curve, as shown in Figure 5.3. Since this situation is more likely to occur at a moderate conversion, $(\partial Q_G/\partial T)_x$ may be considerably greater than (dQ_G/dT), and the stability criterion of Eq. (5.13) will not be satisfied. A slight increase in T will then make $Q_G > Q_R$, and the temperature will increase further. However, since there is no upper stable point, the temperature will eventually decrease and go below the original value. The reactor will operate in cyclic fashion between high and low values of temperature and conversion. The amplitude of the cycles depends on the order of the reaction and the difference between (dQ_R/dT) and $(\partial Q_G/\partial T)_x$ at the average temperature. [6,7,8].

For a zero-order reaction, there is no difference between $(\partial Q_G/\partial T)_x$ and $(\partial Q_G/\partial T)_x$, and Eq. (5.11) is a sufficient criterion for stability. For a second-order reaction, the difference between these derivatives is greater than for a first-order reaction, so Eq. (5.13) is conservative.

Another concept useful in stability analysis is the critical temperature difference, the maximum allowable difference between reactor temperature and jacket temperature to be sure of stable operation. When the kinetics follow the Arrhenius relation, the partial derivative of Q_G is easily calculated:

$$Q_G = r(-\Delta H)V = f(x)Ae^{-E/RT}(-\Delta H)V \tag{5.15}$$

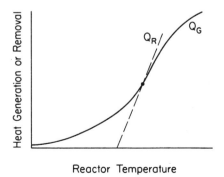

FIGURE 5.3 Heat balance for a CSTR that may exhibit a limit cycle.

$$\left(\frac{\partial Q_G}{\partial T}\right)_x = f(x)(-\Delta H)VAe^{-E/RT}\left(\frac{E}{RT^2}\right)$$

$$= Q_G\left(\frac{E}{RT^2}\right) \tag{5.16}$$

The term $\left(\frac{RT^2}{E}\right)$ has units of temperature and is called the *critical temperature difference*:

$$\Delta T_c = \left(\frac{RT^2}{E}\right) \tag{5.17}$$

The significance of ΔT_c is best shown by considering the case where the sensible heat term is quite small and almost all of the heat of reaction is removed to the jacket. The heat balance for stable operation then becomes

$$Q_R = UA(T - T_j) = Q_G \tag{5.18}$$

For stable operation,

$$\frac{dQ_R}{dT} = UA > \left(\frac{\partial Q_G}{\partial T}\right)_x \tag{5.19}$$

From Eqs. (5.16), (5.17), and (5.19),

$$UA > \frac{Q_G}{\Delta T_c} \tag{5.20}$$

From Eqs. (5.18) and (5.20),

$$UA > \frac{UA(T - T_j)}{\Delta T_c} \tag{5.21}$$

For stable operation,

$$(T - T_j) < \Delta T_c \tag{5.22}$$

Equation (5.22) offers a quick way of estimating the reactor stability. If ΔT_c is 20°C and the preliminary design calls for $T = 80°C$ and $T_j = 30°C$ with a moderate conversion, the reactor would probably be uncontrollable since $(T - T_j) >> \Delta T_c$. With a greater value of UA and $(T - T_j) = 25°C$, the reactor would be unstable according to Eq. (5.22). But if the reaction is first order, a more thorough analysis using Eq. (5.13) might indicate stable operation. However, it would be risky to operate close to unstable conditions unless controls are present to prevent a runaway reaction.

In principle, a reactor can be operated at an unstable point using a feedback control system if the jacket temperature can be changed rapidly enough when the reactor temperature rises. The jacket temperature can be adjusted by blending warm and cold water fed to the jacket or by changing the pressure if the jacket fluid evaporates. Consider what could happen with automatic adjustment of the jacket temperature. If a 1°C increase in reactor temperature leads to a 3°C decrease in jacket temperature, the heat removal rate increases by $4 \times UA$, which means the effective heat removal line is nearly four times steeper than before. Now (dQ_R/dT) may be much greater than $(\partial Q_G/\partial T)_x$, making the system stable. There is a minimum gain for this type of control system as well as the maximum gain that is calculated from stability analysis. If the gains are far enough apart, stable operation at an inherently unstable point is possible [9]. A cascade control system should be used, with the jacket controller receiving a signal from the reactor temperature controller.

Batch Reactions

For batch reactions in a stirred tank, where temperature control is achieved by jacket cooling, the concept of critical temperature difference also applies. If a constant reactor temperature is needed, the fluid might be brought to the desired temperature and then catalyst or a second reagent added to start the reaction. The jacket temperature could be manually or automatically adjusted to allow for decreasing rate as the reaction proceeded. The danger of runaway reaction is greatest at the start, when the reactant concentration is high, and the required $(T - T_j)$ is large.

For semibatch operation, there is little chance of a runaway reaction if the added reactant is always present at low concentrations in the reactor. This is the case when the added reactant is a slightly soluble gas, such as oxygen or hydrogen. The reaction temperature and reaction rate can be controlled by the feed rate of the reactant. However, if the added reactant is a liquid or a very soluble gas, it might accumulate in the reactor, and this could lead to a stability problem if the difference between reactor and jacket temperature becomes large.

Tubular Reactors

To carry out an exothermic reaction in a tubular reactor under nearly isothermal conditions, a small diameter is needed to give a high ratio of surface area to volume. The reactor could be made from sections of jacketed pipe or from a long coil immersed in a cooling bath. The following analysis is for a constant jacket temperature, and the liquid is assumed to be in plug flow, with no radial gradients of temperature or concentration and no axial conduction or diffusion.

If the feed enters at the jacket temperature, the reactor temperature usually rises to a maximum not far from the inlet, as shown in Figure 5.4. At this point, the local rate of heat generation equals the rate of heat removal, and the temperature starts to fall. (For a zero-order reaction, the temperature would remain at the maximum value until the reaction was complete). The reactor temperature continues to decrease with length, and it approaches the jacket temperature if the reaction goes to a high conversion.

If the maximum temperature difference $(T_{max} - T_j)$ is close to the critical value, a slight change in feed concentration or jacket temperature may lead to a large change in T_{max} or to a runaway reaction. The stability analysis differs from that for a CSTR because both temperature and concentration change along the length of the reactor. There are no multiple steady states of the type possible with a CSTR. Instead, there is a unique temperature profile for each jacket temperature and feed condition. What makes the reactor potentially unstable in a practical sense is that very large changes in T_{max} can result from small changes in T_j or in any of the other parameters. This region of very high sensitivity must be avoided to ensure safe operation.

The material balance equation for a first-order homogeneous reaction is written for an incremental length or volume using a volumetric flowrate, F, or a velocity, u:

$$FC_{A_0}\, dx = r\, dV = kC_{A_0}(1 - x)dV \tag{5.23}$$

$$u\left(\frac{\pi D^2}{4}\right)C_{A_0}dx = kC_{A_0}(1 - x)\frac{\pi D^2}{4} dl \tag{5.24}$$

$$\frac{dx}{dl} = \frac{k(1 - x)}{u} \tag{5.25}$$

where $k = Ae^{-E/RT}$

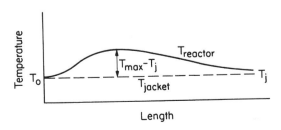

FIGURE 5.4   Temperature profile in a tubular reactor.

The steady-state energy balance shows that the incremental heat released goes partly to change the sensible heat of the fluid, and the rest is transferred to coolant in the jacket.

$$dQ = FC_{A_o} \, dx(-\Delta H) = F\rho c_p \, dT + U \, dA(T - T_j) \qquad (5.26)$$

or

$$u\left(\frac{\pi D^2}{4}\right)C_{A_o} \, dx(-\Delta H) = u\left(\frac{\pi D^2}{4}\right)\rho c_p \, dT + U\pi D \, dl(T - T_j) \qquad (5.27)$$

Combining Eqs. (5.25) and (5.27) and rearranging gives

$$\frac{dT}{dl} = \frac{C_{A_o}k(1-x)(-\Delta H) - U\frac{4}{D}(T - T_j)}{u\rho c_p} \qquad (5.28)$$

Equations (5.25) and (5.28) can be solved numerically to give T and x as functions of l. One approach is to use small increments of l to calculate Δx from Eq (5.25) and then get ΔT from Eq (5.28). At each step, revised values of Δx and ΔT are obtained using k at $(T + \Delta T/2)$ and using $(T + \Delta T/2 - T_j)$ as the heat transfer driving force. Another method is to use small increments of x to get Δl from Eq. (5.25) and then get ΔT. (Don't confuse ΔT, the incremental change in reactor temperature, with $(T - T_j)$, the driving force for heat transfer.)

Example 5.1

Calculate the temperature profiles for the following reaction carried out in a 2-cm-diameter tubular reactor with feed and jacket temperatures of about 350 K.

$$A \rightarrow B + C \qquad \Delta H = -25 \text{ kcal/mol}$$

$$r = kC_A \qquad C_{A_0} = 2M = 0.002 \text{ mol/cm}^3$$

$$k = 0.00142 \text{sec}^{-1} \quad \text{at } 340K$$

$$\frac{E}{R} = 15,000 \text{ K}^{-1}$$

$$\rho = 0.8 \text{ g/cm}^3$$

$$c_p = 0.5 \text{ cal/g, }^\circ\text{C}$$

$$U = 0.025 \text{ cal/sec, cm}^2, {}^\circ\text{C}$$

$$u = 60 \text{ cm/sec}$$

Solution.

$$\ln\left(\frac{k}{0.00142}\right) = 15,000\left[\frac{1}{340} - \frac{1}{T}\right]$$

From Eq. (5.25),

$$\frac{dx}{dl} = \frac{k(1-x)}{60}$$

From Eq. (5.28),

$$\frac{dT}{dl} = \frac{0.002k(1-x)(25,000) - 0.025(2.0)(T - T_j)}{60(0.8)(0.5)}$$

These equations were solved numerically, and the temperature profiles for four feed temperatures are plotted in Figure 5.5. For $T_0 = T_j = 348$ or 349 K, the temperature profiles appear normal in shape, with a maximum temperature at about 25–30% conversion. A 1°C increase in feed temperature results in a 3.5°C increase in maximum temperature, which shows that the reaction has a moderate sensitivity to changes in initial

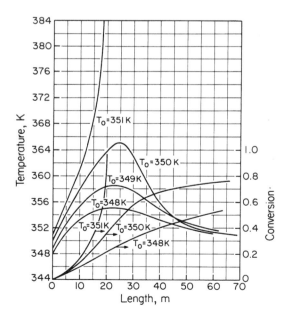

FIGURE 5.5 Temperature profiles and conversion for a jacketed tubular reactor, Example 5.1.

conditions. With $T_0 = 350$ K, the temperature profile still appears normal, but the maximum temperature increases by 6.5°C for the 1°C increase in T_0. This high sensitivity to T_0 is a warning that the reactor is nearing unstable operation. This is confirmed by calculations for $T_0 = 351$ K, which show a runaway reaction with a peak temperature of 419 K (offscale in Fig. 5.5) and complete conversion at $L = 20$ m. Therefore it would not be safe to operate the reactor at $T_0 = 350$ K, and perhaps not even at $T_0 = 349$ K, because a small change in jacket temperature, feed concentration, or heat transfer coefficient could lead to a runaway reaction.

Another way of judging how close a reactor is to unstable operation is to compare the maximum temperature rise with the critical temperature difference, as defined by Eq. (5.17). This ratio is denoted by θ:

$$\theta = \frac{T_{max} - T_j}{\Delta T_c} = \frac{T_{max} - T_j}{RT^2/E} \tag{5.29}$$

A conservative criterion for stability is that θ be less than 1.0. The actual stability limit depends on the reaction kinetics and on α, the ratio of the adiabatic temperature rise to ΔT_c,

$$\alpha = \frac{\Delta T_{ad}}{\Delta T_c} \tag{5.30}$$

Going back to Example 5.1 and Figure 5.5, ΔT_c is about 8.6° C based on $T = 360$ K ($360^2/15{,}000 = 8.64$). For $T_0 = 350$ K, $T_{max} = 365$K, and $\theta = 15/8.6 = 1.74$, and $\alpha = 125/8.6 = 14.5$. The reactor profile appears normal even though $\theta > 1$, because the decrease in reactant concentration with length has a stabilizing effect. If the reaction was second order, the limiting temperature profile would correspond to $\theta \geq 2.0$; but for a zero-order reaction, the stability limit is $\theta = 1.0$ [10].

The reactor stability decreases with increasing values of α, since the fraction converted at the peak temperature is lower when ΔT_{ad} is higher. One study showed that the allowable value of θ for a first-order reaction ranged from 2.4 to 1.1 as α increased from about 7 to 70 [11,12]. There have been many other studies of the stability of tubular reactors and batch reactors, and some complex correlations for the stability limit allowing for changes in coolant temperature with length and the thermal capacity of the reactor wall [13]. However, it is generally not necessary to get the exact stability limit. The conservative criterion that $\theta < 1$ is often used unless calculations for different conditions show that even with $\theta > 1$ the reactor is definitely stable to all likely disturbances.

PACKED-BED TUBULAR REACTORS

Reactions on solid catalysts are often carried out in tubes or pipes packed with spherical, cylindrical, or ring-shaped catalyst particles. For exothermic reactions, small-diameter tubes are used to permit nearly isothermal operation and prevent a temperature runaway. For small-scale operation, a cooling jacket can be placed around each pipe. But for large production, hundreds or thousands of tubes containing catalyst are mounted in a heat exchanger, with coolant circulated on the shell side. The design is more complicated than for homogeneous reactions, since there are radial as well as axial temperature gradients in the tubes, and there may also be significant concentration and temperature gradients near the surface of the individual particles. The effects of internal concentration gradients were discussed in Chapter 4. Here we first treat mass and heat transfer to the external surface of catalyst particles, and then the problem of heat transfer in a bed of particles is discussed, with recommendations for design procedures.

External Mass Transfer

Diffusion of reactants to the external surface is the first step in a solid-catalyzed reaction, and this is followed by simultaneous diffusion and reaction in the pores, as discussed in Chapter 4. In developing the solutions for pore diffusion plus reaction, the surface concentrations of reactants and products are assumed to be known, and in many cases these concentrations are essentially the same as in the bulk fluid. However, for fast reactions, the concentration driving force for external mass transfer may become an appreciable fraction of the bulk concentration, and both external and internal diffusion must be allowed for. There may also be temperature differences to consider; these will be discussed later. Typical concentration profiles near and in a catalyst particle are depicted in Figure 5.6. As a simplification, a linear concentration gradient is shown in the boundary layer, though the actual concentration profile is generally curved.

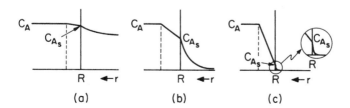

FIGURE **5.6** External and internal concentration gradients for a catalyst pellet: (a) Slow reaction; (b) fast reaction; (c) very fast reaction.

In Figure 5.6a, the surface concentration is almost equal to the bulk value, and external mass transfer has a negligible effect on the reaction rate. Note that the gradient just inside the particle is steeper than that outside, because the effective diffusivity, D_e, is much less than the bulk diffusivity, D_{AB}, and the fluxes to the particle and into the particle are equal:

$$D_{AB}\left(\frac{dC_A}{dr}\right)_{R,outside} = D_e\left(\frac{dC_A}{dr}\right)_{R,inside} \tag{5.31}$$

Figure 5.6b shows the gradient for a fast reaction and a low effectiveness factor. The external gradient becomes steeper, to keep up with the faster reaction in the pellet. Again, the gradient just inside the pellet is several times steeper than the external gradient. In typical cases, D_e would be 0.01–0.2 times D_{AB}, making the internal gradient 5–100 times steeper than that outside the pellet.

With an extremely active catalyst or very high temperature, most of the reactant is consumed very close to the external surface, as shown in Figure 5.6c, and the effectiveness factor is very low. The reactants must still diffuse through the entire boundary layer, which may be a much greater distance that the average diffusion distance inside the pellet. When the surface concentration is almost zero, the reaction rate is controlled by the rate of external mass transfer, and further increases in the kinetic rate constant have almost no effect on rate.

The rate of mass transfer to the surface is expressed using either concentration or partial pressure as a driving force. The term a is the external surface area per unit mass of catalyst:

$$r = k_c a\left(C_A - C_{A_s}\right) \tag{5.32}$$

$$r = k_g a\left(P_A - P_{A_s}\right) \tag{5.33}$$

For a first-order reaction, the mass transfer coefficient can be combined with the effective rate constant to give an overall coefficient, K_O, since we have two first-order steps in series.

$$r = k\eta P_{A_s} \tag{5.34}$$

Combining Eqs. (5.33) and (5.34) gives

$$r = K_O P_A \tag{5.35}$$

where

$$\frac{1}{K_O} = \frac{1}{k_g a} + \frac{1}{k\eta}$$

The importance of external mass transfer can be judged by comparing $k_g a$ and $k\eta$ or their reciprocals, which are resistances. For example, if $k_g a = 10$ and $k\eta = 2$ (any consistent units), $K_O = 1.67$ and external mass transfer contributes $0.1/0.6$ or 17 percent of the overall resistance to the process of mass transfer plus reaction.

If the reaction is not first order, Eq. (5.33) still applies, but there is no general solution for the overall reaction rate. If the reaction follows Langmuir–Hinshelwood kinetics and the effectiveness factor can be estimated, the reaction rate can be calculated for certain values of P_A and P_B:

$$A + B \rightarrow C$$

$$r = k\eta K_A \frac{P_{A_s} K_B P_{B_s}}{\left(1 + K_A P_{A_s} + K_B P_{B_s}\right)^2} \tag{5.36}$$

Equation (5.36) can be used with Eq. (5.33) to permit solving for $\left(P_A - P_{A_s}\right)$ and r if the gradient for B is negligible (B might be in large excess). A graphical solution is shown in Figure 5.7, where the reaction rate is plotted against P_{A_s} for a given value of P_{B_s}. When P_A is high (as with P_{A_1}) and the reaction is nearly zero order to A, the mass transfer driving force, $P_A - P_{A_s}$ is about 20% of P_A, but the reaction rate r is only about 4% less than the ideal rate. However, when P_A is lowered to P_{A_2} and the apparent reaction order is close to 1, external mass transfer has a much greater effect on the overall rate (about a 30% reduction for the second case shown in Fig. 5.7).

Mass and Heat Transfer to a Single Sphere

Because of the analogy between mass transfer by diffusion and heat transfer by conduction in a boundary layer, correlations for mass transfer and heat transfer to particles are similar. For mass transfer to a single isolated sphere,

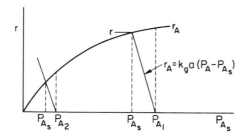

FIGURE 5.7   Effect of external mass transfer with a Langmuir–Hinshelwood reaction.

the Sherwood number depends on the square root of the Reynolds number, but it approaches a theoretical limit of 2.0 as the Reynolds number approaches zero. The following equation is fairly accurate up to a Reynolds number of 1000:

$$\text{Sh} = \frac{k_c d}{D_{AB}} = 2.0 + 0.6 \, \text{Re}^{1/2} \text{Sc}^{1/3} \tag{5.37}$$

For heat transfer, the Nusselt number is

$$\text{Nu} = \frac{hd}{k} = 2.0 + 0.6 \, \text{Re}^{1/2} \text{Pr}^{1/3} \tag{5.38}$$

Equation (5.38) is known as the Frössling [14] equation or the Ranz and Marshall [15] equation. The exponents 1/2 and 1/3 come from boundary-layer theory, and the 0.6 is an empirical constant that gives the average coefficient over the sphere surface. The local coefficient is two to three times higher at the front of the sphere than at the point of boundary-layer separation.

Mass and Heat Transfer in Packed Beds

When spheres or other particles are in a packed bed, the mass and heat transfer coefficients are higher than for isolated particles exposed to the same superficial velocity. There is also more uncertainty in the coefficients, because published data show considerable scatter and significant differences in the Reynolds number exponents. Most of the published data are reported as j-factors, which are dimensionless groups that are functions of the Reynolds number:

$$\text{For mass transfer}: \quad j_m = \frac{k_c}{u} \text{Sc}^{2/3} = f_1(\text{Re}) \tag{5.39}$$

$$\text{For heat transfer}: \quad j_h = \frac{h}{c_p G} \text{Pr}^{2/3} = f_2(\text{Re}) \tag{5.40}$$

Figure 5.8 shows typical reported correlations for j_m and j_h based on tests of water evaporation from porous particles [16,17], sublimation of naphthalene[18], diffusion-controlled decomposition of hydrogen peroxide[19], and electrochemical reduction of ferricyanide ions [20]. In the middle range of Reynolds numbers, Re = 100–1000, the j-factor correlations are in reasonable agreement, considering the difficulties in measuring or estimating the values of surface temperature or surface concentration.

Although early workers reported separate correlations for j_m and j_h, they should be identical unless heat transfer by radiation is important or

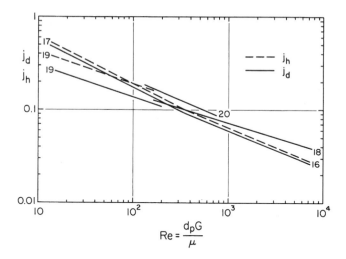

FIGURE 5.8   Mass and heat transfer correlations for packed beds.

unless the net flow to or from the surface is significant. An equation that fits most of the data fairly well is [21]

$$j_m = 1.17(\text{Re})^{-0.42} \tag{5.41}$$

To permit comparison with data for single spheres, the j-factor can be converted to Sherwood or Nusselt numbers by multiplying by Re and $Sc^{1/3}$ or $Pr^{1/3}$:

$$\frac{k_c}{u} Sc^{2/3} \frac{du\rho}{\mu} Sc^{1/3} = \frac{k_c d\rho}{\mu} Sc = \frac{k_c d}{D_{AB}} \tag{5.42}$$

Therefore Eq. (5.41) becomes

$$\text{Sh} = 1.17 \, \text{Re}^{0.58} Sc^{1/3} \tag{5.43}$$

Similarly,

$$\text{Nu} = 1.17 \, \text{Re}^{0.58} Pr^{1/3} \tag{5.44}$$

Figure 5.9 compares Sherwood and Nusselt numbers for single spheres with those for particles in packed beds. For gases ($Sc \cong 1$, $Pr \cong 1$), the transfer coefficients are about three times higher in packed beds, in large part because the maximum gas velocity in the bed is up to several-fold higher than the superficial velocity, which is used in the Reynolds number. Note that for low Reynolds numbers, extrapolation of Eq. (5.43) would give

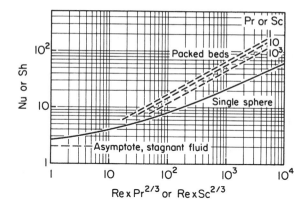

FIGURE 5.9 Comparison of transfer coefficients for a single sphere and particles in packed beds.

values of k_c less than for an isolated sphere, which cannot be correct. The equation for packed beds should be of the form

$$\mathrm{Sh} = \mathrm{Sh}_0 + b\mathrm{Re}^{0.58}\mathrm{Sc}^{1/3} \qquad (5.45)$$

The limiting Sherwood number is probably between 6 and 10; the value will depend on the bed porosity and particle shape.

Internal Temperature Gradients

When an exothermic reaction takes place in a catalyst particle, the heat released by reaction is conducted to the outside of the particle and then transferred to the fluid phase. The temperature gradients for typical cases are sketched in Figure 5.10. The difference between the center temperature, T_C, and the surface temperature, T_s, is usually quite small and can often be

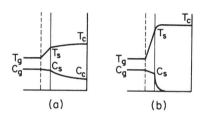

FIGURE 5.10 External and internal temperature and concentration gradients: (a) Slow reaction, high η; (b) fast reaction, low η.

neglected when compared to the external temperature difference $(T_s - T_g)$. The following derivation leads to a simple equation for $(T_C - T_s)$, which can be used when the difference is small.

Consider an exothermic reaction in a spherical pellet with a significant internal concentration gradient for the key reactant, A:

$$A + B \rightarrow C$$
$$r = kP_A$$

The total heat released in the particle is the flux of A into the particle times the heat of reaction:

$$Q_G = 4\pi R^2 D_e \left(\frac{dC_A}{dr}\right)_R (-\Delta H) \tag{5.46}$$

Heat is transferred to the particle surface by conduction, and the thermal conductivity of the catalyst is k_s:

$$Q_t = -4\pi R^2 k_s \left(\frac{dT}{dr}\right)_R \tag{5.47}$$

Similar equations can be written for any radius r within the particle, and the $4\pi r^2$ terms cancel when Q_G and Q_t are equated:

$$D_e \left(\frac{dC_A}{dr}\right)_r (-\Delta H) = -k_s \left(\frac{dT}{dr}\right)_r \tag{5.48}$$

$$-dT = \frac{D_e(-\Delta H)dC_A}{k_s} \tag{5.49}$$

Integration of Eq. (5.49) gives

$$(T_c - T_s) = \frac{D_e(C_{A_s} - C_{A_C})(-\Delta H)}{k_s} \tag{5.50}$$

The maximum temperature difference occurs when the effectiveness factor is low and the concentration at the center is almost zero:

$$(T_C - T_s)_{max} = \frac{D_e C_{A_s}(-\Delta H)}{k_s} \tag{5.51}$$

Example 5.2

The hydrogenation of ethylene is studied at 200°C and 1.2 atm using 5-mm catalyst pellets and a gas with 5% C_2H_4 and 95% H_2. The solid conductivity is about 8×10^{-4} cal/sec,cm^2, °C/cm [10], and D_e for C_2H_4 is 0.02 cm^2/sec. Estimate the maximum internal temperature difference:

$$C_2H_4 + H_2 \rightarrow C_2H_6 \qquad \Delta H = -32.7\,\text{kcal}$$

Solution. Assume

$$C_s = C_g = \frac{0.05}{22,400} \times \left(\frac{1.2}{1}\right) \times \left(\frac{273}{473}\right) = 1.55 \times 10^{-4}\ \text{mole/cm}^3$$

From Eq. (5.51)

$$(T_C - T_s)_{\text{max}} = \frac{0.02\left(1.55 \times 10^{-6}\right)(32,700)}{8 \times 10^{-4}} = 1.27^\circ C$$

A temperature difference of 1.3°C would generally be considered negligible.

It might seem strange that the reaction rate constant and the pellet radius do not affect $(T_C - T_s)_{\text{max}}$, although increasing the radius or the rate constant lowers the effectiveness factor and C_{A_C}, and thus increases $(T_C - T_s)$, according to Eq. (5.50). However, once the effectiveness factor is low enough to make the center concentration almost zero, further increases in the rate constant k shift the reaction zone closer to the external surface, so heat does not have to be transferred as far. The internal temperature gradient becomes steeper as k increases, as shown in Figure 5.10, and $(T_C - T_s)$ remains constant.

If some combination of parameters is suspected to lead to large internal temperature differences, the temperature distribution, effectiveness factor, and reaction rate could be calculated by solving the equations for pore diffusion and reaction simultaneously with the equation for heat conduction. Numerical solutions for some cases of first-order kinetics have been published [22], and the predicted effectiveness factors range from 0.01 to over 100. Effectiveness factors greater than 1.0 result when the effect of higher internal temperature outweighs the effect of lower reactant concentration. However, these solutions are hardly ever applied to real problems, because the kinetics are often more complex than first order, and because the thermal parameters do not match the specific values of the published studies. Furthermore, when the reaction is rapid enough to make $(T_c - T_s)$ significant, the external temperature and concentration differ significantly from the bulk values, and the reaction rate is more dependent on external mass and heat transfer coefficients than on the internal temperature gradient.

External Temperature Difference and Stability Analysis

The temperature gradient in the gas film just outside a catalyst pellet is steeper than the internal gradient at the surface, because the solid conductivity is generally several times greater than the thermal conductivity of the gas, and the heat flux to the surface equals the heat flux away from the surface:

$$Q = 4\pi R^2 k_s \left(-\frac{dT}{dr}\right)_{R,\text{pellet}} = 4\pi R^2 k_g \left(-\frac{dT}{dr}\right)_{R,\text{gas}} \tag{5.52}$$

$$\left(-\frac{dT}{dr}\right)_{R,\text{gas}} = \frac{k_s}{k_g}\left(-\frac{dT}{dr}\right)_{R,\text{pellet}} \tag{5.53}$$

The external temperature difference can be related to the reaction rate and the heat generated in a pellet using the appropriate correlation to predict the heat transfer coefficient for the gas film. At steady state, the rate of heat generation is equal to the rate of heat removal. For a first-order reaction in a spherical pellet, the heat balance is

$$Q = \frac{4}{3}\pi R^3 \rho_s k \eta C_{A_s}(-\Delta H) = 4\pi R^2 h(T_s - T_g) \tag{5.54}$$

Equation (5.54) can be used for a first estimate of $(T_s - T_g)$, but since k and η depend on the average pellet temperature and since C_{A_s} is lower than the bulk concentration C_A, a trial-and-error solution would be needed for exact values of T_s and C_{A_s}. A simpler equation to show the relative values of T_s and C_{A_s} is obtained by relating Q to the rate of mass transfer to the surface and the heat of reaction:

$$Q = 4\pi R^2 k_c(C_A - C_{As})(-\Delta H) = 4\pi R^2 h(T_s - T_g) \tag{5.55}$$

The ratio k_c/h comes from the j-factor correlations for particles in a packed bed, Eqs. (5.39) and (5.40), with $f_1 = f_2$:

$$\frac{k_c}{u}\text{Sc}^{2/3} = \frac{h}{c_p G}\text{Pr}^{2/3} \tag{5.56}$$

$$\frac{k_c}{h} = \frac{u}{c_p G}\left(\frac{\text{Pr}}{\text{Sc}}\right)^{2/3} \tag{5.57}$$

Since $G = u\rho$, combining Eqs. (5.55) and (5.57) gives

$$T_s - T_g = \left(\frac{\text{Pr}}{\text{Sc}}\right)^{2/3}\frac{(C_A - C_{A_s})(-\Delta H)}{\rho c_p} \tag{5.58}$$

The adiabatic temperature rise for the system is a convenient scaling parameter:

$$\Delta T_{\text{ad}} = \frac{C_A(-\Delta H)}{\rho c_p} \tag{5.59}$$

Combining the last two equations shows that the scaled temperature difference is proportional to the relative concentration difference:

$$\frac{T_s - T_g}{\Delta T_{ad}} = \left(\frac{Pr}{Sc}\right)^{2/3} \frac{(C_A - C_{A_s})}{C_A} \tag{5.60}$$

Since the Schmidt and Prandtl numbers for gases are not far from 1.0 and since ΔT_{ad} is often quite large (200–500°C), an external concentration difference of only 5–10% is accompanied by an appreciable external temperature difference. A 10°C rise in surface temperature would have a much greater effect on the reaction rate than a 5–10% decrease in surface concentration. When a reaction is rapid enough for external gradients to be important, attention is generally focused on the external temperature difference. If this difference is too large, the pellet may become unstable and jump to a much higher temperature.

The stability problem for an exothermic reaction in a catalyst particle is similar to that for a reaction in a CSTR, in that multiple solutions of the heat and mass balance equations are possible. A typical plot of heat generation and removal rates is shown in Figure 5.11. The values of Q_G and Q_R are in cal/sec, g, and a is the external area in cm^2/g. The plot differs from the one for a CSTR (Fig. 5.2) in that the highest possible value for Q_G is a mass transfer limit corresponding to $C_s = 0$ and not to complete conversion. The mass transfer limit increases with temperature because of the increase in diffusivity, and the limit also increases with gas velocity. The heat removal

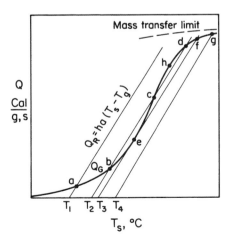

FIGURE 5.11   Stable and unstable operating points for an exothermic reaction in a catalyst particle.

lines in Figure 5.11 have a constant slope, but a more accurate plot over a wide temperature range would show increases in slope because of radiant heat transfer at high temperatures.

The heat removal lines in Figure 5.11 are for different bulk gas temperatures, and the lines start at $Q_R = 0$, where $T_s = T_g$. For the lowest gas temperature, $T_G = T_1$, there is only one intersection, point a, which is a stable operating point, where T_s is only slightly greater than T_g. For $T_g = T_4$, there is also only one intersection, point g, which is a stable surface temperature at a large concentration difference. For $T_g = T_2$, there are three intersections, and point c is an unstable operating point. A slight increase in T_s would make $Q_G > Q_R$, and the temperature would rise to the upper stable point. If the gas temperature was gradually increased starting at T_1, the surface temperature would increase, and the pellet would be stable until the Q_R curve became tangent to the Q_G curve at point e. This is an unstable point, and the temperature would rise to point f. The corresponding value of $(T_s - T_g)$ at point e is the critical temperature difference for stable operation.

A conservative estimate of the critical temperature difference can be obtained by equating ha, the rate of change of Q_R with T_s, to the partial derivative of Q_G with T_s, assuming an exponential increase in reaction rate with temperature. The result is the same equation that was derived for a CSTR, Eq. (5.17):

$$\Delta T_c = \frac{RT_s^2}{E} \tag{5.61}$$

The true allowable temperature difference is somewhat higher than the value from Eq. (5.61), because C_s decreases as T_s increases. The stability depends on the order of the reaction and the relative concentration change. Another stabilizing effect is the decrease in apparent activation energy with temperature if the reaction is in the region of moderate to strong pore diffusion limitations.

When a catalyst is operating with a temperature difference close to the critical value, a slight change in reactant concentration, gas temperature, or gas flow rate can lead to a rapid rise in pellet temperature. This jump might be called a runaway if it leads to greatly increased byproduct formation or other undesired effects. However, if a high reaction temperature is desired, as in incinerators and catalytic mufflers, the jump is called *ignition*, and the ignition temperature is the gas temperature needed to initiate a jump to an upper stable state.

The transition from stable operating conditions at low temperatures to an unstable point and then stable high-temperature operation can be illustrated on an Arrhenius plot of ln(rate) versus reciprocal gas temperature,

Figure 5.12. In the low-temperature region, the plot may appear linear, implying a constant activation energy. However, as the rate increases, the difference between surface and gas temperature also increases, so the slope of the line based on T_g indicates an apparent activation energy greater than the value based on surface temperature. The rate plotted as a function of the estimated surface temperature is shown as a dashed line in Figure 5.12. When the critical ΔT is reached, the plot of $\ln r$ vs $1/T_g$ becomes discontinuous, since there is no steady-state solution until the upper stable point is reached. Further increases in temperature give small increases in rate because the mass transfer coefficient increases slightly with increasing temperature.

If the gas temperature is gradually decreased, the rate remains high as the temperature passes the ignition point and then drops abruptly at the extinction point, which occurs when the heat removal line becomes tangent to the heat generation curve, as at point h in Figure 5.11. The heat removal line is not shown for this situation, but the gas temperature at extinction would be slightly less than T_2. The width of the hysteresis zone in Figure 5.12 depends on the shape of the Q_G curve and could be only a few degrees

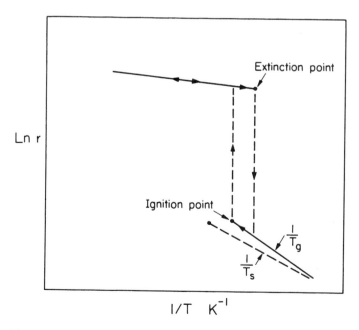

FIGURE 5.12 Ignition and extinction points for an exothermic reactions in a catalyst pellet.

Centigrade or more than 100°C. Some systems with a low activation energy and a high heat transfer coefficient may not have multiple solutions and show only a smooth rise in surface temperature and reaction rate as the gas temperature increases.

There are many examples of ignition phenomena in the literature. The burning of carbon particles in air is controlled by kinetics at moderate temperatures and has a high activation energy. Once ignition occurs, the particles become red hot, and the reaction rate is limited by mass transfer. Combustion of single carbon spheres was studied by Tu, Davis, and Hottel [23], who found a high activation energy and a rate independent of gas velocity up to temperatures of about 1000 K. At high temperatures, combustion was diffusion controlled with a low activation energy, and the rate increased with the 0.5 power of the gas velocity. There was a jump in surface temperature on ignition, but it was not as large as predicted by Eq. (5.60), because radiant heat transfer was the dominant mode of heat removal.

The catalytic hydrogenation of olefins provides an example of ignition at relativity low temperatures. Figure 5.13 shows rate data for hydrogenation of ethylene with 0.14-in. spheres of Ni/Al_2O_3 [24]. For temperatures of 34°C to 73°C, the reaction rate nearly doubled for each 12°C increase in temperature and then increased ninefold for an 8°C increase in temperature. Analysis shows that the external temperature difference $(T_s - T_g)$ was about 12°C at 73°C and exceeded ΔT_c at the next value of T_g.

The ignition point can also be reached by changing other parameters, such as reactant concentration, gas velocity, and particle size. The effect of

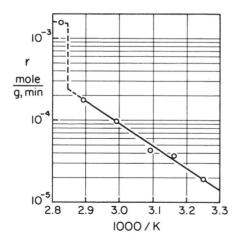

FIGURE 5.13   Instability observed in ethylene hydrogenation. (Data from Ref. 24).

particle size on ethylene hydrogenation is shown in Figure 5.14 [25]. The initial decline in rate is due to a decrease in effectiveness factor, and the large increase in rate at about 0.04 cm occurs because the critical temperature difference was exceeded. For large particles, the surface temperature is high and the reaction is diffusion controlled. The decrease in rate with particle size is due mainly to the decrease in external area.

RADIAL HEAT TRANSFER IN PACKED BEDS

When an exothermic reaction is carried out in a packed tube and heat is removed at the wall, the radial temperature profile is approximately parabolic, with a maximum at the center, as shown in Figure 5.15. The shape of the profile is similar to that for a laminar flow reactor, but the radial heat flux for a given gradient is much greater than with laminar flow of gas, because the particles contribute to an enhanced thermal conductivity. Near the wall, the profile becomes much steeper, because heat transfer is mainly by conduction through the gas boundary layer. At any point in the bed, the solid temperature is at least slightly higher than the gas temperature, as indicated by the dashed line in Figure 5.15. However, the local difference $(T_s - T_g)$ is much smaller than the overall temperature drop in the bed, and in the model used here, called a *homogeneous* model, T_s and T_g are assumed to be the same. The reactor temperature is T_g, with T_{ave} referring to a cross-sectional average temperature and T_C to the center or maximum temperature.

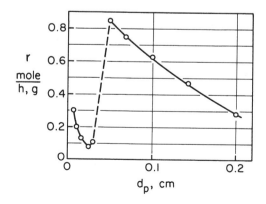

FIGURE 5.14 Effect of particle size on ethylene hydrogenation. (Data from Ref. 25).

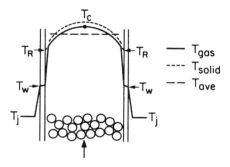

FIGURE 5.15   Radial temperature profile in a packed-bed tubular reactor.

In a one-dimensional model of a tubular reactor, the radial heat flux is expressed using an overall coefficient and an average driving force:

$$Q = UA(T_{\text{ave}} - T_j) \qquad (5.62)$$

An alternate approach might be based on the maximum driving force and a different coefficient, U'.

$$Q = U'A(T_C - T_j) \qquad (5.63)$$

Use of the average driving force, $(T_{\text{ave}} - T_j)$ and Eq. (5.62) is preferred to simplify the analysis. In a stepwise calculation to get the concentration and temperature as a function of reactor length, T_{ave} is used to evaluate the rate constant k and the rate of heat transfer. The centerline temperature, T_C, can be calculated later if necessary [see Eq. (5.88)].

The overall coefficient U is obtained by summing the resistances in the jacket, the metal wall, the gas film at the wall, and the catalyst bed:

$$\frac{1}{U} = \frac{1}{h_j} + r_{\text{wall}} + \frac{1}{h_w} + \frac{1}{h_{\text{bed}}} \qquad (5.64)$$

Determining the values of h_w and h_{bed} from experiments is a challenging task, and a great many empirical correlations have been presented. Most of the data are for heat transfer without reaction, such as for heating air in a steam-jacketed pipe packed with spheres. For these tests, Q is measured from the change in sensible heat of the air, and U is calculated from the usual equation, $Q = UA\overline{\Delta T}_L$. The small steam–film and metal–wall resistances can be subtracted from the overall resistance to obtain an overall bed coefficient, h_o:

$$\frac{1}{U} - \frac{1}{h_j} - r_w = \frac{1}{h_w} + \frac{1}{h_{bed}} = \frac{1}{h_o} \tag{5.65}$$

Early attempts to get a correlation for h_o similar to the correlations for pipe flow showed wide variations in the exponents for the dimensionless groups and even differences in the equations for heating and cooling [26,27]. Such variations are understandable when we consider the different mechanisms of heat transfer in the bed itself and in the gas film at the wall.

Mechanisms of Heat Transfer in Packed Beds

The principle mechanisms of heat transfer in packed beds are (1) conduction through the gas and solid phases, (2) radiation between particles combined with conduction, and (3) convective flow of fluid elements. Conduction of heat in a two-phase system is analogous to conduction of electricity or diffusion, and rigorous solutions for the effective conductivity or diffusivity are available for regular arrays, such as uniform spheres in cubic or hexagonal packing. For a packed-bed or a tubular reactor, where particles are just dumped in the bed, the random arrangement of the packing makes a theoretical approach too difficult. Instead we rely on measurements of effective thermal conductivity that are made for rectangular beds of dumped packing with no fluid flow. Some typical results are shown in Figure 5.16, where the effective thermal conductivity relative to that of the gas, k_e^0/k_g, is

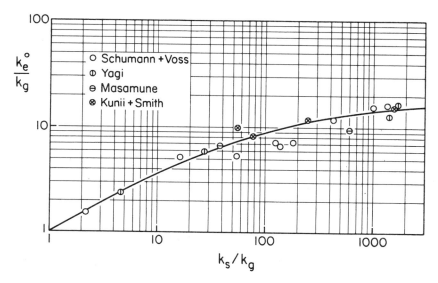

FIGURE 5.16 Effective thermal conductivity of packed beds.

plotted as a function of the conductivity ratio for the individual phases [28–31]. When k_s/k_g is about 10, a typical value for a porous catalyst in air, the effective bed conductivity k_e^0 is about $4k_g$. The highest values of k_s/k_g are for quartz or metal particles, and k_e^0/k_g may be 10–15. With solids of very high conductivity, the limiting resistance is in the thin film of gas where the particles touch or almost touch, so changes in k_s have little effect on k_e^0 if $k_s/k_g > 1000$. The zero superscript in k_e^0 is a reminder that these are zero-flow values. The conductivity of a bed or a suspension of particles depends on the packing density or the void fraction, which may explain some of the scatter in Figure 5.16. The typical void fraction for a bed of spheres or stubby cylinders is 0.4, but higher values are found when the tube diameter is only a few times the particle size.

For tests at 100°C, radiation is probably negligible, but for catalysts beds at 200°C or higher, radiation may have a significant effect on the bed conductivity. The radiant flux between two surfaces is proportional to the fourth-power temperature differences:

$$\frac{Q_{\text{rad}}}{A} = \sigma\epsilon(T_1^4 - T_2^4) \tag{5.66}$$

The temperature term can be expanded to

$$(T_1^4 - T_2^4) = (T_1^2 + T_2^2)(T_1 - T_2)(T_1 + T_2)$$

Since $T_1 \cong T_2$ for adjacent particles in the bed, a radiation heat transfer coefficient can be calculated:

$$\frac{Q_{\text{rad}}}{A} = \sigma\epsilon \times 4T_1^3(T_1 - T_2) \equiv h_r(T_1 - T_2) \tag{5.67}$$

$$h_r = 4\sigma\epsilon T_1^3 \tag{5.68}$$

To convert the radiant heat transfer coefficient to a contribution to the effective bed conductivity, the particle diameter must be included, because this affects the path length for radiant energy transfer. Radiation between particles and conduction through the solid are accounted for in the model of Schotte [32]:

$$k_r = \frac{1 - \epsilon}{\dfrac{1}{k_s} + \dfrac{1}{h_r d_p}} + \epsilon h_r d_p \tag{5.69}$$

For 3-mm catalyst particles in air, k_r is about $1.2k_g$ at 300°C and $2.0k_g$ at 500°C.

The major mechanism of heat transfer at high flow rates is the random sideways motion of fluid elements as they pass through successive layers of particles. A tracer introduced at the center of the bed will spread in a conical plume and reach the wall after a certain number of layers. The radial spread can be described by a turbulent diffusivity and presented as a Peclet number for mass transfer:

$$\text{Pe}_m = \frac{d_p u_0}{D_{td}} \tag{5.70}$$

Hot gas at the center will also move radially carrying heat to the walls. The Peclet number for heat transfer includes a turbulent diffusion conductivity:

$$\text{Pe}_h = \frac{d_p u_0 \rho c_p}{k_{td}} \tag{5.71}$$

Random walk theory shows that both Pe_m and Pe_h should be about 8–10, in agreement with diffusion measurements [33]. Taking the Peclet number as 10, the contribution to bed conductivity is normalized using gas conductivity:

$$\frac{k_{td}}{k_g} = 0.1 \frac{d_p u_0 \rho c_p}{k_g} \tag{5.72}$$

Multiplying by μ/μ gives the useful form

$$\frac{k_{td}}{k_g} = 0.1 \text{Re} \times \text{Pr} \tag{5.73}$$

The final equation for bed conductivity is

$$\frac{k_e}{k_g} = \frac{k_e^0}{k_g} + \frac{k_r}{k_g} + 0.1 \text{Re} \times \text{Pr} \tag{5.74}$$

Note that k_e is a linear function of Re, and for high Reynolds number, k_e is almost proportional to Re. To relate k_e to h_{bed}, the bed geometry and shape of the temperature profile are needed. If q, the rate of heat generation per unit volume of bed, is assumed to be independent of the bed radius, a heat balance for a unit length of cylindrical bed is:

$$\pi r^2 q = 2\pi r k_e \left(-\frac{dT}{dr} \right) \tag{5.75}$$

$$q \int_0^R r \, dr = -2 k_e \int_{T_c}^{T_R} dT \tag{5.76}$$

$$T_C - T_R = q \frac{R^2}{4 k_e} \tag{5.77}$$

The temperature profile in the bed is parabolic, and T_R is the temperature that would be reached by extending the parabola to the wall, as shown in Figure 5.15. The total heat generated is equated to the heat conducted from the bed, using the driving force $(T_{ave} - T_R)$:

$$Q = \pi R^2 q = h_{bed}(2\pi R)(T_{ave} - T_R) \tag{5.78}$$

Combining Eqs. (5.77) and (5.78) gives

$$h_{bed} = \frac{2k_e}{R}\left(\frac{T_C - T_R}{T_{ave} - T_R}\right) \tag{5.79}$$

Since

$$T_{ave} = \frac{\int_0^R 2\pi r T\, dr}{\pi R^2}, \qquad \text{and} \qquad T = T_C - (T_C - T_R)\left(\frac{r}{R}\right)^2$$

integration leads to

$$T_{ave} = \frac{(T_C + T_R)}{2} \qquad \text{or} \qquad (T_C - T_R) = 2(T_{ave} - T_R).$$

As a result,

$$h_{bed} = 4\frac{k_e}{R} \tag{5.80}$$

The assumption that q is constant may seem unrealistic, but allowing for the change in q usually has only a slight effect on the temperature profile and the effective value of h_{bed}.

Heat Transfer at the Wall

The main mechanism of heat transfer at the wall of a packed bed is conduction through a boundary layer whose thickness depends on the gas velocity, the particle size, and the arrangement of particles near the wall. A correlation similar to that for heat transfer to particles in the bed might be expected, but a rigorous theory has not yet been developed. There are many empirical correlations for h_w or for Nu_w, but the values and exponents differ widely, reflecting the difficulty in separating $1/h_0$, the overall bed resistance, into its two parts, $1/h_w$ and $1/h_{bed}$. Some workers have measured radial temperature profiles inside or just above the bed in an attempt to get a direct determination of h_w. However, the unusual velocity profile in a small-diameter packed bed makes rigorous calculations very difficult. As sketched in Figure 5.17, the gas velocity goes from zero at the wall to a maximum about

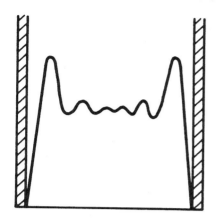

Figure **5.17** Velocity profile in a packed tubular reactor.

half a particle diameter from the wall [35–37]. The maximum velocity may be as high as three times the centerline velocity. Several maxima and minima may be found in a radial traverse due to variations in local void fraction. A circular traverse just above the bed near the wall will also show variations in velocity and temperature as the probe passes over particle and void spaces.

The correlation for Nu_w used here is based on tests by Peters and Schiffino [38], who studied heat transfer from steam to air in 1-, 2-, and 4-inch pipes packed with spheres or cylinders of several sizes. Gases from the bed were passed through a mixing cup to get the average exit temperature, which was used to calculate the overall coefficient, U. A correction for the small resistance of the metal wall and condensate film gave values of h_o, the overall bed coefficient. The wall coefficient h_w was calculated from h_o using predicted values of h_{bed}:

$$\frac{1}{h_w} = \frac{1}{h_o} - \frac{1}{h_{bed}} \tag{5.81}$$

The simplest correlation was based on an Eq. (5.80) for h_{bed} and an assumed Peclet number of 10 to get k_e. The zero-flow term was estimated to be 5:

$$\frac{k_e}{k_g} = 5 + 0.1 \mathrm{Re} \times \mathrm{Pr} \tag{5.82}$$

The resulting Nusselt numbers for spheres are shown in Figure 5.18, and a single line fits all the data except for $\frac{3}{4}$-inch spheres in a 1-inch pipe. The

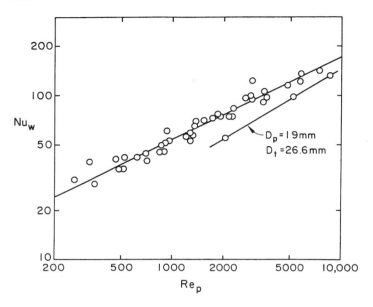

Figure 5.18 Nusselt numbers for spheres, constant approach. (From Ref. 38 = Pe_m with permission of the American Chemical Society.)

effect of gas properties was not investigated, and a 0.33-power dependence on Prandtl number was assumed to give the following correlation:

$$Nu_w = 1.94Re_p^{0.5}Pr^{0.33} \qquad (5.83)$$

The Nusselt numbers for cylinders and rings showed more variation with particle size, but the differences from the results for spheres were not enough to justify a separate correlation.

An alternate correlation for Nu_w was developed allowing for a linear increase in Pe with (d_p/d_t) [38], but the fit to the data was not improved, and the simpler correlation given by Eq. (5.83) is recommended.

Increasing d_p decreases h_w but increases h_{bed}, and h_o goes through a broad maximum as d_p increases. From the standpoint of heat transfer, the optimum (d_p/d_t) ratio is in the range 0.1–0.2. However, increasing d_p decreases the pressure drop through the bed and often reduces the effectiveness factor, so there are many factors to consider in choosing the particle size. Both h_w and h_{bed} increase with gas velocity, and quite high velocities may be used to get high values of U and decrease the chance of a runaway reactor. When the Reynolds number is very high (Re > 2000) and $d_p/d_t = 0.1–0.2$, the major resistance to heat transfer is in the wall film, and the temperature drop in the bed is relatively small. Under these conditions, the assumption of a parabolic temperature gradient is more realistic then when most of the temperature change is in the bed.

Example 5.3

The partial oxidation of ethylene (E) with air is carried out in a tubular reactor with 0.4-cm catalyst pellets packed in 3.8-cm tubes. The feed gas contains 4% E, 6% CO_2, 7% O_2, and 83% N_2 and enters the reactor at 230°C, a total pressure of 5.0 atm, and a superficial velocity of 1.5 ft/sec. The reaction rate for these pellets at 230°C is

$$r = \frac{0.076 P_{O_2} P_E}{1 + 2P_E + 15P_{CO_2}} \quad \text{moles E consumed/hr, g cat}$$

$$P_{O_2}, P_E, P_{CO_2} = \text{atm}$$

About 70% of the ethylene reacting forms ethylene oxide, and 30% forms CO_2 and H_2O:

$$\left(\begin{array}{c} \Delta H \\ (\text{kcal/mol}) \end{array} \right)$$

$$E + \frac{1}{2}O_2 \longrightarrow EO \qquad\qquad -29.9$$

$$E + 3O_2 \longrightarrow 2CO_2 + 2H_2O \quad -317$$

The apparent activation energy is 18 kcal, and the selectivity may be assumed constant for moderate changes in temperature. The particle density is 2.5 g/cm^3, and $k_s \cong 8 \times 10^{-4}$ cal/sec, cm°C.

a. Calculate the overall heat transfer coefficient for velocities of 1.5 ft/sec and 3.0 ft/sec (cal/cm^2, sec,°C or W/m^2, K).

b. For a constant jacket temperature of 230°C, estimate the peak radial average bed temperature for both velocities, assuming the ethylene conversion is 10% at this point. Compare these temperature differences with the critical temperature difference.

See Table 5.1.

Solution.

$M_{\text{ave}} = 29.24, \qquad \bar{c}_p = 7.91$ cal/mol, °C $= 0.270$ cal/g, °C

$$\rho = \frac{29.24}{22,400} \times \frac{5}{1} \times \frac{273}{503} = 3.54 \times 10^{-3} \text{g/cm}^3$$

$u = 1.5$ ft/sec $= 45.8$ cm/sec

$\mu \cong \mu_{\text{air}} = 0.026$ cp

TABLE 5.1   Data for Example 5.3

Gas	M	Mole fraction	c_p cal/mol, °C at 230°C (= 503 K)
C_2H_4	28	0.04	15.3
O_2	32	0.07	7.4
CO_2	44	0.06	10.7
N_2	28	0.83	7.4

$$Re_p = \frac{0.4(45.8)(3.5 \times 10^{-3})}{2.6 \times 10^{-4}} = 249$$

$$k \cong k_{air} = 9.72 \times 10^{-5} \text{cal/sec, cm, °C}$$

$$Pr = \frac{0.270(2.6 \times 10^{-4})}{9.72 \times 10^{-5}} = 0.722$$

$$\frac{k_s}{k_g} = \frac{8 \times 10^{-4}}{9.72 \times 10^{-5}} = 8.2$$

From Figure 5.16,

$$\frac{k_e^0}{k_g} \cong 3.5$$

From Eqs. (5.68) and (5.69),

$$\frac{k_r}{k_g} = 2.5$$

From Eq. (5.74),

$$\frac{k_e}{k_g} = 3.5 + 2.5 + 0.1(249)(0.722) = 24.0$$

$$k_e = 24(9.72 \times 10^{-5}) = 2.33x10^{-3}$$

$$h_{bed} = 4\frac{k_e}{R} = \frac{4(9.72 \times 10^{-3})}{1.9} = 4.91 \times 10^{-3} \text{ cal/sec, cm}^2, K$$

From Eq. (5.83),

$$Nu_w = 1.94(249)^{0.5}(0.722)^{0.33} = 27.5$$

$$h_w = \frac{27.5(9.722 \times 10^{-5})}{0.4} = 6.68 \times 10^{-3} \text{ cal/sec, cm}^2, K$$

Assume $h_j \cong 100 \times 10^{-3}$cal/sec, cm^{2}K—boiling water. Neglecting r_{wall},

$$\frac{1}{U} = \frac{1}{100 \times 10^{-3}} + \frac{1}{6.68 \times 10^{-3}} + \frac{1}{4.91 \times 10^{-3}}$$

$$U = 2.75 \times 10^{-3} \text{cal/sec, cm}^2, \text{ K}$$

$$-\Delta H_{ave} = 0.7(29.9) + 0.3(317) = 116 \text{ kcal/mol E}$$

$$\Delta T_c = \frac{RT^2}{E}$$

If $T_{max} \cong 230 + 20 = 250°C = 523K$,

$$\Delta T_c = \frac{1.987(523)^2}{18,000} = 30.2°C$$

Estimate r and Q at $T = 250°C$ at 10% conversion:

$$\ln\left(\frac{k_{250}}{k_{230}}\right) = \frac{18,000}{1.987}\left(\frac{1}{503} - \frac{1}{525}\right) = 0.689$$

$$\frac{k_{250}}{k_{230}} = 1.99$$

At $x = 0.1$ and $1 - x = 0.9$,

$$P_E = 0.04(5)(0.9) = 0.18 \text{ atm}$$
$$P_{O_2} = 0.33\text{atm}, \qquad P_{CO_2} = 0.31 \text{ atm}$$
$$r = 1.99 \times \frac{0.076(0.33)(0.18)}{1 + 2(0.18) + 15(0.31)} = 1.49 \times 10^{-3}\text{mol/hr, g}$$
$$Q' = 1.49 \times 10^{-3}\frac{(116,000)}{3600} = 48.2 \times 10^{-3}\text{cal/sec, g}$$

For 1-cm^3 bed,

$$\epsilon = 0.4, \qquad \rho_{bed} = 2.5(0.6) = 1.5 \text{ g/cm}^3$$
$$A = \frac{4}{D} = \frac{4}{2.8} = 1.05 \text{ cm}^2/\text{cm}^3$$
$$Q = Q'\rho_{bed} = UA \ \Delta T$$
$$\Delta T = \frac{48.2 \times 10^{-3}(1.5)}{2.75 \times 10^{-3}(1.05)} = 25°C$$

Since $\Delta T < \Delta T_c$, the reactor should be stable at $T_{max} = 250°C$, but the margin of safety would not be very large. At the higher velocity of 3.0 ft/sec, Re$_p$ is doubled to 498 and U is considerably increased.

$$\frac{k_e}{k_g} = 6 + 0.1(498)(0.722) = 42$$

$$k_e = 4.08 \times 10^{-3}$$

$$h_{\text{bed}} = 4\frac{(4.08 \times 10^{-3})}{1.9} = 8.59 \times 10^{-3} \text{ cal/sec, cm}^2, \text{K}$$

Since $\mathrm{Nu}_w \propto \mathrm{Re}_p^{0.5}$,

$$h_w = 6.68 \times 10^{-3}(2)^{0.5} = 9.45 \times 10^{-3} \text{ cal/sec, cm}^2, \text{K}$$

$$U = 4.31 \times 10^{-3} \text{ cal/sec, cm}^2, \text{K}$$

For $T = 250°C$ and $x = 0.1$, $Q = 48.2 \times 10^{-3}$, as before, and

$$\Delta T = \frac{48.2 \times 10^{-3}(1.5)}{4.31 \times 10^{-3}(1.05)} = 16°C$$

The maximum ΔT is about half ΔT_c, and the reactor could be operated with a good margin of safety. The peak temperature and sensitivity to disturbances should be checked by numerical integration.

The axial temperature profile for a packed tubular reactor with a constant jacket temperature is similar to that for a pipeline reactor, as shown in Figure 5.5. If flowing water, oil, or molten salt is used for heat removal, a nearly constant jacket temperature could be obtained with a very high flow rate of coolant. Figure 5.19a shows the reactor and jacket temperatures for parallel flow of reactant and coolant with high coolant flow rate. The peak temperature occurs early in the reactor, and the final reactor temperature is a little above the inlet temperature. Better performance might be obtained with parallel flow and a moderate coolant rate to give a profile like that in Figure 5.19b. The larger change in jacket temperature makes the reactor exit temperature closer to the peak temperature, and the conversion would be higher than for the case of nearly constant jacket temperature.

A third method of operation is to use counterflow of coolant and reactant, which results in a profile of the type shown in Figure 5.19c. By adjusting the coolant rate, the peak reactor temperature can be kept about the same, but the final reactor temperature is close to the coolant inlet temperature. The low concentration and temperature make the final reaction rate quite low, and a larger reactor would be needed to reach the same conversion. Unlike normal fluid–fluid heat exchange, there is no inherent advantage, and perhaps a penalty for countercurrent operation. However, for practical reasons, countercurrent flow is often used with upflow in the shell and downflow in the tubes to prevent fluidization.

In designing a tubular reactor, the approximate temperature may be selected as a compromise that gives good selectivity and reasonable reaction rate. The selectivity and catalyst life often decrease with increasing tempera-

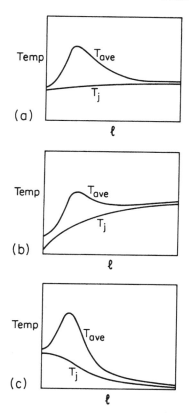

FIGURE 5.19 Temperature profiles for a tubular reactor with jacket cooling: (a) Parallel flow, high coolant rate; (b) parallel flow, moderate coolant rate; (c) countercurrent flow, moderate coolant rate.

ture. To obtain the desired temperature without risking a runaway reaction means choosing the best combination of particle size, tube size, flow rate, and method of cooling. Small particles have higher effectiveness factors but higher pressure drop, and heat transfer in the bed is poor when d_p/d_t is small. The optimum particle size is generally large enough to have some rate limitation due to pore diffusion. The choice of tube diameter is influenced by heat transfer rates and reactor costs. A reactor with 4000 1-inch tubes costs more than one with 1000 2-inch tubes and the same amount of catalyst. To permit the use of larger tubes, the mass velocity can be increased so that the higher overall coefficient compensates for the decrease in area, though this requires longer tubes for the same conversion. With larger tables, the feed concentration may be reduced to give about the same temperature profile, as

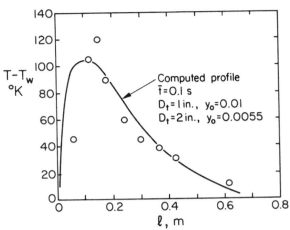

FIGURE 5.20 Temperature profiles for o-xylene oxidation in tubular reactors. (From Ref. 44.)

shown in Figure 5.20. With a very exothermic reaction, it may be worthwhile to dilute the catalyst with inert particles in the first part of the reactor or use multiple coolant zones.

ALTERNATE MODELS

The 1-D homogeneous model can be used to predict the effects of particle and tube size, flow rate, and gas properties on the temperature profile, the rate of heat removal, and the overall conversion. However, it is not expected to be very accurate because of the simplifying assumptions made. Several more complex models have been proposed, and the features of some of these are reviewed here.

    1. Two-dimensional (2-D) models allow for the change in temperature and reaction rate constant with tube radius. Most 2-D homogeneous models still assume plug flow of the gas and uniform radial concentration. (Calculations show that radial mixing is rapid enough to minimize the concentration differences.) Several radial increments are used for the computations, and the heat flux is set proportional to the radial temperature gradient and the local conductivity, k_e. The conductivity can be taken as a constant or as a function of radial position.

    For a partial oxidation example, when the maximum temperature difference was about 10°C, there was almost no difference in the temperature profiles for 1-D and 2-D models [37]. When ΔT reached 20–30°C, the

2-D model gave center temperatures 2–3°C higher than the 1-D model, and a runaway reaction occurred at a slightly lower feed temperature than with the 1-D model. However, the 1-D model could be modified to be conservative by decreasing h_{bed} to $3k_e/R$ to compensate for the higher reaction rate in the center. In any case, it is unlikely that a tubular reactor would be designed to operate close to the stability limit because of uncertainties in the kinetic and heat transfer parameters.

2. More rigorous 2-D models allow for the radial velocity profile instead of assuming plug flow. The peak velocity near the wall Figure (5.17) is especially pronounced for large values of d_p/d_t, where over half the total flow passes through the region less than one particle diameter from the wall. The model can include several radial increments with a different axial velocity for each increment. However, this makes the calculations very lengthy, and some have suggested dividing the bed into an annular zone and a core with different average velocities. Care must be taken in choosing values for k_e, since heat transfer correlations derived assuming plug flow of gas would not be appropriate here.

3. Heterogeneous 2-D models allow for the difference between local surface and gas temperatures. Usually $(T_s - T_g)$ is quite small relative to the overall radial ΔT and might be 2–3°C when $\Delta T = 30$°C. In a laboratory study of ethylene oxidation [39], where catalyst and gas temperature were directly measured, the differences were only a few degrees Centigrade and were about half the predicted values. When $(T_s - T_g)$ is large, a more rigorous analysis could consider the asymmetric temperature profile in the pellet and the variations in heat transfer coefficient around the surface. A local instability might be found where none would be expected based on average values of h.

A final point to consider is whether heat transfer in a tubular reactor is inherently different from heat transfer without reaction. Several workers have claimed that special correlations for the heat transfer parameters are needed for reliable reactor models. In a number of reactor studies [40], the maximum temperature rise was greatly overpredicted by the models used, but others have reported the opposite effect. These differences may have been caused by inaccurate kinetic equations or errors in the heat transfer parameters, but they were probably not due to a fundamental difference between heat transfer with and without reaction. It is true that in steady-state heat transfer tests, the local gas and solid temperatures are equal, whereas with an exothermic reaction, T_s is greater than T_g. If the reactor analysis is based on T_g, the heat transfer rate will be slightly underestimated because of increased radiation and conduction. However, for moderate or high Reynolds numbers, convection is the major contributor to radial heat transfer, and the effect of a higher surface temperature should be quite small. Separate correlations for heat transfer with reaction are not justified except perhaps for studies of ignition phenomenon.

More tests are needed comparing measured conversions and temperature profiles with model predictions for tubular reactors. Comparisons will be easier for reactions with simple kinetics than for complex reactions such as partial oxidations. Tests should be made over a wide range of Reynolds numbers, which may require high velocities and long reactors. If kinetic data are uncertain or unavailable, the overall heat transfer coefficient for the 1-D model can be obtained from the axial temperature profile and the total heat removal [41]:

$$Q_R = \int U \, dA (T_{ave} - T_j) \tag{5.84}$$

Remember that U in Eq. (5.84) is based on the radial average temperature, T_{ave}, and not on the centerline temperature, T_C. If only T_C is known, T_{ave} can be estimated using the relative resistances:

$$h_{bed}(T_{ave} - T_R) = U(T_{ave} - T_j) \tag{5.85}$$

Since $T_{ave} - T_R = T_C - T_{ave}$,

$$T_C - T_{ave} = \frac{U}{h_{bed}}(T_{ave} - T_j) \tag{5.86}$$

or

$$T_C - T_{ave} = f(T_{ave} - T_j) \tag{5.87}$$

where f = fraction of total resistance in the bed. Solving for T_{ave} and then for $T_C - T_{ave}$ gives an alternate equation:

$$T_C - T_{ave} = \frac{f}{1+f}(T_c - T_j) \tag{5.88}$$

For example, if half the total resistance is in the bed and $(T_C - T_j) = 30°C$, the average bed temperature is 10°C below T_C, and the driving force for heat transfer $(T_{ave} - T_j)$ is 20°C.

Using the foregoing equations and the ethylene oxidation data in Ref. 39, overall coefficients were obtained that were within 20% of the U values predicted following the procedure of Example 5.3, and the changes in U with flow rate agreed with predictions.

NOMENCLATURE

Symbols

A	heat transfer area, constant in Arrhenius equation
a	external area per unit mass of catalyst
C	concentration
C_A	molar concentration of A
c_p	heat capacity
D	diameter
D_a	diameter of agitator
D_c	diameter of coil
D_{He}	diameter of helix
D_t	diameter of tank
D_{AB}	bulk diffusivity
D_e	effective diffusivity
D_{td}	turbulent diffusivity
d_p	particle diameter
d_t	tube diameter
E	activation energy
F	volumetric feed rate
F_A	molar feed rate of A
f	fraction of resistance in the bed
G	mass velocity
h	heat transfer coefficient
h_{bed}	heat transfer coefficient of bed
h_c	heat transfer coefficient of coil
h_i	inside heat transfer coefficient
h_j	heat transfer coefficient of jacket
h_o	outside heat transfer coefficient
h_r	heat transfer coefficient for radiation
h_w	heat transfer coefficient of wall film
h_o	heat transfer coefficient for packed bed, including h_{bed} and h_w
j_d	mass transfer factor
j_m	mass transfer factor
j_h	heat transfer factor
K_A	adsorption constant
K_B	adsorption constant
K_o	overall mass transfer coefficient
k	reaction rate constant
k	thermal conductivity
k_e	effective thermal conductivity
k_e^0	effective thermal conductivity at zero flow

k_g	thermal conductivity of gas
k_r	thermal conductivity due to radiation
k_s	thermal conductivity of solid
k_{td}	turbulent diffusion conductivity
k_c	mass transfer coefficient
l	length
Nu	Nusselt number
Nu_w	Nusselt number for wall film
n	stirrer speed, rps
P	pressure
P_A, P_B	pressure of gases A and B
Pe_h	Peclet number for heat transfer
Pe_m	Peclet number for mass transfer
Pr	Prandtl number
Q	Heat rate
Q_G	heat generation rate
Q_R	heat removal rate
Q_j	heat to jacket
Q_{rad}	radiant heat transfer rate
Q_t	total heat transfer rate
q	heat generation rate per unit bed volume
R	gas constant, pellet radius, tube radius
Re	Reynolds number
Re_p	Reynolds number based on particle size
r	reaction rate, radius of pellet or tube
r_w	heat transfer resistance of reactor wall
Sc	Schmidt number
Sh	Sherwood number
T	Absolute temperature
T_o or T_f	feed temperature
T_j	average jacket temperature
$\bar{T}$ or T_{ave}	average temperature
T_c	center temperature
T_R	temperature near the wall
T_s	surface temperature
U	overall heat transfer coefficient
U'	alternate definition for Eq. (5.63)
u	linear velocity
u_o	superficial velocity
V	reactor volume
x	fraction converted

Greek Letters

α	temperature ratio, Eq. (5.30)
ΔH	heat of reaction
ΔT_c	critical temperature difference
ΔT_{ad}	adiabatic temperature rise
ΔT_{ave}	average driving force for heat transfer
ΔT_L	log-mean temperature difference
ϵ	void fraction
ϵ	emissivity, Eq. (5.66)
η	effectiveness factor
θ	temperature change ratio, Eq. (5.29)
μ	viscosity
μ_w	viscocity at wall
π	pi
ρ	density
ρ_s	density of solid
ρ	Stefan–Boltzman constant, 5.67×10^{-8} W/m^2-K^4 or 0.1713×10^{-8} Btu/ft^2-hr-$^\circ$R^4

PROBLEMS

5.1 In large jacketed reactors used for suspension polymerization of vinyl chloride, the reaction rate is limited by the rate of heat removal [42,43]. Thick-walled reactors are needed because of the high pressure, and a very smooth surface is desired to minimize fouling. Compare the overall heat transfer coefficients for the following conditions, assuming $h_i = 300$ and $h_j = 500$ Btu/hr · ft^2 · $^\circ$F and a metal wall thickness of $\frac{3}{4}$ inch.

    a. Stainless steel wall
    b. Steel wall with $\frac{1}{8}$-inch cladding of polished stainless steel
    c. Steel wall with $\frac{1}{16}$-inch glass lining
    d. All of the above with a 0.02-inch layer of polymer on the wall

5.2 In the scaleup of a stirred reactor at constant power per unit volume, how do the following parameters change with vessel diameter if geometric similarity is maintained?

    a. Stirrer speed
    b. Inside coefficient for the jacket
    c. Coil coefficient, same tubing size
    d. Coil coefficient, scaled tubing size

5.3 The design conditions for a continuous stirred-tank reactor are as given here. Would the reactor be stable with a constant jacket temperature?

Feed = 1000 kg/hr at 20°C, containing 50% A

$c_p = 0.75$ cal/g°C   3.1 J/g°C

$W = 1200$ kg holdup

$A =$ jacket area = 4.5 m^2

$U = 850$ W/m$^2 \cdot$ K

$\Delta H = -1400$ J/g (exothermic)

Kinetics:

$r = kC_A$

$k = 1.3$ hr^{-1} at 70°C

$E = 70,000$J/g $\cdot$ mol

Desired reaction conditions: 50% conversion at 65–75°C.

5.4 The combustion of 1-inch carbon spheres was studied by Tu, Davis, and Hottel [23]. The sphere was suspended from one arm of a balance, and the reaction rate was determined from the change in weight. The gas flow was great enough so that changes in gas composition and temperature were negligible, and the fraction carbon burned was very small. An optical pyrometer was used to measure the surface temperature. Some results are shown in Table 5.2.

    a. Plot the data and determine the activation energy for the low- and high-temperature regions.
    b. What is the effect of gas velocity at high temperature? Does this agree with theory?
    c. Compare the maximum measured rate with that predicted if external mass transfer controls.

5.5 The air oxidation of o-xylene was studied using 0.6-cm particles of V$_2$O$_5$/SiC catalyst [44]. At 450°C, the overall rate constant was given as

$k = 0.8$ g $-$ mol/hr $\cdot$ g $\cdot$ atm

    About 75% of the o-xylene reacted forms phthallic anhydride, and the rest burns to water and carbon dioxide. The average heat

TABLE 5.2   Data for Problem 5.4

Carbon surface temp (°K)	Furnace temp. (°K)	Gas velocity at S.T.P. (cm/sec)	Combustion rate (g/sec, cm²) $\times 10^3$
Air used to oxidize the carbon			
947	953	3.51	0.011
1091	1015	3.51	0.0569
986	972	3.51	0.0126
1251	1172	3.51	0.122
1348	1268	3.51	0.128
1659	1633	3.51	0.146
1027	1004	7.52	0.022
1137	1044	7.52	0.102
1341	1207	7.52	0.169
1607	1550	7.52	0.224
1193	1074	7.52	0.132
1205	1084	27.4	0.174
1488	1355	27.4	0.323
1622	1530	27.4	0.393
Feed gas contained 2.98% O_2			
1025	985	3.51	.00518
1081	1050	3.51	.0138
1113	1088	3.51	.0167
1527	1542	3.51	.0225

of reaction is −470 kcal/mol. For 1% o-xylene in air, estimate the internal and external temperature differences for $Re_p = 50$ and $k_s = 10^{-3}$ cal/sec · cm · °C.

5.6   Calculate the effect of flow rate on the overall heat transfer co-efficient for a 2-inch-diameter reactor packed with $\frac{1}{4}$-inch catalyst spheres and operating with air at 400°C.

  a.   Present the results as a graph of U versus Re_p for a wide range of flow rates. Neglect the metal wall and jacket resistances.
  b.   What fraction of the overall resistance is in the wall film for different flow rates?
  c.   How much does radiation contribute to the overall coefficient?

5.7   An exothermic catalytic reaction was studied at 2 atm and 280°C in a 1-inch-diameter 4-ft-long jacketed reactor. The catalyst was $\frac{1}{8}" \times \frac{1}{8}"$

cylinders. The maximum difference between average bed temperature and the jacket temperature was 15°C. For a commercial reactor, tubes 2 inch × 24 ft are suggested with the same d_p/d_t ratio.

    a.   What change in flow rate per tube would be chosen to keep the conversion about the same? What is the new Reynolds number if Re_p was 130 for the 1-inch reactor?

    b.   Will the maximum ΔT in the 2-inch tubes be higher or lower than 15°C if the jacket temperature is adjusted to make T_{max} about the same?

5.8   Study the temperature profile reported by Calderbank [44] for the oxidation of o-xylene in 1-inch and 2-inch reactors. See Figure 5.20.

    a.   Predict the change in ΔT_{max} in going from a 1-inch to a 2-inch tube while decreasing the feed concentration. How does your prediction compare with the data?

    b.   How close is the reactor to a runaway, or has runaway already happened? The activation energy is reported to be 27.4 kcal/mol up to 440°C and 8 kcal/mol at higher temperatures.

5.9   The dehydrogenation of ethyl benzene (EB) to styrene (S) is carried out in 3-inch tubes packed with $\frac{3}{16}$-inch catalyst pellets. The feed contains 15 moles H_2O/mole EB and enters the reactor at 1.3 atm and 600°C. The particle Reynolds number is 800 and $Pr \cong 1.0$. The heat of reaction is 33 kcal/mol, and $E = 21$ kcal/mol.

    a.   Sketch the radial and axial temperature profiles, assuming that the maximum temperature difference is about 50°C. Is there any danger of a runaway reaction?

    b.   The use of $\frac{3}{8}$-inch pellets at the same mass flow rate was suggested to improve radial mixing. By what factor would the overall heat transfer coefficient be changed?

5.10   A reaction with hundreds of 1.5-inch tubes packed with $\frac{1}{8}$-inch catalyst pellets has axial thermocouples in a few of the tubes. The thermocouple diameter is $\frac{3}{8}$-inch, and the maximum indicated temperature is 30°C above the wall temperature.

    a.   Assuming the same average rate of heat generation, about how much higher would the maximum temperature be in tubes with no thermocouple?

    b.   If the external void fraction is 0.45 in the tubes with thermocouples and 0.40 in the other tubes, what is the difference in peak temperatures?

5.11   To control the temperature rise in the very exothermic methanation reaction, a porous nickel catalyst is deposited in a thin layer ($\sim \frac{1}{8}$ inch) on the inside of 2-inch-diameter reactor tubes. Reacting gases pass in turbulent flow through the tubes, and steam is generated on the outside:

$$CO + 3H_2 \rightleftharpoons CH_4 + H_2O \qquad \Delta H = -52 \text{ kcal}$$

a.   Sketch the radial temperature and concentration gradients that might exist near the inlet and near the exit of the reactor. How do these differ in shape from those for a conventional packed-tube reactor?

b.   Derive an equation that gives the approximate shape of the temperature profile in the catalyst layer.

c.   Discuss the stability of this reactor relative to a packed-bed reactor containing $\frac{1}{4}$-inch particles of the same porous nickel catalyst. Do you think the tube-wall reactor will always be stable, very much more stable, somewhat more stable, or about as stable as the packed bed? Give an intuitive answer, and try to justify it by rough calculations.

5.12   An exothermic reaction is studied in a $\frac{3}{4}$-inch laboratory reactor equipped with a $\frac{1}{16}$-inch thermocouple. The catalyst is 0.10-inch particles diluted with 50% inert porous alumina, but the reactor is not as close to isothermal as desired.

a.   How much would it help to dilute the bed with quartz particles instead of Al_2O_3?

b.   Would it help to use He rather than N_2 as the carrier gas, which is 95% of the total flow?

REFERENCES

1.   WL McCabe, JC Smith, P Harriott. Unit Operations of Chemical Engineering. 6th ed. New York: McGraw-Hill, 2000, pp 460–462.
2.   RM Fogg, VW Uhl. Chem Eng Prog 69(7):76, 1973.
3.   WH McAdams. Heat Transmission. 3rd ed. New York: McGraw-Hill, 1954, p 228.
4.   P Harriott. Process Control. New York: McGraw-Hill, 1964, p 312.
5.   O Bilous, NR Amundson. AIChE J 1:513, 1955.
6.   R Aris, NR Amundson. Chem Eng Sci 7:121, 1958.
7.   GP Baccaro, NY Gaitonde, JM Douglas. AIChE J 16:249, 1970.
8.   JM Douglas, DWT Rippin. Chem Eng Sci 21:305, 1966.
9.   P Harriott. Process Control. New York: McGraw-Hill, 1964, p 315.
10.   JMH Fortuin, JJ Heiszwolf, CS Bildea. AIChE J 47:920, 2001.

11. DE Boynton, WB Nichols, HM Spurlin. Ind Eng Chem 51:489, 1959.
12. P Harriott. Process Control. New York: McGraw-Hill, 1964, p 320.
13. M Morbidelli, A Varma. AIChE J 28:705, 1982.
14. N Frössling. Beitr Geophys 52:170, 1938.
15. WE Ranz, WR Marshall Jr. Chem Eng Prog 48:141, 173, 1952.
16. BW Gamson, G Thodos, OA Hougen. Trans Am Inst Chem Engrs 39:1, 1943.
17. CR Wilke, OA Hougen. Trans Am Inst Chem Engrs 41:445, 1945.
18. RD Bradshaw, CO Bennett. AIChE J 7:48, 1961.
19. CN Satterfield, H Resnick. Chem Eng Progr 50:504, 1954.
20. KR Jolls, TJ Hanratty. AIChE J 15, 199, 1969.
21. TK Sherwood, RL Pigford, CR Wilke. Mass Transfer. New York: McGraw-Hill, 1975, p 242.
22. PB Weisz, JS Hicks. Chem Eng Sci 17:265, 1962.
23. CM Tu, H Davis, HC Hottel. Ind Eng Chem 26:753, 1934.
24. AC Pauls, EW Comings, JM Smith. AIChE J 5:453, 1959.
25. JW Fulton, OK Crosser. AIChE J 11:513, 1965.
26. CG Hill Jr. An Introduction to Chemical Engineering Kinetics and Reactor Design. New York: Wiley, 1977, p 495.
27. M Leva. Ind Eng Chem 42:2498, 1950.
28. TE Schumann, V Voss. Fuel 13:249, 1934.
29. S Yagi, D Kunii. AIChE J 3:373, 1957.
30. S Masamune, JM Smith. IEC Fund 2:136, 1963.
31. D Kunii, JM Smith. AIChE J 6:71, 1960.
32. W Schotte. AIChE J 6:63, 1960.
33. JM Smith. Chemical Engineering Kinetics. 3rd ed. New York: McGraw-Hill, 1981, p 557.
34. JF Wehner, RH Wilhelm. Chem Eng Sci 6:89 1956.
35. H Delmas, GF Froment. Chem Eng Sci 43:2281, 1988.
36. D Vortmeyer, J Schuster. Chem Eng Sci 38:1691, 1983.
37. GF Froment, KB Bischoff. Chemical Reactor Analysis and Design. 2nd ed. New York: Wiley, 1990, pp 457–467.
38. PE Peters, RS Schiffino, P Harriott. Ind Eng Chem Res 27: 226, 1988.
39. EPS Schouten, PC Borman, KR Westerterp. Chem Eng Sci 49:4725, 1994.
40. WR Patterson, JJ Carberry. Chem Eng Sci 38:175, 1983.
41. KB Wilson. Ind Eng Chem 23: 910 1931.
42. LF Albright, CG Build. Chem Eng (Sept. 15), 1977, p 121.
43. RE Markovitz. Hydrocarbon Process (Aug. 1973), p 117.
44. PH Calderbank. ACS Adv in Chem 133: 646, 1974.

6

Nonideal Flow

In the previous chapters, the reactors were assumed to have ideal flow, which means perfect back-mixing for a stirred-tank reactor and plug flow (no mixing) for a pipeline or packed-bed reactor. This chapter deals first with stirred reactors that have concentration gradients because of mixing delays and later with tubular reactors that have a distribution of residence times because of axial mixing or channeling. The effect of nonideal flow on conversion in stirred tanks is small and can usually be neglected. The problem with mixing delays is that the selectivity may be decreased when the reactions are so fast that significant conversion occurs before complete mixing is achieved.

MIXING TIMES

In a stirred reactor, it takes time for liquid added at the surface or at any point in the tank to become blended with the bulk of the liquid. The mixing process can be followed by observing the color change after a basic solution with an indicator is neutralized by suddenly adding a slight excess of acid. If this test is done in a 2-liter vessel, most of the solution will appear free of base in less than a second, but wisps of color may persist for 2–3 seconds until the mixing is complete. If the same test is carried out in a 5000-liter tank, it may take a few seconds for most of the reaction to take place and 10–20 seconds for complete neutralization. Many studies confirm that mixing is much slower in large tanks under normal agitation conditions, though it is often difficult to get reproducible values of the mixing time by visual observation.

Another type of test that gives a continuous record of the mixing process is adding a pulse of salt solution to a tank of water and monitoring the change in conductivity at some point in the tank. A typical response is sketched in Figure 6.1. There is a time delay, or lag, L, before any concentration change is noticed, because of the time it takes for fluid to travel from the injection point to the conductivity probe. The concentration then rises rapidly and overshoots the final value, which is reached after a few cycles of oscillation. The total mixing time, t_t, which is also called the *blending time*, can be defined as the time for the concentration to settle within ±1% or ± 5% of the final concentration change. The mixing time depends mainly on the pumping capacity of the impeller and the volume of the tank. Theory and experiments show that for low-viscosity fluids, the 99% mixing time is approximately the time needed to circulate the tank contents five times [1]. The mixing time does not change much with the location of the sensor or the injection point, though the initial delay does depend on the position.

Many correlations have been presented for the mixing time in stirred tanks with different types and sizes of impellers. For low-viscosity fluids, the mixing time (t_t) varies inversely with the stirrer speed. With a standard turbine in a baffled tank and $Re > 5000 (Re = nD_a^2\rho/\mu)$,

$$nt_t \cong 4\left(\frac{D_t}{D_a}\right)^2\left(\frac{H}{D_t}\right)$$

(6.1)

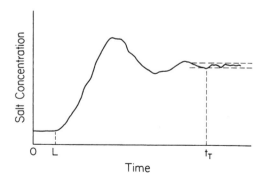

FIGURE **6.1** Determination of total mixing time.

When $D_t/D_a = 3$ and $H/D_t = 1$, nt_t is about 36. For a large tank ($D_t \cong 2m$) with the stirrer at 120 rpm, or 2 $\sec^{-1}$, the predicted mixing time is $36/2 = 18$ seconds. Mixing times with other impellers and for viscous fluids may be much longer.

A blending time of 10–20 seconds in a large stirred reactor might seem unimportant if the reaction time is many minutes or a few hours, and imperfect mixing does have little effect on the conversion in systems where only one reaction is taking place. However, when there are multiple fast reactions, the amount of byproduct formed may increase if the feed streams are not immediately blended with the bulk liquid, and a mixing delay of only a few seconds can significantly lower the selectivity.

Parallel Reactions

Consider a system of parallel reactions, where A and B combine to form the desired product C and where A can react with itself to form byproduct D:

$$A + B \xrightarrow{k_1} C$$

$$A + A \xrightarrow{k_2} D$$

For simple kinetics, the selectivity depends on the ratio of rate constants and the reactant ratio:

$$\frac{r_1}{r_2} = \frac{k_1 C_A C_B}{k_2 C_A^2} = \left(\frac{k_1}{k_2}\right)\left(\frac{C_B}{C_A}\right) \tag{6.2}$$

$$S = \frac{r_1}{r_1 + r_2} = \frac{r_1/r_2}{1 + r_1/r_2} \tag{6.3}$$

To get a high selectivity, a semibatch reactor could be used, with A fed continuously to an initial charge of B, as described in Chapter 3. Because of imperfect mixing, there will be regions near the feed pipe where C_A is greater than the bulk concentration and C_B is somewhat depleted. These differences lower the ratio (C_B/C_A) and decrease the local selectivity. The extent of the decrease depends on the relative rates of mixing and reaction.

An industrial example of fast parallel reactions is the alkylation of isobutane with butenes to form isooctane, which is accompanied by the oligomerization of butenes to form C-8 and C-12 olefins, which have a lower octane rating. The reactions take place in a suspension of hydrocarbon droplets in sulfuric acid, and the octane number of the product increases with stirrer speed because of more rapid droplet coalescence and mixing of the hydrocarbons [2].

Another type of parallel reaction system is the neutralization of a process solution that contains a labile substrate. For example, when strong caustic B was added to an acid solution A at the end of a batch reaction, some hydrolysis of the product C occurred, because there were small regions of high caustic concentration [3]. The competing reactions are

$$B + A \xrightarrow{k_1} H_2O + salt$$
$$B + C \xrightarrow{k_2} Q$$

The reaction of A and B to form water is very fast and limited only by the rate of mixing. The degradation of product C to byproduct Q is much slower but still fast enough to be affected by imperfect mixing. In laboratory tests, byproduct formation was decreased by going to higher agitator speeds, as shown in Figure 6.2. In a large plant reactor, byproduct formation was also noted, and higher stirrer speeds were not feasible. The yield was improved by using a weaker basic solution for neutralization.

Consecutive Reactions

Imperfect mixing can also affect the selectivity for consecutive reactions of the following type, which are sometime called *consecutive-parallel reactions*:

$$A + B \xrightarrow{k_1} C$$
$$A + C \xrightarrow{k_2} D$$

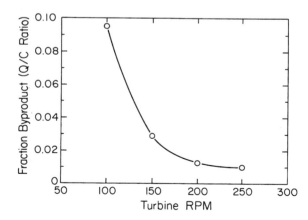

FIGURE 6.2 Effect of stirrer speed on byproduct formation. (After Ref. 3.)

If C is the desired product, the selectivity based on consumption of A depends on the reactant/product ratio:

$$\frac{r_1}{r_2} = \frac{k_1 C_A C_B}{k_2 C_A C_C} = \left(\frac{k_1}{k_2}\right)\left(\frac{C_B}{C_C}\right) \tag{6.4}$$

$$S = \frac{r_1}{r_1 + r_2}$$

If A is fed continuously to a reactor containing B, the high concentration of A close to the feed pipe results in a local decrease in C_B and an increase in C_C, both adversely affecting the selectivity. Several test systems of this type have been developed to study mixing effects in laboratory reactors, including the successive addition of iodine or bromine to aromatic compounds and the reaction of diazotized sulphanilic acid with 1-naphthol [4]. Typical results are given in Figure 6.3, where the segregation index x_s, defined as the fraction of A reacting that forms C, is shown to depend on the stirrer speed and the feed location.

Bioreactors

The previous examples dealt with the effect of mixing delays on fast parallel or consecutive reactions. Mixing effects are also found in bioreactors, even though the cell growth and product formation reactions are relatively slow [5]. Because of the low solubility of oxygen in water, the dissolved

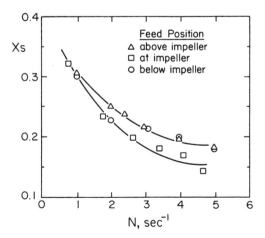

FIGURE 6.3 Effects of stirrer speed and feed location on segregation index in a 20-L reactor. (From Ref. 4 with permission from Pergammon Press.)

oxygen concentration near the top of a large fermentor may be much lower than in the impeller region, where the air is admitted. There may also be gradients in nutrient concentration, with a high value near the feed pipe and low values in other regions. Cells exposed briefly to a different environment may undergo changes in metabolism that are not reversed when the cells are swept into other regions of the reactor [6]. These effects of imperfect mixing add to the difficulty of scaling up bioreactors, where oxygen transfer limitations, foaming, and heat removal rates are often major problems.

Macromixing, Micromixing, and Mesomixing

Predicting the effect of imperfect mixing on the selectivity is a difficult challenge because of the complexity of the mixing process and because reactions and mixing are taking place simultaneously. There have been many theoretical and experimental studies of mixing plus reaction, and most have characterized stages of the mixing process with mixing times that are quite different from the blending time.

The terms *macromixing* and *micromixing* have often been used in discussing different stages of the mixing process. Macromixing is the blending of major flow streams by the interaction of large eddies. Some researchers arbitrarily define the *macromixing time*, t_{ma}, as the time to circulate the tank contents once, which is V/q, where V is the tank volume and q is the flow induced by the impeller [7]. For a standard Rushton turbine and $D_a/D_t = 1/3$, this leads to

$$t_{ma} = \frac{V}{q} \cong \frac{8-10}{n} \tag{6.5}$$

The uncertainty in Eq. (6.5) arises because the induced flow is about twice the direct discharge from the turbine, and it is difficult to measure or predict the exact value. Since the blending time t_t is about $36/n$ (Eq. 6.1), the macromixing time is roughly one-fourth the blending time, or about 2–10 seconds for a large tank.

The *micromixing time*, t_{mi}, is the time required for equilibration in the smallest eddies by a process of stretching, engulfment, and molecular diffusion. For most liquids, the thinning of fluid elements by stretching and engulfment is the limiting factor, and the micromixing time then depends on the kinematic viscosity, μ/ρ, and the local rate of energy dissipation, $\phi\bar{\varepsilon}$ [8]:

$$t_{mi} \cong 17\left(\frac{\mu/\rho}{\phi\bar{\varepsilon}}\right)^{1/2} \tag{6.6}$$

For $\mu/\rho = 10^{-6}$ m^2/sec and $\phi\bar{\varepsilon} = 1.0$ W/kg (5 HP/1000 gal), Eq. (6.6) gives $t_{mi} = 0.017$sec. However, the local energy dissipation rate varies greatly with position in the tank. Near the tip of a turbine impeller, ϕ may be 50–70, and far from the impeller ϕ is about 0.1–0.2. Therefore, values of t_{mi} for $\bar{\varepsilon} = 1.0$W/kg might range from 0.002 to 0.05 sec, one or two orders of magnitude less than the typical macromixing time.

In recent studies [7,10], a three-stage mixing process is described; the first stage is called *mesomixing*, because the mesomixing time usually falls between the micromixing time and the macromixing time. The *mesomixing time*, t_{me}, is the time for "significant mixing" of the incoming jet of feed liquid with the surrounding fluid. During this time an appreciable amount of the feed may engage in side reactions that lower the selectivity. One formula for t_{me} comes from estimating the time for turbulent diffusion to transport matter a distance equal to d_o, the feed pipe diameter [7]:

$$t_{me} \cong \frac{5.3d_o^2}{(\phi\bar{\varepsilon})^{1/3}D_a^{4/3}} \tag{6.7}$$

Another approach leads to a somewhat similar equation that includes n_f, the number of feed points and the ratio of feed velocity to solution velocity [11]:

$$t_{me} \cong 1.26\left(\frac{d_o^2}{\phi\bar{\varepsilon}n_f}\frac{v_f}{u}\right)^{1/3} \tag{6.8}$$

If geometric similarity is maintained and d_o is proportional to D_a, both equations predict that t_{me} increases with $d_o^{2/3}$, but the values of t_{me} are quite different. For $\phi\bar{\varepsilon} = 1.0$ W/kg, $d_o = 0.05$ m, and $D_a = 0.5$ m, Eq. (6.7) gives $t_{me} = 0.033$ sec. For the same parameters with $n_f = 1$ and $v_f/u = 1.0$, Eq. (6.8) gives $t_{me} = 0.17$sec. Note that t_{me} is quite dependent on feed pipe location, because of the wide range of local energy dissipation rates.

With several types of multiple reaction systems to consider and four different mixing times that can be estimated, there is no simple approach to predicting the effect of imperfect mixing on selectivity. If all the mixing times are much less than the characteristic reaction time, mixing should have little effect. However, it is not always clear how the reaction time should be defined and which of the mixing times is most important. Several authors have focused on reactions influenced or controlled by micromixing, but mesomixing may be more important in many cases, since it is the first step in the mixing process.

To understand the effect of mesomixing on reactor performance, consider what happens when a feed solution containing reactant A enters a semibatch reactor at high velocity through a submerged feed pipe. As the feed jet entrains and mixes with the bulk liquid, it forms a plume of increasing cross section, as shown in Figure 6.4. The expanding plume has radial gradients of both reactants, with high concentrations of A near the centerline and high concentrations of reactant B or product C near the outer edge. In parts of the plume, the concentration products $C_A C_B$, C_A^2, and $C_A C_C$ may be much greater than in the bulk solution, leading to changes in relative reaction rates and decreased selectivity. If the first reaction is an instantaneous acid–base reaction, A and B will not coexist, and this reaction will take place at a mixing front with a rate limited by diffusion. However, slower secondary reactions may occur in zones of increasing volume as the plume expands.

Detailed models for mixing plus reaction have been presented, and some use computational fluid mechanics to calculate velocities and the local reaction rates throughout the tank [12,13]. Others have developed cell models, with four to six interacting zones to account for different reaction rates [11,14]. These approaches require extensive computations and detailed kinetic data, which may not be available or completely reliable. In industry, multiple reaction systems are generally scaled up from laboratory or pilot-plant data. Mixing theories offer some guidance, but often there is still uncertainty about the correct procedure.

Reactor Scaleup

In scaling up stirred reactors, geometric similarity is usually maintained so that D_a/D_t and H/D_t are kept constant. The stirrer speed is chosen to meet some criterion such as constant power per unit volume, constant mixing time, or constant impeller tip speed. Only one of these criteria can be satisfied if geometric similarity is maintained. In a study of consecutive iodination reactions, Paul and Treybal [15] found that equal tip speed gave the same product distribution in 5- and 30-liter reactors when the feed was introduced at the turbine discharge. They said the fluctuating velocity near the feed point was the key parameter, and u' is proportional to

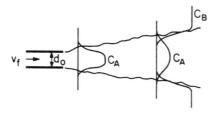

FIGURE 6.4   Expansion of the plume from a reactant feed pipe.

$n\pi D_a$, the tip speed. Constant tip speed means less power per unit volume on scaleup, and most other workers have found that equal or greater P/V is needed for larger reactors.

Power per unit volume is frequently used as a scaleup criterion for gas–liquid reactions, but there is disagreement about its merit for homogeneous reactions with imperfect mixing. If micromixing effects are most important, the selectivity should not change at scaleup at constant P/V, since $\bar{\varepsilon}$ would be constant and ϕ should be the same at comparable positions in the reactors, giving the same value of $\phi\bar{\varepsilon}$ and t_{me} [Eq. (6.6)]. In a study of parallel reactions in 1- and 20-liter tanks, Fournier and coworkers [4] reported that the local rate of energy dissipation was a good parameter for scaleup and could also explain the effect of feed location on product distribution. However, in a similar study of parallel reactions using 2.3-, 19-, and 71-liter reactors, Bourne and Yu [14] found that using constant P/V did not give the same product distribution. The amount of byproduct formed increased with tank size at constant $\bar{\varepsilon}$, as shown in Figure 6.5. Note that to get the same product distribution the power dissipation rate in the largest tank would have had to be about 10 times the value for the smallest tank.

Large-scale tests of consecutive reactions were carried out by Rice and Baud [16], who used five Pfaudler reactors from 3 to 2680 liters in size. For all reactors, the amount of secondary product decreased with increasing P/V; but at equal P/V, it was four times greater for the largest reactor than for the smallest reactor. They also found less secondary product when D_a/D_t was increased at constant P/V. No single criterion for satisfactory scaleup was suggested.

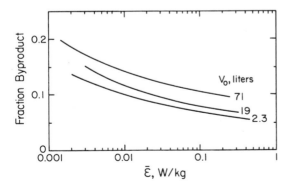

FIGURE 6.5 Effects of energy dissipation and reactor size on byproduct formation with parallel reactions. (After Ref. 14.]

Constant blending time was recommended by Fasano and Penney [17] as the scaleup criterion for fast reactions. This corresponds to constant stirrer speed, as shown by Eq. (6.1). This recommendation is supported by the work of Paul [18], who found that keeping n constant gave the same product distribution after a 10-fold increase in reactor volume. In another study [13], slightly lower selectivity was obtained in the larger reactor at the same n, but the difference in selectivity at constant P/V was much greater. Constant blending time seems to be the most reliable simple scaleup criterion, though it might be conservative for some cases and slightly underestimate byproduct formation in other cases.

A theoretical justification for using the blending time as a scaleup criterion is that t_{me} is usually proportional to t_l, and t_{me} is the critical mixing time if most of the extra byproduct formation occurs in or near the feed plume. If the ratio d_o/D_a is kept the same, with $d_o = \alpha D_a$, and the average rate of energy dissipation is $\bar{\varepsilon} = \gamma n^3 D_a^2$, then Eq. (6.7) becomes

$$t_{me} = \frac{5.3(\alpha D_a)^2}{\left(\phi \gamma n^3 D_a^2\right)^{1/3} D_a^{4/3}} = \frac{1}{n}\text{(constant terms)} \qquad (6.9)$$

Since t_{me} varies with n^{-1}, as does n_t, a constant blending time means a constant mesomixing time if the geometric ratios are maintained.

The main problem with keeping the blending time constant is that power per unit volume increases greatly on scaleup. The agitator power increases with $n^3 D_a^5$ for turbulent flow, since the power number, N_p, is constant:

$$P = N_p \rho n^3 D_a^5 \qquad (6.10)$$

Since the tank volume is proportional to D_t^3, and if H/D_t is constant,

$$\frac{P}{V} \propto n^3 D_a^2 \left(\frac{D_a}{D_t}\right)^3 \qquad (6.11)$$

At constant n, P/V increases with D_a^2, and if a several-fold increase in diameter is planned, it may be impossible to scale up with the same n. A power consumption of 10–15 HP/1000 gal (2–3 kW/m^3) is considered very vigorous agitation, and higher values are likely to cause vibration problems. The power needed for the large reactor might be decreased by changing the D_a/D_t ratio or by using a different type of impeller that gives faster blending for the same power input. These changes could be tested in the laboratory reactor and might lead to satisfactory performance at lower stirrer speeds and lower P/V.

One variable that usually changes on scaleup is the velocity of the feed jet. If the feed time is not changed, the flow rate of the feed solution increases with D_a^3, but the feed pipe area changes with d_o^2. If d_o is proportional to D_a, the feed velocity increases with $D_a^{1.0}$. At constant n, the impeller tip speed also increases with $D_a^{1.0}$, so the ratio v_f/u does not change. A high feed velocity is needed to prevent bulk fluid from entering the feed pipe, particularly when feeding near the impeller, where velocity fluctuations are quite large. In small tanks, the d_o/D_a ratio may be large and the feed velocity quite low, and this has led to decreased selectivity in some laboratory tests [19].

When studying systems with multiple reactions in small stirred tanks, the possible effects of imperfect mixing may not be apparent. Perhaps tests at 1000 and 1500 rpm show no change in product distribution, and the yield and selectivity seem to depend only on kinetic parameters. However, a large reactor would have to be operated at a much lower stirrer speed, where mixing effects could be critical. The scale-down approach should be used and the lab unit operated with agitation conditions similar to those achievable in the large reactor. Tests at low stirrer speed and with different feed locations might show an increase in byproduct formation. Once the boundaries for satisfactory performance are determined, scaleup to a large reactor can be considered.

If direct scaleup at constant stirrer speed is impractical, other schemes can be considered. If the feed was introduced just below the surface, changing to a feed location in the impeller discharge stream would give faster mixing and might permit operation at lower stirrer speed. Multiple feed nozzles could be used with smaller-diameter feed pipes, which would decrease the mesomixing time. Larger-diameter impellers would give faster mixing for the same power input, as would some of the newer "high-efficiency" impellers.

When very fast blending of reactants is needed, static mixers should be considered. These mixers have no moving parts, and they contain several short elements placed in series in a straight pipe. Units for laminar flow have 6–18 helical elements, and each element divides the stream in two and gives it a 180° twist. For turbulent flow, the elements have short tabs that protrude at an angle into the pipe. For both types, the energy dissipation rate is very high, and the mixing is more rapid than in a stirred tank. In a typical installation with a semibatch reactor, solution from the stirred tank is pumped through the static mixer, and the other reactant is added through a side port near the front of the mixer. The combined streams are returned to the tank or are sent to a heat exchanger and then to the tank. Methods of predicting the pressure drop, heat transfer rate, and the performance as a chemical reactor are described in recent papers [20–22].

Example 6.1

Tests in a turbine-stirred reactor with $D_a = 0.1$m and $D_t = H = 0.3$ m showed some product degradation when the final neutralization step was carried out at stirrer speeds less than 300 rpm.

    a. What was the power consumption per unit volume at 300 rpm, and what would it be at the same speed in a similar reactor 1.8 m in diameter?

    b. If P/V is limited to 10 HP/1000 gal, what would be the stirrer speed in the 1.8-m reactor? How much would the blending time be increased by using the lower speed?

    c. If D_a/D_t is increased to 0.5 at the same power input, what would be the new stirrer speed and blending time?

Solution.

    a. $N_p = 5.5$ for a standard six-blade turbine. Assume $\rho = 1000$ kg/m^3 and use Eq. (6.10) with $n = 5$ sec^{-1}:

$$\frac{P}{V} = \frac{5.5(1000)(5)^3(0.1)^5}{\frac{\pi}{4}(0.3)^2(0.3)} = 324 \, \text{W/m}^3$$

$$\frac{P}{V} = 0.324 \, \text{kW/m}^3 \times 5.076 = 1.64 \, \text{HP/1000 gal}$$

From Eq. (6.1), $t_t = 36/5 = 7.2$ sec.

Scaling up sixfold in diameter means 6^3 increase in volume and 6^5 increase in power or 6^2 increase in P/V:

$$\frac{P}{V} = 36 \times 1.64 = 59 \, \text{HP/1000 gal} \quad \text{impractical}$$

    b. Since $\dfrac{P}{V} \propto n^3$,

$$\left(\frac{n}{5}\right)^3 = \frac{10}{59} = 0.169$$

$$n = 0.553 \times 5 = 2.76 \, \text{sec}^{-1}, \quad \text{or } 166 \, \text{rpm}$$

t_t increases by the factor $5/2.76 = 1.81$,

    c. If $D_a/D_t = 0.5$ instead of $1/3$ but $n^3 D_a^5$ is the same

$$\left(\frac{n}{2.76}\right) = \left(\frac{1/3}{1/2}\right)^{5/3} = 0.509$$

$$n = 0.509(2.76) = 1.40 \, \text{sec}^{-1}$$

Use Eq. (6.1). At 2.76 sec^{-1}, $D_a/D_t = 1/3$, $t_t = 36/2.76 = 13.0$ sec.

With $D_a/D_t = 1/2$, $nt_1 = 4(2)^2(1) = 16$, so $t_t = 16/1.4 = 11.1$ sec.

The larger impeller would reduce the blending time by 22%, but t_t would still be greater than in the lab test by the factor $11.1/7.2 = 1.54$.

PIPELINE REACTORS

Reactors that consist of long open pipes or of tubes packed with catalyst pellets are usually designed assuming plug flow, which means every element of the fluid spends the same time in the reactor. In practice, all such reactors have a distribution of residence times, which may be a narrow and nearly normal distribution or may be strongly skewed, with some residence times much greater than the average. The measurement and interpretation of residence time distribution are discussed later when dealing with packed beds and multiphase reactions. For homogeneous systems, the most obvious departure from plug flow is when the flow is laminar, but deviations are also significant for some cases of turbulent flow.

Laminar-Flow Reactors

If the Reynolds number in a pipeline reactor is less than 2100, the flow will be laminar; and if there are no radial temperature gradients, the velocity profile will be parabolic.

$$u = 2\bar{u}\left[1 - \left(\frac{r}{R}\right)^2\right] \tag{6.12}$$

The velocity at the wall is zero, and the centerline velocity is twice the average, so the residence time, ignoring diffusion, varies from half the average time to infinity. This distribution of residence times lowers the average conversion, since the higher conversion for elements near the wall is not enough to compensate for the lower conversion near the center. The difference between the actual and the ideal or plug-flow performance depends on the reaction rates, the conversion, and the effect of diffusion, which acts to reduce the radial concentration gradients. In a simple analysis, radial diffusion is neglected to give a conservative solution for the average conversion with a parabolic velocity profile.

Consider a first-order homogeneous reaction taking place isothermally with no volume change in a laminar-flow reactor. Assuming zero diffusivity, the exit concentration for each element of fluid is given by

$$C = C_0 e^{-kt} = C_0 e^{-kL/u} \tag{6.13}$$

The average concentration at the reactor exit is found from the following integral, which takes into account the amount of fluid passing though each area element $(2\pi r \, dr)$ as well as the exit concentration for that element:

$$C_{ave} = \frac{\int_0^R uC(2\pi r)dr}{\bar{u}(\pi R^2)} \tag{6.14}$$

or

$$C_{ave} = 2 \int_0^1 \frac{u}{\bar{u}} C\left(\frac{r}{R}\right) d\left(\frac{r}{R}\right) \tag{6.15}$$

The integral can be evaluated numerically using Eq. (6.12) for u and Eq. (6.13) for C. The average exit concentration corresponds to an apparent rate constant k', which is less than the true rate constant k:

$$\ln\left(\frac{C_0}{C_{ave}}\right) = k't \tag{6.16}$$

If k'/k is 0.8, the laminar-flow reactor would have to be $1/0.8$, or 1.25, times longer than a plug flow reactor with the same conversion. Values of k'/k are given in Table 6.1.

The effect of laminar flow on conversion becomes more important as the required conversion increases, but it is still not very great at 90% conversion. For reactions other than first order, Eq. (6.15) could be used, with the appropriate kinetic equation in place of Eq. (6.13). The effect of laminar flow is greater for second-order kinetics and less for half-order kinetics.

TABLE 6.1 Effect of Laminar Flow on First-Order Reaction

Conversion	k'/k
0.02	0.96
0.17	0.91
0.56	0.81
0.78	0.76
0.94	0.70

Diffusion in Laminar-Flow Reactors

The effect of molecular diffusion is to decrease the radial concentration differences and increase the conversion. Cleland and Wilhelm [23] showed that the conversion depends on a diffusion parameter $\alpha = D_m/kR^2$ and a kinetic parameter $kL/\bar{u}$ or $k\bar{t}$. The values in Table 6.1 correspond to $\alpha = 0$, and the plug-flow case is $\alpha = \infty$. The conversions for $\alpha \leq 0.01$ are so close to those for $\alpha = 0$, that radial diffusion can be neglected. For $\alpha \geq 1.0$, radial diffusion makes the conversion almost the same as for plug flow. The conversions for $\alpha = 0.1$ fall about midway between these extremes, as shown in Figure 6.6.

Laminar flow of gases can occur in small laboratory reactors, but the diffusivity is usually high enough to nearly eliminate the radial concentration differences, so the reactor is almost equivalent to a plug-flow unit. In Example 6.2, the rate constant and diameter are arbitrarily chosen, and the diffusivity is typical for gases in air at 1 atmosphere and high temperature.

Example 6.2

Predict the effect of diffusion on the conversion for a laminar-flow gas-phase reactor under the following conditions:

$$D = 1.0 \text{ cm} \qquad D_m = 10^{-4} \text{ m}^2/\text{sec} \qquad k = 1.0 \text{ sec}^{-1}$$

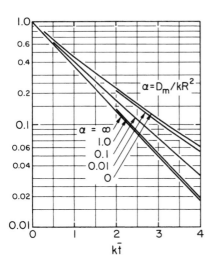

FIGURE 6.6 Effect of diffusion on conversion for first-order kinetics and a laminar-flow reactor.

Solution.

$$\alpha = \frac{D_m}{kR^2} = \frac{10^{-4}}{1.0(5 \times 10^{-3})^2} = 4.0$$

Since the conversion for $\alpha = 1.0$ is quite close to that for plug flow, as shown in Figure 6.6, the conversion for $\alpha = 4.0$ can be assumed the same as for plug flow. If the diameter was 2.0 cm, α would be 1.0 and the conversion would still be very close to the plug-flow value. For larger diameters, the flow would probably no longer be in the laminar range.

When reacting liquids in a pipeline reactor, the low diffusivity makes it less likely that radial diffusion has a significant effect. For example, if $D = 10^{-9}$ m/sec^2, $R = 10^{-2}$ m, and $k = 10^{-3}$ sec^{-1}, then $\alpha = 10^{-9} / (10^{-3} \times 10^{-4}) = 10^{-2}$ and the effect of diffusion would be negligible. For a slower reaction in a smaller tube, there might be a small effect of molecular diffusion. However, even when the effect of diffusion can be neglected, the classic solution for laminar flow is often not applicable because of changes in the velocity profile. When carrying out an exothermic reaction in a jacketed pipe, the lower temperature near the wall increases the viscosity and decreases the local velocity gradient. If the reaction is a solution or bulk polymerization, the viscosity is further increased by the higher polymer concentration near the wall, and the velocity profile differs greatly from parabolic flow, as sketched in Figure 6.7.

In a detailed model for tubular polymerization reactors, Hamer and Ray [24] showed that the viscosity and monomer conversion at the wall

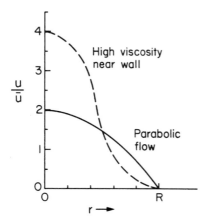

FIGURE 6.7 Typical velocity profile for a laminar-flow polymerizer.

could be much greater than the values at the center of the reactor. The velocity at the center might be several times the average, and because of the low residence time for fluid near the center, a much longer reactor is needed than for parabolic flow. The model allowed for radial flow and radial diffusion of monomer and polymer; but in a 1-inch-diameter reactor, diffusion is predicted to have little effect on conversion or polymer properties.

Turbulent-Flow Reactors

With turbulent flow in a straight pipe, the velocity profile is blunter than with laminar flow but still quite different from the flat profile assumed for plug flow. The ratio of maximum velocity to average velocity is about 1.3 at $Re = 10^4$, and this ratio slowly decreases to 1.15 at $Re = 10^6$. A pulse of tracer introduced at the inlet gradually expands, but the distribution of residence times at the exit is fairly narrow. The effect of the axial velocity profile is largely offset by rapid radial mixing due to the turbulent velocity fluctuations.

The pulse spreading has been characterized by an axial diffusion coefficient and an axial Peclet number:

$$Pe_a = \frac{uD}{D_{ea}} \tag{6.17}$$

Axial Peclet numbers range from 2 to 5, increasing slowly with Reynolds number [25]. For $Pe_a = 2$ and $L/D = 50$, the effect of axial diffusion on conversion is very small, as will be shown later when discussing reaction with axial dispersion in catalyst beds.

More important departures from plug flow can occur when gases are flowing in large-diameter reactors, such as combustion chambers or flue-gas treatment systems. If the L/D ratio is relatively small, the end effects due to changes in diameter or flow direction may be significant. One example is the waste gas incinerator shown in Figure 6.8. A large-diameter chamber with a moderate L/D is used to provide sufficient residence time in a relatively compact unit. The inlet and exit velocities are much higher than the average velocity in the chamber, which makes the flow patterns quite different from

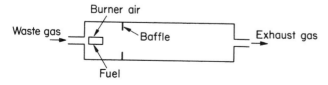

FIGURE 6.8   Turbulent-flow gas incinerator.

plug flow. Actually, vigorous mixing at the inlet is desirable to bring the waste gas rapidly to combustion temperature. A baffle may be used to promote mixing in the first part of the incinerator, but plug flow is desirable in the rest of the chamber to get a high destruction efficiency. A simple model for the incinerator is a CSTR in series with a PFR. This combination gives a higher conversion than either type of ideal reactor alone when a strongly exothermic reaction is carried out in an adiabatic reactor.

The incinerator in Figure 6.8 has an auxiliary burner that is supplied with natural gas or other fuel and sometimes with extra air. If the waste gas has enough oxygen for combustion, has a moderate heating value, and is preheated in a heat recovery exchanger, the supplemental fuel may be needed only for startup. The preheated gas then mixes rapidly with hot combustion gases in the first part of the incinerator, where 80–90% of the reaction takes place. The rest of the conversion occurs under nearly plug-flow conditions in the long section of the incinerator. However, when very high conversion is needed, the temperature and velocity near the wall cause deviations from plug-flow performance.

In a large incinerator operating at high Reynolds number, about 1% of the gas flows in the laminar boundary layer near the wall, where the average velocity and temperature are much lower than the midstream values. The conversion in the boundary layer is decreased, because the temperature effect is more important than the increase in residence time. The predicted effect of boundary-layer flow on toluene destruction in a large incinerator is shown in Figure 6.9 [26]. There is little effect at 99% conversion, but for $x \geq 0.999$, the nonideal reactor requires more than twice the residence time of an ideal plug-flow reactor.

Another example of nonideal gas flows in a large reactor is the system for selective noncatalytic reduction of nitric oxide (SNCR) in flue gas by reaction with ammonia [27]:

$$4NO + 4NH_3 + O_2 \rightarrow 4N_2 + 6H_2O$$

In laboratory reactors, nearly complete NO conversion is obtained with a stoichiometric feed, temperatures of 900–1000°C, and a short residence time. In commercial installations, ammonia is injected through multiple nozzles into hot combustion gases flowing at high velocity in large ducts. Rapid mixing is difficult, even with many feed ports, because of the low flow rate of ammonia relative to the flue gas, which may have only 200 ppm NO. Near the feed jets there are local variations in NH_3 and NO concentration, which lead to side reactions, including ammonia cracking. The overall NO conversion is typically 50–70%, even with careful design of the ammonia injection system.

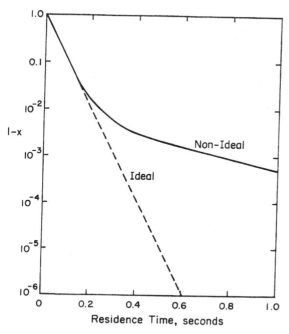

FIGURE 6.9 Predicted effect of boundary layer on toluene destruction in a large incinerator. (From Ref. 26 with permission from Elsevier Science.)

PACKED-BED REACTORS

Packed-bed reactors include multitube reactors, which use many small tubes to get good heat transfer, and large-diameter adiabatic reactors. In tubular reactors, the d_t/d_p ratio is often only 5–10, and the velocity profile is far from plug flow, as was shown in Figure 5.17. The maximum velocity, which occurs near the wall, may be several times the centerline velocity. The dispersion effect of this velocity distribution is moderated by radial mixing, so the measured residence time distribution and conversion in a long reactor may not be much different than for ideal flow. With a large adiabatic reactor, the maximum velocity still occurs near the wall, but only a small fraction of the flow is in that region. However, in a large-diameter bed, it is more difficult to get uniform flow distribution over the entire cross section, particularly if the feed enters through a single central pipe.

In addition to the effect of nonuniform flow distribution, packed beds have variations in local velocity that also cause departures from plug flow. The average interstitial velocity is u_0/ϵ, or $2.5\,u_0$ for a typical bed of spheres with $\epsilon = 0.4$. However, some of the void volume is in spaces behind particles or in channels not aligned with the central axis. The maximum local velocity is estimated to be 5–10 u_0, based on pressure-drop and heat-transfer data. As gas passes at high velocity through constricted areas, the emerging jet mixes with slow moving gas in pockets between particles. The combined effects of flow maldistribution and local mixing contribute to dispersion that is characterized by an effective axial diffusivity, D_{ea}. Values of D_{ea} are calculated from the response of the bed to a step change in tracer concentration or by the response to pulse or sinusoidal inputs.

Tests show that D_{ea} is approximately proportional to the product of the flow rate and the particle size, and the data are usually presented as axial Peclet numbers:

$$\text{Pe}_a = \frac{u_0 d_p}{D_{ea}} \tag{6.18}$$

There have been dozens of studies of axial dispersion in packed beds, and there is considerable scatter in the results. Figure 6.10 shows Peclet numbers for gases and liquid as a function of the particle Reynolds number. Both Re and Pe_a are based on the superficial fluid velocity, but in some references the average interstitial velocity, u_0/ϵ, is used for one or both of these numbers. For gases, the Peclet number is 2.0 ± 0.5. A theory that treats the bed as a series of n perfect mixers, where $n \cong L/d_p$, gives $\text{Pe}_a \cong 2$, in agreement with the data [28]. This suggests that the high velocity near the wall does not have a major effect on the dispersion.

Peclet numbers for liquids are much lower than for gases, with values of about 0.2 ± 0.1 at Re $= 1 - 10$ slowly increasing to 0.6 ± 0.2 at Re $=$

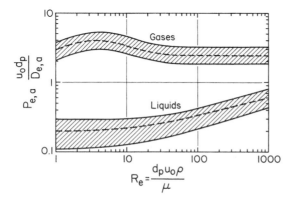

FIGURE 6.10 Axial dispersion of gases and liquid in packed beds.

1000 [29]. Schmidt numbers for liquids are about 10^3, compared to ~ 1 for gases, so molecular diffusion may be the key to the difference. The effect of high velocity near the tube wall might be offset by molecular diffusion in gas tests but not in tests with liquids. Another explanation is that pockets of fluid act as side capacities throughout the bed, and diffusion from these pockets is quite slow in liquids, leading to greater tracer dispersion [30].

Effect of Axial Dispersion on Conversion

For a first-order irreversible reaction in a packed bed where axial dispersion is significant, the concentration profile has the shape shown in Figure 6.11. The material balance for a differential element is given in Eq. (6.19), where D is the same as D_{ea} the effective axial dispersion coefficient, and u is the superficial velocity. The temperature, pressure, and molar flow rate are assumed not to change, so u and k are constant, and the equation is written for a unit cross section of the reactor:

$$\text{flow in} - \text{flow out} + \text{diffusion in} - \text{diffusion out} = \text{amount reacted}$$

$$uC - u(C + dC) - D\frac{dC}{dl} + D\left(\frac{dC}{dl} + \frac{d^2C}{dl^2}dl\right) = k\rho_b C \, dl \tag{6.19}$$

Canceling terms and dividing by u gives

$$-\frac{dC}{dl} + \frac{D}{u}\frac{d^2C}{dl^2} = \frac{k\rho_b C}{u} \tag{6.20}$$

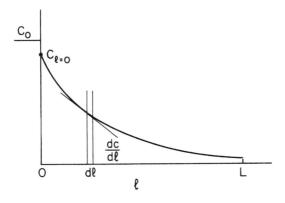

FIGURE 6.11 Concentration profile for a packed-bed reactor with axial dispersion.

The selection of appropriate boundary conditions has received much attention, and different sets of conditions have been proposed for alternate reactor configurations.

When there is no diffusion across the inlet and exit planes, the reactor is considered closed, and the boundary conditions given by Danckwerts [31] should be used. A bed of catalyst particles held between porous or perforated support plates is considered a closed reactor, because diffusion in the support plates would be negligible. However, a reactor with downflow of fluid might be closed at the bottom but have several inches of open space between the top of the bed and the gas inlet. A similar reactor with upflow of gas would be closed at the inlet and open at the exit. Boundary conditions for open–closed and closed–open configurations have been proposed [32], but for fixed beds nearly everyone uses the closed–closed conditions of Danckwerts, which are:

1. At $l = 0$, $\qquad uC_0 = uC_{l=0} + D\left(-\dfrac{dC}{dl}\right)_{l=0}$ (6.21)

2. At $l = L$, $\qquad \dfrac{dC}{dl} = 0$ (6.22)

The inlet boundary condition states that the feed to the reactor, uC_0, is equal to the reactant flux just inside the reactor, which is the sum of a convection term and a diffusion term. Since the diffusion term is finite and positive, the concentration just inside the reactor must be less than the feed concentration, and there is a step decrease in concentration at the inlet, as shown in Figure 6.11. The magnitude of the step change depends on the dispersion coefficient. As D increases, the change at the inlet becomes greater, and the profile in the reactor becomes flatter.

The exit boundary condition is zero gradient in the reactor at $l = L$. If this gradient were finite, using an equation like that for the inlet would make the exit concentration greater than the concentration just inside the reactor, which is impossible. In an ideal plug-flow reactor, where $D = 0$, there is a finite concentration gradient at the reactor exit and no step change in concentration at the inlet.

The solution of Eq. (6.20), assuming zero conversion at the inlet, is:

$$1 - x = \frac{4\beta}{(1+\beta)^2\exp\left(-\dfrac{\mathrm{Pe}'(1-\beta)}{2}\right) - (1-\beta)^2\exp\left(-\dfrac{\mathrm{Pe}'(1+\beta)}{2}\right)}$$

(6.23)

where

$$\text{Pe}' = \frac{uL}{D} = \text{Pe}_a\left(\frac{L}{d_p}\right)$$

$$\beta = \left(1 + \frac{4k\rho_b L}{u\text{Pe}'}\right)^{1/2}$$

The term $\text{Pe}' = uL/D$ is a modified Peclet number based on the reactor length instead of the particle diameter, and so Pe' may have quite large values. Figure 6.12 shows the fraction unconverted as a function of $k\rho_b L/u$ for several values of Pe'. For $\text{Pe}' > 40$, the conversion is very close to that for plug flow. For $\text{Pe}' \leq 1.0$, the perfect mixing curve is approached, but such low values of Pe' would not occur for fixed-bed reactors. The dimensionless group $k\rho_b L/u$ corresponds to kW/F in previous examples and to kt for a homogeneous reaction.

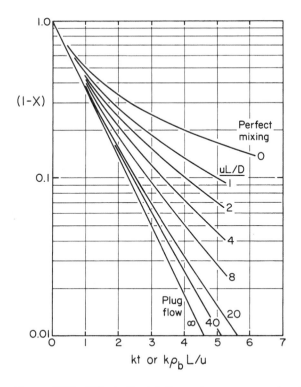

FIGURE 6.12 Effect of axial dispersion on conversion for a first-order reaction in a pipe or packed bed.

For gas flow in packed beds, where $Pe_a = 2$, the effect of axial dispersion is significant only if the bed is quite short. For example, consider a 5-inch bed of $\frac{1}{4}$-inch catalyst pellets that is predicted to give 95% conversion with ideal plug flow based on $k\rho_b L/u = 3.0$. Since $L/d_p = 20$, $Pe' = 40$, and Figure 6.12 shows $(1 - x) = 0.06$, compared to 0.05 for the ideal reactor. Rather than stating the dispersion effect as a 1% decrease in conversion, the additional amount of catalyst needed can be expressed as a correction factor. Figure 6.12 shows that about 10% more catalyst is needed to reach 95% conversion for this example. The correction factor increases slightly with increasing conversion because of the curvature of the lines for constant Pe' and the straight line for plug flow.

Most gas-phase reactors have $L/d_p > 50$, so axial dispersion can be neglected. However, there are some cases where short beds are used for mass transfer studies or for very fast reactions, such as catalytic incineration, and correction for axial dispersion may be justified.

Another way of assessing the effect of axial dispersion is to use the model of n perfectly mixed reactors in series. Since the theory for dispersion in packed beds predicts $Pe_a = 2$ if the gas is mixed between each layer of particles, $n \cong L/d_p$. For $n \geq 10$, the conversion is almost the same as for a plug-flow reactor, as was shown in Figure 3.9.

For liquids reacting in packed beds, the lower values of Pe_a might seem to make dispersion effects more important than for gases at the same L/d_p. However, even for very fast chemical reactions, high conversion of liquid cannot be obtained in short beds because of mass transfer limitations. Large values of L/d_p are needed at to get high conversion, and the effect of axial dispersion is small.

Example 6.3

A reaction catalyzed by ion-exchange resin is carried out in a 48-cm-long bed of 3-mm beads. The superficial velocity is 1 cm/sec, and the molecular diffusivity is $2 \times 10^{-5} cm^2/sec$, giving $Re = 30$ and $Sc = 500$. If the reaction rate is limited by external mass transfer, the predicted conversion is 93%.

    a.   Estimate the effect of axial dispersion on the conversion.
    b.   Would the effect be greater or smaller for a bed 24 cm long?

Solution.

    a.   For $Re = 30$, $Pe_a \cong 0.25$ from Figure 6.10:

$$Pe' = Pe_a \frac{L}{d_p} = 0.25\left(\frac{48}{0.3}\right) = 40$$

From Figure 6.12, for $1 - x = 0.07$ with plug flow, $k\rho_b \, L/u = 2.65$. With $\text{Pe}' = 40$, $1 - x \cong 0.085$. To get $1 - x = 0.07$, we need $k\rho_b L/u \cong 2.85$, or 8% more catalyst is needed.

b. If $L = 24$ cm instead of 48, then

$$\frac{k\rho_b L}{u} = \frac{2.65}{2} = 1.32$$

With plug flow, $1 - x = e^{-1.32} = 0.267$, $\text{Pe}' = 40/2 = 20$, $1 - x \cong 0.28$, to get $1 - x = 0.267$; about 6% more catalyst is needed.

The effect of axial dispersion is slightly less for the shorter bed, in spite of the lower modified Peclet number, because of the lower conversion level.

The examples just presented show that axial dispersion has only a small effect on reactor performance for single-phase flow through a packed bed of particles. With two-phase flow through a packed bed or with gas flow in a fluidized bed, dispersion effects can be quite important because of the complex flow patterns; these cases are discussed in later chapters. Complex flow patterns can also occur for various reasons with single-phase flow in fixed beds, and then reactor performance may be worse than predicted allowing for normal axial dispersion. When abnormal flow patterns are suspected because of poor reactor performance, the residence time distribution should be investigated.

Residence Time Studies

The residence time distribution in a reactor can be measured by adding a pulse of tracer to the feed or making a step change in feed concentration while continuously monitoring the outlet concentration. Frequency response techniques can also be used, but control of the feed concentration is more difficult. When the feed is distributed uniformly and the bed has no abnormal features, the response to a pulse or step change is usually like that shown in Figure 6.13 or 6.14. For beds with large L/d_p and large values of Pe', the pulse response curves show a narrow distribution of residence times, and the step response curves are almost symmetrical. Either type of curve could be compared to theoretical response curves to get the best value of Pe' or D_{ea}. However, if $\text{Pe}' \geq 40$, the effect of axial dispersion on the conversion is very small and the exact value of D_{ea} is not important.

The problem comes when the response curves are very nonsymmetrical, such as those in Figure 6.15. There may be an early breakthrough, with some elements of flow having much less than the average residence time and some having several times the average residence time. Curves of this type are not well fitted by the axial dispersion model; and if a "best-fit" value of D_{ea}

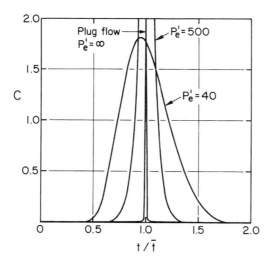

FIGURE 6.13 Pulse response curves for a dispersion model.

is obtained, it will probably not lead to correct values of conversion when used in a reactor model. Other models, such as the tanks-in-series model, could be tried, but they might not fit any better. What should be done is to consider possible causes for the abnormal response curves and to try to correct the problem.

Three situations that could lead to abnormal flow patterns and poor reactor performance are sketched in Figure 6.16. The bed in Figure 6.16a is uneven because the high-velocity inlet stream has blown particles away from the center. The difference in local velocities will lead to a greater spread of

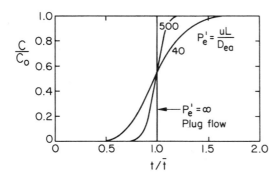

FIGURE 6.14 Step response curves for a dispersion model.

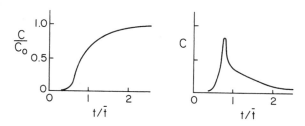

FIGURE 6.15   Response to step and pulse inputs for a nonuniform bed.

residence times and a lower average conversion. The problem could be corrected by having a feed manifold with multiple discharge pipes, by using a set of baffles to divert flow from the center, or by covering the level bed with a layer of larger and heavier particles.

Figure 6.15b shows a bed partially plugged with dirt, scale, or perhaps coke deposits. Most of the flow goes through part of the bed at a velocity much higher than normal, and the average conversion is decreased. A slow flow through the fouled section gives a long tail to the residence time curve and may lower the average selectivity because of the long residence time. Evidence for partial plugging might come from an increase in pressure drop across the bed or changes in the temperature profile.

An abnormal flow pattern can also be caused by a plugged support plate, as shown in Figure 6.16c. Less gas will pass through bed in the region above the plugged section, and the uneven flow distribution will lower the conversion. Measuring the pressure drop across the support plate and across the bed would help in diagnosis of the problem.

Example 6.4

Show the possible effect of uneven catalyst distribution in a shallow packed bed such as that in Figure 6.16a. Assume that the bed depth is 3.0 ft over half the cross section and 2.0 ft in the other half.

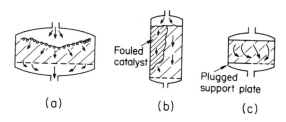

FIGURE 6.16   Nonuniform flow in packed beds.

a.   If a uniform bed would give 95% conversion, what con-
     version would result if each section had the same superficial
     velocity?

b.   If the pressure drop increases with $u_0^{1.7}$ and both sections have the
     same ΔP, what are the relative velocities? What is the overall
     conversion, allowing for the different velocities?

Solution.

a.   For first-order reaction in an ideal bed with $L = 2.5$ ft:

$$\ln\frac{1}{1-x} = \frac{k\rho_b L}{\bar{u}_0} = \ln\frac{1}{0.05} = 3$$

In section a, $L_a = 3.0$ ft instead of 2.5 ft:

$$\frac{k\rho_b L}{u_0} = 3\left(\frac{3.0}{2.5}\right) = 3.6$$

$$1 - x_a = e^{-3.6} = 0.027$$

$$x_a = 0.973$$

In section b, $L_b = 2.0$ ft:

$$\frac{k\rho_b L}{u_0} = 3\left(\frac{2}{2.5}\right) = 2.4$$

$$1 - x_b = e^{-2.4} = 0.091$$

$$x_b = 0.909$$

$$x_{\text{ave}} = \frac{0.909 + 0.973}{2} = 0.941$$

b.   The flows in sections a and b will adjust to make the pressure
     drops equal. Find the split by trial. Assume $\Delta P \propto L\bar{u}_0^{1.7}$ [see Eq.
     (3.64)].

If $u_{oa} = 0.88\bar{u}_0$,

$$\Delta P_a = \alpha(3.0)(0.88\bar{u}_o)^{1.7} = 2.41\alpha\bar{u}_0^{1.7}$$

and if $u_{ob} = 1.12\bar{u}_0$,

$$\Delta P_a = \alpha(2.0)(1.12\bar{u}_o)^{1.7} = 2.42\alpha\bar{u}_0^{1.7}$$

In section a,

$$\frac{k\rho_b L}{u_0} = 3\left(\frac{3}{2.5}\right)\frac{1}{0.88} = 4.09 \qquad x_a = 0.983$$

In section b,

$$\frac{k\rho_b L}{u_0} = 3\left(\frac{2}{2.5}\right)\frac{1}{1.12} = 2.14 \qquad x_a = 0.883$$

Using a weighted average,

$$x = \frac{0.88(0.983) + 1.12(0.883)}{2} = 0.927$$

The higher flow rate in the shallower section of the bed increases the effect of the difference in bed lengths. The amount unconverted is 7.3% compared to 5.0% for an ideal reactor.

NOMENCLATURE

Symbols

C	concentration
$\overline{C}$	average concentration
D	diameter, axial dispersion coefficient
D_a	diameter of agitator
D_t	diameter of tank
D_m	molecular diffusivity
D_{ea}	effective axial diffusivity
d_o	feed pipe diameter
d_p	particle diameter
d_t	tube diameter
H	depth of liquid
k	reaction rate constant
k'	apparent rate constant
L	time delay, length
N_p	power number for agitator
n	stirrer speed
n_f	number of feed points
P	power
Pe_a	axial Peclet number
Pe'	modified Peclet number $= uL/D$
q	flow from impeller, including induced flow
R	pipe radius
r	reaction rate, radius
Re	Reynolds number
S	selectivity
t	time

$\bar{t}$	average residence time
t_{ma}	macromixing time
t_{me}	mesomixing time
t_{mi}	micromixing time
t_t	blending time, total mixing time
u	velocity
u'	fluctuating velocity
$\bar{u}$	average velocity
u_o	superficial velocity
V	volume of reactor
v_f	velocity of feed stream
x_s	segregation index

Greek Letters

α	proportionality factor, diffusion parameter D_m/kR^2
β	parameter in Eq. (6.23)
γ	parameter in Eq. (6.9)
ε	void fraction
ε	energy dissipation rate, w/kg
$\bar{\epsilon}$	average energy dissipation rate
μ	viscosity
ρ	density
ρ_b	density of catalyst bed
π	pi
ϕ	relative rate of energy dissipation

PROBLEMS

6.1   An organic synthesis will be carried out in a jacketed baffled reactor 1.5 m in diameter with an average liquid depth of 1.8 m. The reactor has a standard six-blade turbine 0.45 m in diameter, and the maximum power input is 1.8 kW/m³.

   a.   Pilot-plant tests are planned in a similar 0.2-m reactor to see if agitation conditions affect the selectivity. What stirrer speed should be used in the small reactor to match the blending time achievable in the large reactor?

   b.   How much would the blending time be changed by making $D_a/D_t = 0.4$ instead of 0.3?

6.2   If a stirred reactor is scaled up, maintaining geometric similarity and keeping the same maximum shear rate, what would be the changes in n, P/V, and t_t for a 1000-fold increase in reactor volume?

6.3 A gas-phase reaction is carried out at 250°C in a pipeline reactor with $D = 5.0$ cm and $L = 20$ m. For a plug-flow reactor, the expected conversion is 98%.

    a.   What effect would axial dispersion have on the conversion?

    b.   If the reactor diameter was increased to 15 cm and the residence time kept the same, what conversion would be expected?

6.4 An irreversible exothermic reaction is carried out in an adiabatic reactor with a feed temperature of 500°C. The rate constant at 500°C is $0.18\sec^{-1}$, and the activation energy is 45 kcal/mol. The adiabatic temperature rise is 180°C. The residence time based on an average temperature of 600°C is 1.5 sec.

    a.   Calculate the conversion for a completely back-mixed reactor.

    b.   Show that a combination of a CSTR in series with a PFR will give a higher conversion than either a CSTR or a PFR. About what residence time should be used in the CSTR?

6.5 A catalytic oxidation is carried out in a shallow bed of 5-mm spherical catalyst particles. The bed depth is only 4 cm, yet the conversion is 97%.

    a.   What would the conversion be for a plug-flow reactor under these conditions?

    b.   What would be the conversion for a plug-flow reactor if external mass transfer in the rate-limiting step?

6.6 For a gas-phase reaction in a packed bed with $L/d_P = 10$, the predicted conversion for a PFR is 0.90.

    a.   What conversion is expected if the axial Peclet number is 2.0? Compare this with the conversion predicted for n stirred reactors in series.

    b.   Repeat the comparison if the PFR conversion for $L/d_P = 10$ is 0.99.

REFERENCES

1. WL McCabe, JC Smith, P Harriott. Unit Operations of Chemical Engineering. 6th ed. New York: McGraw-Hill, 2000, p 260.
2. L Lee, P Harriott. Ind Eng Chem Proc Des Dev 16:282, 1977.

3. EL Paul, H Mahadevan, J Foster, M Kennedy, M Midler. Chem Eng Sci 47:2837, 1992.
4. MC Fournier, L Falk, J Villermaux. Chem Eng Sci 51:5053, 1996.
5. A Humphrey. Biotechnol Prog 14:3, 1998.
6. PK Namdev, PK Yegneswaran, BG Thompson, MR Gray. Can J Chem Eng 69:513, 1991.
7. D Thoenes. Chemical Reactor Development. Dordrecht: Netherlands Klumer Academic, 1994.
8. J Baldyga, FR Bourne. Chem Eng Sci 47:1839, 1992.
9. LA Cutter. AIChE J 12:35, 1966.
10. J Baldyga, R Pohorecki. Chem Eng Journal 58:183, 1995.
11. KD Samant, KM Ng. AIChE J 45:2371, 1999.
12. RA Bakker, HE VandenAkker. I Chem E Symposium Ser 136:259, 1994.
13. JC Middleton, F Pierce, PM Lynch. Chem Eng Res Des 64:18, 1986.
14. JR Bourne, S Yu. Ind Eng Chem Res 33:41, 1994.
15. EL Paul, RE Treybal. AIChE J 17:718, 1971.
16. RW Rice, RE Baud. AIChE J 36:293, 1990.
17. JB Fasano, WR Penney. Chem Eng Progr 87(12):46, 1991.
18. EL Paul. Chem Eng Sci 43:1773, 1998.
19. JR Bourne, F Kozicki, U Moergelli, P Rys. Chem Eng Sci 36:1655, 1981.
20. CS Knight, WR Penney, JB Fasano. Paper presented at AIChE Winter Annual Meeting, 1995.
21. KJ Myers, A Bakker, D Ryan. Chem Eng Progr 93(6):28, 1997.
22. RA Taylor, WR Penney HX Vo. Paper presented at AIChE Winter Annual Meeting, 1998.
23. FA Cleland, RH Wilhelm. AIChE J 2:489, 1956.
24. JW Hamer, WH Ray, Chem Eng Sci 41:3083, 1986.
25. H Kramers, KR Westerterp. Elements of Chemical Reactor Design and Operation. New York: Academic Press, 1963.
26. P Harriott, J Ellet. J Hazard Mat 45:233, 1996.
27. RK Lyon. Environ Sci Technol 21:231, 1987.
28. KW McHenry Jr, RH Wilhelm. AIChE J 3:83, 1957.
29. SF Chung, CY Wen. AIChE J 14:856, 1968.
30. H Kramers, KR Westerterp. Elements of Chemical Reactor Design and Operation. New York: Academic Press, 1963, p 95.
31. PV Danckwerts. Chem Eng Sci 2:1, 1953.
32. JB Butt. Reaction Kinetics and Reactor Design. 2nd ed. New York: Marcel Dekker, 2000, p 371.

7

Gas–Liquid Reactions

In many processes, a gas-phase reactant dissolves in a liquid to react with the liquid or with other substances present in the solution. Examples of partial oxidation of organic liquids are the oxidation of cyclohexane to cyclohexanone and the oxidation of p-xylene to terephthalic acid, processes that are carried out by bubbling air through the liquid in a stirred tank. Reactions of oxygen in aqueous solution include aerobic fermentations and destruction of organic contaminants in polluted water. Reactions of other gases such as chlorine, hydrogen, carbon monoxide, and ethylene, with organic compounds are often carried out in the liquid phase, and in all these examples the gas must dissolve before reaction takes place. The dissolving of a gaseous reactant is a mass transfer step that may have a slight or a large effect on the rate of reaction, depending on the gas solubility, the mass transfer coefficient, and the intrinsic kinetics of the reaction.

In this chapter, the theories for gas absorption plus reaction are presented first for relatively slow reactions and then for fast reactions and for instantaneous reactions. Performance data and scaleup criteria for several types of gas–liquid reactors are then reviewed. For a given reaction, the intrinsic kinetics of the liquid-phase reaction are the same for all types of reactors, including stirred tanks, packed columns, bubble columns, and spray contactors, but the mass transfer coefficients differ greatly, and the selection of reactor type and reaction conditions are strongly influenced by the mass transfer characteristics.

CONSECUTIVE MASS TRANSFER AND REACTION

When the reaction rate in the liquid is slow compared to the maximum possible rate of mass transfer of the gaseous reactant, the processes of mass transfer and chemical reaction can be considered to take place in series. Consider an irreversible reaction that takes place when gas A is bubbled into a tank containing B:

$$A_g + nB \rightarrow C \tag{7.1}$$

Assume the reaction in the liquid is first order to both A and B. The rate per unit volume of liquid is

$$r' = k_2 C_A C_B \tag{7.2}$$

The concentration gradients near the gas–liquid interface are shown in Figure 7.1. The figure shows a small driving force $(P_A - P_{Ai})$ for diffusion of A through the gas film near the interface, a discontinuity at the interface, where C_{Ai} is the concentration of A on the liquid side of the interface, and a modest driving force $(C_{Ai} - C_A)$ for diffusion of A into the bulk liquid, where reaction occurs. No gradient is shown for B, because B is already present in the bulk liquid, where all the reaction is assumed to take place. Often C_B is one or two orders of magnitude greater than C_A because of the low solubility of the gas. The gradients for other types of gas–liquid reactors, such as packed columns and spray absorbers, are similar to those in Figure 7.1, but of course the mass transfer coefficients and relative driving forces might be different.

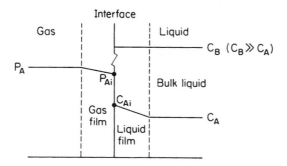

FIGURE 7.1   Concentration gradients for slow reaction of A with B.

The reaction rate can be defined in different ways using a unit volume of liquid, a unit volume of reactor, or unit interfacial area as a basis. Since mass transfer coefficients for tanks and packed columns are often based on the volume of the apparatus, the overall reaction rate will be expressed in units such as lb-mol-A/hr, ft^3 or kg mol A/hr, m^3. The volume is the active reactor volume, which is the volume of the gas–liquid mixture in a stirred tank or bubble column, the packed column volume, or the chamber volume for a spray reactor. The mass transfer of A to the interface is the first step:

$$r_1 = k_g a (P_A - P_{Ai}) \tag{7.3}$$

If the gas follows Henry's law, then

$$P_{A_i} = H C_{Ai} \tag{7.4}$$

The diffusion of A through the liquid film is the second step. The a in $k_L a$ and $k_g a$ is the interfacial area per unit volume:

$$r_2 = k_L a (C_{Ai} - C_A) \tag{7.5}$$

The third step is the reaction of A and B in the bulk liquid. For a stirred tank or bubble column, the amount of liquid per unit reactor volume is $(1 - \epsilon)$, where ϵ is the void fraction or bubble fraction:

$$r_3 = k_2 C_B C_A (1 - \epsilon) \tag{7.6}$$

Continuous Back-Mixed Reactor

If the reaction is taking place in an ideal continuous stirred-tank reactor, the concentrations C_B and C_A are uniform throughout the tank and do not change with time. Then the rate of mass transfer of A to and into the liquid must equal the rate of reaction of A:

$$r_1 = r_2 = r_3 = r \tag{7.7}$$

Since all the steps are first order to A, Eqs. (7.3)–(7.7) can be combined to give an overall rate expression. The overall driving force $(P_A - 0)$ is split into its parts, and each part is divided by the appropriate rate expression:

$$P_A = P_A - P_{Ai} + P_{Ai} = P_A - P_{Ai} + H(C_{Ai} - C_A + C_A) \tag{7.8}$$

$$\frac{P_A}{r} = \frac{P_A - P_{Ai}}{r_1} + \frac{H(C_{Ai} - C_A)}{r_2} + \frac{HC_A}{r_3} \tag{7.9}$$

Substituting for r_1, r_2, and r_3 gives

$$\frac{P_A}{r} = \frac{1}{k_g a} + \frac{H}{k_L a} + \frac{H}{k_2 C_B (1 - \epsilon)} = \frac{1}{K_g a} \tag{7.10}$$

$$r = K_g a P_A \tag{7.11}$$

Often the gas-film resistance is negligible because $H/k_L a \gg 1/k_g a$, and the overall coefficient becomes

$$\frac{1}{K_g a} \cong H\left[\frac{1}{k_L a} + \frac{1}{k_2 C_B(1-\epsilon)}\right] \tag{7.12}$$

For example, the Henry's law coefficient for oxygen in water is $H_{25°C} \cong 4.3 \times 10^4$ atm/mole fraction, and the gas-film resistance can be neglected for most air oxidations. The gas-film resistance is zero if a pure gas is used and the solvent partial pressure is negligible.

For gas absorption in a large stirred tank, the concentration in the liquid phase may be the same throughout the tank, but there may be appreciable differences in the partial pressure of A in the gas phase because of the hydraulic pressure gradient and because of incomplete mixing of the gas. Equation (7.11) would still apply, but P_A would vary from the bottom to the top of the tank, and an integrated form of the rate expression or an appropriate average partial pressure would have to be used. A log-mean average is sometimes used; but for vigorous agitation, the effective average may be closer to the exit mole fraction times the average pressure.

Example 7.1

An aqueous catalyst solution is continuously regenerated by air oxidation in a stirred tank, where monovalent metal ions are converted to the active divalent form:

$$4B^+ + O_2 + 4H^+ \rightarrow 4B^{++} + 2H_2O$$

$$r = k_2 C_{O_2} C_{B^+}$$

$$k_2 = 8.5 \text{ L/mol-sec at } 50°C$$

The reaction conditions are:

$T = 50°C$, $H_{O_2} = 8 \times 10^4$ atm/mole fraction
$P = 2$ atm (top of tank)
$F = 17,000$ L/hr solution feed rate
$C_{B^+} = 1.6$ M feed, 0.8 M product

100% excess air

The estimated mass transfer coefficients are

$k_L a = 900 \text{ hr}^{-1}$

$k_g a = 80 \text{ mol/hr-L-atm}$

a. Calculate the overall reaction rate coefficient and the percent resistance due to gas-phase mass transfer, liquid-phase mass transfer, and chemical reaction.
b. What reaction volume is needed and what reactor size should be chosen?

Solution.

a. To get $K_g a$ in mole/hr-L-atm, convert all rate terms to the same units:

$$k_g a = 80 \text{ mol/L-hr-atm}$$

convert H assuming $1000/18 = 55.5 \text{ mol/L}$:

$$H = \frac{8 \times 10^4 \text{atm}}{\text{mole } O_2/\text{total moles}} \times \frac{1.0 \text{ 1}}{55.5 \text{ mol}} = 1440 \frac{\text{atm-L}}{\text{mole } O_2}$$

$$\frac{k_L a}{H} = \frac{900}{1440} = \frac{0.625 \text{ mol}}{\text{atm-L-hr}}$$

assume $\epsilon = 0.1, (1 - \epsilon) = 0.9$:

$$C_{B^+} = 0.8 \text{ M}$$

$$\frac{k_2 C_{B^+}(1 - \epsilon)}{H} = \frac{8.5(0.8)(0.9)(3600)}{1440} = 15.3\text{-mol/L-hr-atm}$$

$$\frac{1}{K_g a} = \frac{1}{80} + \frac{1}{0.625} + \frac{1}{15.3} = 1.678$$

$$K_g a = 0.596 \text{ mol/L-hr-atm}$$

$$\text{gas-film resistance} = \frac{1/80}{1.678} \times 100 = 0.7\%$$

$$\text{liquid-film resistance} = \frac{1/0.625}{1.678} \times 100 = 95.4\%$$

$$\text{reaction resistance} = \frac{1/15.3}{1.678} \times 100 = 3.9\%$$

The gas-film resistance could have been neglected for this example.

b. moles O_2 needed $= F(\Delta C_{B^+})\dfrac{1}{4} = 17{,}000\left(\dfrac{0.8}{4}\right) = 3400$ moles/hr

per 100 moles air, O_2 fed $= 20.9$,

$$\begin{aligned} O_2 \text{ out} &= 10.45 \quad \text{(half of } O_2 \text{ fed)}\\ N_2 \text{ out} &= 79.10\\ \hline \text{Gas out} &= 89.55 \end{aligned}$$

$$P_{O_2\text{out}} \cong 2 \text{ atm}\left(\frac{10.45}{89.45}\right) = 0.234 \text{ atm} \quad \text{(neglects } P_{H_2O})$$

$$P_{O_2\text{in}} \cong 2(0.209) = 0.418 \quad \text{(neglects depth factor)}$$

$$\overline{P_{O_2}} = \text{log mean } P_{O_2} = 0.317 \text{ atm}$$

$$\text{volume} = \frac{O_2 \text{ needed}}{K_g a \overline{P_{O_2}}} = \frac{3400}{0.596(0.317)} = 18{,}000\text{L} \quad (4750 \text{ gal})$$

The total reactor volume would be at least 20% greater than the volume of aerated liquid, to allow for disengagement space, so a 6000-gal tank might be chosen.

Semibatch Reactor

If a gas–liquid reaction is carried out by passing gas continuously through a batch of liquid in a stirred tank, the concentration of liquid-phase reactant changes with time, and this usually changes the reaction rate and the relative importance of the mass transfer and reaction steps. The change in reaction rate also leads to gradual accumulation of gas A in the bulk liquid, which means that the rate of absorption of A is no longer equal to the rate of reaction, as was assumed in Eq. (7.7).

Consider the reaction $A_g + nB \rightarrow C$ taking place in a semibatch reactor with a concentration C_B and an active volume V. Assume the reaction is first order to A and to B and that the gas-phase resistance is negligible. The unsteady-state material balance for A is

$$\text{accumulation} = \text{input} - \text{reaction} \tag{7.13}$$

$$V(1 - \epsilon)\frac{dC_A}{dt} = k_L a V(C_{Ai} - C_A) - k_2 C_A C_B (1 - \epsilon)V \tag{7.14}$$

The volume terms cancel, and the equation is rearranged to

$$\frac{dC_A}{dt} + C_A\left(\frac{k_L a}{1 - \epsilon} + k_2 C_B\right) = \frac{k_L a}{1 - \epsilon} C_{Ai} \tag{7.15}$$

Although C_B decreases with time because of reaction, the change takes place over many minutes or a few hours, whereas the rate of change of C_A is quite rapid when gas is first introduced to the reactor. Assuming C_B is practically constant for a short time permits integration of Eq. (7.15):

For $C_A = 0$ at $t = 0$,

$$C_A = C_{Ai}\left(\frac{k_L a'}{k_L a' + k_2 C_B}\right)\left(1 - e^{-t\left(k_L a' + k_2 C_B\right)}\right) \tag{7.16}$$

where

$$k_L a' = \frac{k_L a}{1 - \epsilon} \qquad a' = cm^2/cm^3 \text{ liquid}$$

The reciprocal of the term $(k_L a' + k_2 C_B)$ is the time constant τ for changes in C_A. For a first-order system, 95% of steady state is reached when $t = 3\tau$. In a typical stirred-tank reactor with good gas dispersion, $k_L a' \cong 0.1 \text{ sec}^{-1}$, and if the reaction rate term $k_2 C_B$ is also about 0.1 sec^{-1}, the time constant is 5 sec. The solution would reach 95% of the final value of C_A in about 15 seconds. The final value (99.7% of equilibrium), which is really a pseudo-steady-state value is reached after $t \cong 5\tau$:

$$C_{A_{ss}} = C_{Ai}\left(\frac{k_L a'}{k_L a' + k_2 C_B}\right) \tag{7.17}$$

Note that the steady-state bulk concentration of A is half the saturation value when the mass transfer and reaction terms are equal. As C_B decreases slowly because of reaction, C_A gradually increases and approaches C_{Ai} as the reaction nears completion. Since the time constant for changes in C_A is usually so much smaller than the time required for appreciable changes in C_B, Eq. (7.17) can be used for C_A in the reaction rate and mass transfer equations. The accumulation term $\left(\frac{dC_A}{dt}\right)$ in Eq. (7.14) is assumed negligible, and the rate of consumption of B is set equal to the corresponding rate of absorption of A:

$$V(1 - \epsilon)\frac{dC_B}{dt} = -\frac{1}{n}k_L aV(C_{Ai} - C_A) \tag{7.18}$$

Substituting for C_A using Eq. (7.17) yields

$$V(1 - \epsilon)\frac{dC_B}{dt} = -\frac{1}{n}k_L aV C_{Ai}\left(1 - \frac{k_L a'}{k_L a' + k_2 C_B}\right) \tag{7.19}$$

Integration of Eq. (7.19) gives the time required for a specified conversion of B. Since $k_L a' = k_L a/(1 - \epsilon)$, Eq. (7.19) can also be written

$$\frac{dC_B}{dt} = -\frac{k_L a' C_{Ai}}{n}\left(\frac{k_2 C_B}{k_L a' + k_2 C_B}\right) \tag{7.20}$$

When $k_L a' \gg k_2 C_B$, the solution is almost saturated with A, and the mass transfer terms cancel, giving

$$-\frac{dC_B}{dt} = \frac{k_2 C_{Ai} C_B}{n} \tag{7.21}$$

With C_{Ai} constant, the change in C_B follows pseudo-first-order kinetics:

$$\ln\frac{C_{Bo}}{C_B} = \frac{k_2 C_{Ai} t}{n} \tag{7.22}$$

The n appears in Eqs. (7.18)–(7.22) because the reaction rate was defined for A rather than for B.

When the reaction rate term $k_2 C_B$ is much greater than the mass transfer coefficient $k_L a'$, the reaction becomes mass transfer controlled, and Eq. (7.20) becomes

$$-\frac{dC_B}{dt} = \frac{k_L a' C_{Ai}}{n} \tag{7.23}$$

The concentration of B then decreases at a constant rate, as if the reaction was zero order to B. The changes in C_B with time for reaction control and mass transfer control are sketched in Figure 7.2.

When $k_2 C_B$ and $k_L a'$ are comparable in magnitude, the reaction rate decreases as B is converted, but not as much as when the reaction step controls the overall rate. For very high conversions of B, the last stages of the batch reaction follow pseudo-first-order kinetics.

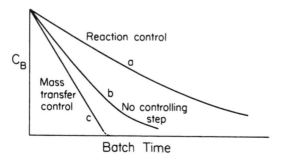

FIGURE 7.2 Concentration curves for a semibatch reaction: (a) $k_L a \gg k_2 C_B$; (b) $k_L a \cong k_2 C_B$; (c) $k_L a \ll k_2 C_B$.

The batch time required for a given conversion of B depends on C_{Ai} for all the cases considered so far, but C_{Ai} may not be constant if the gas fed is a mixture. For example, if an oxidation is carried out using air bubbled into a tank, an average partial pressure of oxygen should be used with the Henry's law constant to get C_{Ai}. At the start of the reaction, an appreciable fraction of the oxygen fed might react, and the partial pressure of oxygen in the exit gas could be considerably less than that in the feed gas. The appropriate average partial pressure could be the log mean of the two values; or if the tank is very vigorously agitated, the exit partial pressure would be used. If mass transfer of oxygen is not controlling and the reaction slows down, a smaller fraction of the oxygen will be used and the average partial pressure will increase with time. A numerical or graphical integration of Eq. (7.20) would be needed, with the appropriate values of P_A and C_{Ai} used.

Continuous Reaction in an Unstirred Reactor

When a gas–liquid reaction is carried out continuously in a packed absorber or in a bubble column, and if the solubility of gas A is low compared to the concentration of reactant B, the pseudo-steady-state approach can be used and Eq. (7.20) applied to give the reaction rate at any point in the column. If plug flow of liquid and gas is assumed, which is reasonable for a packed column, and if the ratio of flows is specified, the change in P_A can be related to the change in C_B and the required tower size calculated by integration or by using an average P_A, if the change in P_A is not very large.

SIMULTANEOUS MASS TRANSFER AND REACTION

A key assumption made in the previous section on consecutive mass transfer and reaction was that essentially all reaction takes place in the bulk liquid. However, when the reaction is rapid enough to make the concentration of reactant A almost zero in the bulk liquid, significant reaction may take place in the liquid film. The amount of liquid in the film region is very much less than the amount of bulk liquid, but the average concentration of A in the film is much higher than the bulk value. When reaction in the film region is significant, the amount reacting must be determined using theories for simultaneous diffusion and reaction. The case of a stagnant liquid film is considered first, since it is easier to picture the gradients, and then results based on the penetration theory are presented.

When some reaction occurs in a stagnant liquid film, the steady-state gradient for A is steeper at the gas–liquid interface than at the other boundary of the film, since the flux into the film is greater than the flux into the bulk liquid. Therefore the concentration profile for A is curved, as shown in

Figure 7.3 for a typical case where about half the reaction occurs in the film. The gas-film resistance is neglected for this example, and the concentration of B is constant throughout the film, as was assumed in the previous section. The concentration of A in the bulk liquid is C_{AL}, and C_A is the (variable) concentration in the liquid film. The distance from the interface is x, and x_L is the film thickness.

At steady state, the difference between the diffusion fluxes into and out of the differential element dx for a unit interfacial area is equal to the amount consumed in the element:

$$-D_A\left(\frac{dC_A}{dx}\right) + D_A\left(\frac{dC_A}{dx} + \frac{d^2C_A}{dx^2}dx\right) = k_2C_BC_A\ dx \qquad (7.24)$$

Since C_B does not vary, k_2C_B is replaced with a first-order rate constant:

$$D_A\frac{d^2C_A}{dx^2} = k_2C_BC_A = kC_A \qquad (7.25)$$

The boundary conditions are

$$C_A = C_{Ai} \qquad \text{at } x = 0$$
$$C_A = C_{AL} \qquad \text{at } x = x_L$$

Note that Eq. (7.25) has the same form as Eq. (4.75) for pore diffusion and reaction in a flat slab, but the boundary conditions are different, since the gradient for A is not zero at the edge of the liquid film.

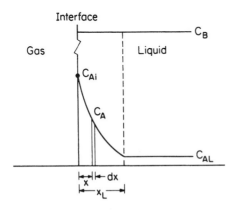

FIGURE 7.3 Concentration gradients for simultaneous diffusion and reaction of gas A in the liquid film.

The solution of Eq. (7.25) is given in terms of a dimensionless modulus sometimes called the *square-root modulus* [1] and sometimes the Hatta number [2]:

$$C_A = \frac{C_{AL} \sinh\left(\frac{x}{x_L}\sqrt{M}\right) + C_{Ai} \sinh\left(\frac{x_L - x}{x_L}\sqrt{M}\right)}{\sinh(\sqrt{M})} \qquad (7.26)$$

where

$$\sqrt{M} = x_L \sqrt{\frac{k}{D_A}} = \text{square-root modulus} \qquad (7.27)$$

An alternate equation for the square-root modulus is based on the normal value of the film coefficient, k_L^*, rather than on the film thickness:

$$k_L^* = \frac{D_A}{x_L} \qquad (7.28)$$

$$\sqrt{M} = \frac{D_A}{k_L^*}\sqrt{\frac{k}{D_A}} = \frac{\sqrt{kD_A}}{k_L^*} \qquad (7.29)$$

The flux of A is obtained by differentiating Eq. (7.26) and evaluating the derivative at $x = 0$:

$$N_A = D_A\left(\frac{-dC_A}{dx}\right)_0 = \frac{\sqrt{kD_A}\left(C_{Ai}\cosh\sqrt{M} - C_{AL}\right)}{\sinh\sqrt{M}} \qquad (7.30)$$

The concentration of A in the bulk can be calculated by setting the flux at $x = x_L$ equal to the amount consumed in the bulk liquid per unit area of interface:

$$D_A\left(\frac{-dC_A}{dx}\right)_{x_L} = \frac{kC_{AL}}{a} \qquad (7.31)$$

In almost all cases where the reaction is fast enough to cause significant reaction in the film, the value of C_{AL} is very low and can be taken as zero. This simplifies the analysis and Eq. (7.30) becomes

$$N_A = \frac{\sqrt{kD_A}C_{Ai}\cosh\sqrt{M}}{\sinh\sqrt{M}} = \frac{C_{Ai}\sqrt{kD_A}}{\tanh\sqrt{M}} \qquad (7.32)$$

The flux of A is proportional to C_{Ai}, just as it is for consecutive mass transfer and reaction, since both mass transfer and reaction are first-order processes. The subsequent analysis focuses on mass transfer of A using the concept of an effective mass transfer coefficient. In Eq. (7.32), the driving force is C_{Ai} (or $C_{Ai} - 0$), and since N_A is proportional to C_{Ai}, the proportionality factor is the effective mass transfer coefficient, k_L:

$$k_L = \frac{N_A}{C_{Ai}} = \frac{\sqrt{kD_A}}{\tanh\sqrt{M}} \tag{7.33}$$

The ratio of the effective mass transfer coefficient to the normal value is always greater than 1.0 and is called the *enhancement factor*, ϕ:

$$\phi \equiv \frac{k_L}{k_L^*} = \frac{\sqrt{kD_A}}{k_L^* \tanh\sqrt{M}} = \frac{\sqrt{M}}{\tanh\sqrt{M}} \tag{7.34}$$

The reaction rate per unit volume is the normal rate times the enhancement factor

$$r = k_L^* a \phi C_{Ai} \tag{7.35}$$

For low values of $\sqrt{M}$, $\tanh\sqrt{M} \cong \sqrt{M}$, giving $\phi = 1.0$; and the equations for consecutive absorption and reaction can be used. For $\sqrt{M} = 1.0$, $\tanh 1.0 = 0.76$ and $\phi = 1.31$, which means that the rate of absorption (and reaction) of A is 31% faster than the rate of physical absorption with the same driving force. For $\sqrt{M} \geq 3$, $\tanh\sqrt{M} \cong 1$ and the enhancement factor is equal to $\sqrt{M}$.

To understand how chemical reaction enhances the mass transfer process, consider the concentration gradients sketched in Figure 7.4. For $\sqrt{M} = 3$, the gradient at the interface is about three times steeper than that for physical absorption, which is shown as a dashed line. Much of the A reacts quite close to the interface, and a negligible amount diffuses into the bulk liquid. For $\sqrt{M} = 10$, the rate of absorption of A is enhanced 10-fold, as if the effective film thickness was only one-tenth the actual value.

For high values of $\sqrt{M}$, the rate of absorption per unit area is the modulus times the normal rate. For $\sqrt{M} \geq 3$, $\phi = \sqrt{M}$,

$$N_A = \sqrt{M} k_L^* C_{Ai} = \sqrt{kD_A} C_{Ai} \tag{7.36}$$

$$r = a\sqrt{kD_A} C_{Ai} \tag{7.37}$$

The absorption rate varies with the square root of the rate constant and the molecular diffusivity. Both mass transfer and chemical reaction influence the overall process, but neither can be said to "control" the rate. Note that the square-root dependence on a rate constant and a diffusivity matches that found for diffusion and reaction in porous catalysts at high values of the Thiele modulus [Eq. (4.34)].

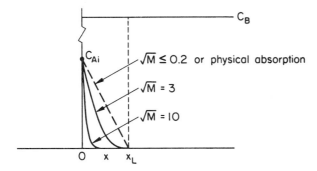

FIGURE 7.4 Concentration profiles for absorption plus slow, moderate, or fast reaction.

Allowing for Gradients of B

When the enhancement factor is large, indicating that most of the reaction takes place quite close the interface, the assumption of constant concentration of B, which leads to pseudo-first-order kinetics, may no longer be valid. Based on the stagnant film model, B has to diffuse most of the way through the film if the reaction takes place in a narrow zone very close to the interface. Typical gradients for this situation are shown in Figure 7.5. The gradient for B is linear through most of the liquid film, where no reaction is taking place. In the reaction zone, the gradient for B decreases as B is consumed, and the gradient goes to zero at the interface, since no B leaves the liquid. The average concentration of B in the reaction zone, $\bar{C}_B$, is

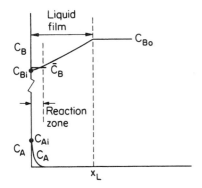

FIGURE 7.5 Gradients for a fast reaction very near the interface.

slightly greater than the concentration at the interface, C_{Bi}. If $\bar{C}_B$ is significantly less than C_{Bo}, the enhancement factor is smaller than that given by Eq. (7.34) with $k = k_2 C_{Bo}$. To calculate the correct enhancement factor allowing for the change in C_B would require simultaneous solution of the equations for diffusion and reaction for both A and B in the liquid film. The exact solution is not given here, since solutions based on the more realistic penetration theory are presented later. However, the following example shows how film theory can be used for an approximate calculation to show when the gradient for B becomes important.

Example 7.2

For the reaction $A + B \to C$ with $C_{Bo} = 40 C_{Ai}$, $D_A = 1.2 D_B$, and $\sqrt{M} = 10$, is the gradient for B in the liquid film significant?

Solution. Assume the gradient for A is the same as when the gradient for B is negligible, which means $\phi = \sqrt{M} = 10$:

$$N_A \cong \frac{10 D_A C_{Ai}}{x_L} = C_{Ai} \frac{D_A}{0.1 x_L}$$

The effective diffusion distance for A is $0.1 x_L$. Assume

$$N_B = \frac{D_B(C_{Bo} - \overline{C_B})}{0.9 x_L} = N_A = \frac{C_{Ai} D_A}{0.1 x_L}$$

$$C_{Bo} - \overline{C_B} = \frac{D_A}{D_B} C_{Ai} \left(\frac{0.9}{0.1}\right) = 9 C_{Ai} \frac{D_A}{D_B}$$

$$\frac{C_{Bo} - \overline{C_B}}{C_{Bo}} = 9(1.2) \frac{C_{Ai}}{C_{Bo}} = 9 \frac{(1.2)}{40} = 0.27$$

$$\overline{C_B} = 0.73 \, C_{Bo}$$

If r varies with $\sqrt{k C_B}$, r decreases by $\sqrt{0.73} = 0.85$. The corrected $\phi = 10(0.85) = 8.5$. A 15% decrease in rate is marginally significant, considering probable uncertainties in k_L^* and $\sqrt{M}$. For higher values of $\sqrt{M}$ or a lower C_{Bo}/C_{Ai} ratio, the change in C_B should certainly be taken into account.

INSTANTANEOUS REACTION

When A and B react instantaneously, the reaction occurs at a plane parallel to the interface, or at the gas–liquid interface if the gas-film resistance controls. For the case where C_{Ai} is not zero, the gradients for reaction in a stagnant film are shown in Figure 7.6. The steepness of the gradients reflects the diffusivities, and if $D_A \cong D_B$ and $C_{Bo} >> C_{Ai}$, then the reaction plane is quite close to the interface.

Based on the gradients shown in Figure 7.6, the fluxes of A and B are

$$N_A = \frac{C_{Ai}D_A}{\alpha}, N_B = \frac{C_{Bo}D_B}{x_L - \alpha} \tag{7.38}$$

For $A + nB \rightarrow C$,

$$N_B = nN_A$$

$$\frac{C_{Ai}D_A}{\alpha} = \frac{C_{Bo}D_B}{n(x_L - \alpha)} \tag{7.39}$$

$$C_{Ai}D_A nx_L - C_{Ai}D_A n\alpha = C_{Bo}D_B\alpha \tag{7.40}$$

$$\alpha = \frac{C_{Ai}D_A nx_L}{C_{Bo}D_B + nC_{Ai}D_A} \tag{7.41}$$

$$N_A = \frac{C_{Ai}D_A(C_{Bo}D_B + nC_{Ai}D_A)}{nC_{Ai}D_A x_L} \tag{7.42}$$

$$N_A = \frac{nC_{Ai}D_A + C_{Bo}D_B}{nx_L} \tag{7.43}$$

$$k_L = \frac{N_A}{C_{Ai}} = \frac{D_A}{x_L} + \frac{C_{Bo}}{nC_{Ai}}\frac{D_B}{x_L} \tag{7.44}$$

$$\phi = \frac{k_L}{k_L^*} = \frac{k_L}{(D_A/x_L)} = 1 + \left(\frac{C_{Bo}}{nC_{Ai}}\right)\left(\frac{D_B}{D_A}\right) \tag{7.45}$$

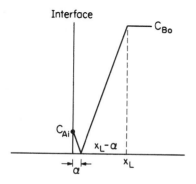

FIGURE 7.6 Concentration profiles for instantaneous reaction of A and B.

The enhancement factor for instantaneous reaction depends on the ratio of concentrations and the ratio of diffusivities. A limiting case is a very high ratio of C_{Bo}/C_{Ai}. When $C_{Bo} >> nC_{Ai}$,

$$\phi \cong \frac{C_{Bo}}{nC_{Ai}}\left(\frac{D_B}{D_A}\right) \tag{7.46}$$

The reaction rate per unit interfacial area is

$$N_A = \phi k_L^* C_{Ai} = \frac{C_{Bo}}{nC_{Ai}}\frac{D_B}{D_A}\frac{D_A}{x_L}C_{Ai} = \frac{C_{Bo}}{n}\left(\frac{D_B}{x_L}\right) \tag{7.47}$$

The reaction rate per unit volume is

$$r_A = a\frac{C_{Bo}}{n}\left(\frac{D_B}{x_L}\right) \tag{7.48}$$

Since $\frac{D_B}{x_L}$ is the mass transfer coefficient for diffusion of B through the film, the rate of absorption of A is then limited by the rate of diffusion of B to the gas–liquid interface, where the reaction occurs. Increases in C_{Ai} or P_A have no effect on the rate and make the apparent overall coefficient for A decrease. This has been observed for some fast gas–liquid reactions [3].

PENETRATION THEORY

In the previous sections, stagnant films were assumed to exist on each side of the interface, and the normal mass transfer coefficients were assumed proportional to the first power of the molecular diffusivity. In many mass transfer operations, the rate of transfer varies with only a fractional power of the diffusivity because of flow in the boundary layer or because of the short lifetime of surface elements. The penetration theory is a model for short contact times that has often been applied to mass transfer from bubbles, drops, or moving liquid films. The equations for unsteady-state diffusion show that the concentration profile near a newly created interface becomes less steep with time, and the average coefficient varies with the square root of (D/t) [4]:

$$k_L^* = 2\sqrt{\frac{D}{\pi t}} = 1.13\sqrt{\frac{D}{t}} \tag{7.49}$$

When a pseudo-first-order reaction takes place in the liquid, the concentration profile for reactant A is initially very steep and becomes less steep as A diffuses further into the liquid. However, with a fast chemical reaction, the concentration profile reaches a steady state after a relatively short time, and the absorption rate is given by

$$N_A = C_{Ai}\sqrt{kD_A} \tag{7.50}$$

where $k = k_2 C_{Bo}$. This equation is identical to that derived for steady-state diffusion plus reaction in a stagnant liquid film for the case of constant C_{Bo}. [Eq. (7.36)].

When the reaction is fast enough so that diffusion of B toward the reaction zone is important, the penetration theory does not give the same results as the stagnant-film model. Numerical solutions of the equations for simultaneous diffusion and reaction of A and B based on the penetration theory have been reported [5]; the enhancement factors are shown in Figure 7.7. The parameter ϕ_a is the asymptotic value of ϕ, which is reached when rate constant k becomes infinite (instantaneous chemical reaction). The diffusivity ratio D_B/D_A and the concentration ratio influence ϕ_a in a complex

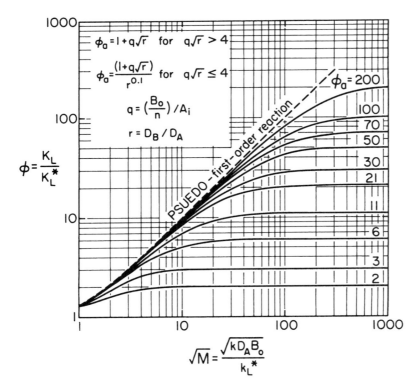

FIGURE 7.7 Enhancement factor for a second-order irreversible reaction. (Adapted from Ref. 5.)

fashion; but for a high concentration ratio, which is often the case, the following simple equation applies:

$$\phi_a = 1 + \frac{C_{Bo}}{nC_{Ai}} \left(\frac{D_B}{D_A}\right)^{1/2} \tag{7.51}$$

Note that Eq. (7.51) differs from Eq. (7.45) only in the exponent for the diffusivity ratio.

Figure 7.7 shows that when $\sqrt{M}$ is less than about $0.1\phi_a$, the gradient for B is small and the enhancement factor is almost the same as for a first-order reaction. When $\sqrt{M} \cong \phi_a$, the change in C_B is significant and the enhancement factor is about 35% less than for a first-order reaction. Since the enhancement factor depends on C_{Bo} and C_{Ai}, which are generally not constant with time or position, the enhancement factor should be evaluated at several points in a flow reactor or at different times of a batch cycle.

Example 7.3

A column packed with 0.5-in. rings is used to scrub CO_2 from air using an aqueous solution of NaOH at 25°C. The initial and final partial pressures of CO_2 are 0.04 and 0.004 atm, respectively, and the caustic concentration is 0.75 M at the top and 0.5 M at the bottom. The first step in the reaction is

$$CO_2 + OH^- \rightarrow HCO_3^-, \qquad k_2 = 8500 L/mol\text{-sec at } 25°C$$

The second step is instantaneous

$$HCO_3^- + OH^- \rightarrow CO_3^= + H_2O$$

Estimated mass transfer coefficients and other properties are:

$$k_g a = 7.4 \text{ mol/hr ft}^3 \text{ atm}$$

$$k_L^* a = 32 \text{ hr}^{-1}$$

$$a \cong 34 \text{ ft}^2/\text{ft}^3$$

$$H_{CO_2} = 1.9 \times 10^3 \text{ atm/m.f.}$$

$$D_{CO_2} = 2 \times 10^{-5} \text{ cm}^2/\text{sec}$$

$$D_{OH^-} = 2.8 \times 10^{-5} \text{ cm}^2/\text{sec}$$

Predict the overall mass transfer coefficient at both ends of the column and the percent resistance in the gas phase.

Solution.

At the gas inlet, $C_{Ai} = P_{Ai}/H$. If $P_{Ai} \cong P_A = 0.04$ atm, then

$$C_{Ai} \cong \frac{0.04}{1.9 \times 10^3} \times 55.5 \text{ mol/L} = 1.17 \times 10^{-3} \text{ mol/L}$$

Since the second step is instantaneous, the overall reaction is

$$CO_2 + 2OH^- \rightarrow CO_3^= + H_2O$$

So

$$n = 2$$

$$\phi_a = 1 + \frac{0.5/2}{1.17 \times 10^{-3}} \left(\frac{2.8 \times 10^{-5}}{2.0 \times 10^{-5}}\right)^{0.5} = 253$$

$$k_L^* = \frac{k_L^* a}{a} = \frac{32}{3600} \times \frac{1}{34} \times \frac{30.5 \text{ cm}}{\text{ft}} = 7.97 \times 10^{-3} \text{ cm/sec}$$

$$\sqrt{M} = \frac{\sqrt{8500 \times (0.5) \times (2 \times 10^{-5})}}{7.97 \times 10^{-3}} = 36.6$$

From Figure 7.7, $\phi \cong \sqrt{M} = 36.6$.
  H is converted to H' to be consistent with $k_g a$ units:

$$\frac{62.3}{18} = 3.46 \text{ mol } H_2O/ft^3$$

$$H' = \frac{1.9 \times 10^3 \text{ atm}}{\text{mol } CO_2/\text{mol } H_2O} \times \frac{1}{3.46 \text{ mol } H_2O/ft^3} = \frac{549 \text{ atm}}{\text{mol}/ft^3}$$

$$\frac{1}{K_g a} = \frac{1}{k_g a} + \frac{H'}{\phi k_L^* a} = \frac{1}{7.4} + \frac{549}{36.6(32)} = 0.604$$

$$K_g a = 1.66 \text{ mol/hr-ft}^3\text{-atm}$$

$$\frac{1.66}{7.4} = 0.22, \qquad 22\% \text{ resistance in gas film}$$

P_{Ai} is 22% less than P_A, which increases ϕ_a, but since $\phi_a \gg \sqrt{M}$, ϕ is not changed.
  At the top of the column,

$$\sqrt{M} = \frac{\sqrt{8500(0.75)(2 \times 10^{-5})}}{7.97 \times 10^{-3}} = 44.8$$

$$C_{Ai} \cong \frac{0.004}{1.9 \times 10^3} = 55.5 = 1.17 \times 10^{-4} \ \text{mol/L}$$

$$\phi_a = 1 + \frac{0.75/2}{1.17 \times 10^{-4}} \left(\frac{2.8}{2.0}\right)^{0.5} = 3792$$

$$\phi = \sqrt{M} = 44.8$$

$$\frac{1}{K_g a} = \frac{1}{7.4} + \frac{549}{44.8(32)} = 0.518$$

$$K_g a = 1.93 \ \text{mol/hr-ft}^3\text{-atm}$$

$$\frac{1.93}{7.4} = 0.26, \quad 26\% \ \text{resistance in gas film}$$

GAS-FILM CONTROL

Example 7.3 showed that a very fast reaction in the liquid can make the gas-film resistance important even for a slightly soluble gas such as CO_2. For an instantaneous reaction in the liquid and a high value of C_{Bo}, diffusion in the gas film may be the controlling step. The gradients for this case are shown in Figure 7.8, which is based on the film theory for simplicity.

To determine whether gas-film control is likely, the maximum flux of A through a unit area of the gas film is compared with the maximum flux of B through the liquid film:

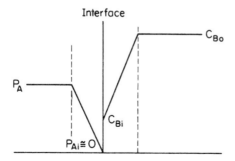

FIGURE 7.8 Concentration gradients for gas-film control and reaction at the interface.

For the gas film: $N_{A,\max} = k_g P_A$ (7.52)

For the liquid: $N_{B,\max} = k_{L,B} C_{Bo} = k_{L,A} \left(\dfrac{D_B}{D_A} \right)^{0.5} C_{Bo}$ (7.53)

For the reaction $A + nB \rightarrow C$, if $N_{A,\max}$ is less than $N_{B,\max}/n$, C_{Ai} will approach zero and C_{Bi} will adjust until the flux of B to the interface matches the flux of A. When $N_{A,\max}$ is greater than $N_{B,\max}/n$, C_{Ai} is finite and the reaction takes place at a plane near the interface where the fluxes of A and B are matched, as illustrated in Figure 7.6.

EFFECT OF MASS TRANSFER ON SELECTIVITY

When consecutive or parallel reactions are carried out between a gas and a liquid, the concentration gradients near the interface may influence the selectivity as well as the overall rate of reaction. For chlorination or partial oxidation of hydrocarbons, several workers have reported that the yield of intermediate products was influenced by agitation variables [6,7] and was less than predicted from the kinetic constants. Rigorous analysis of multiple reactions is complex, but film theory can be used to show when mass transfer effects are likely to change the selectivity [8].

Consider a consecutive reaction system where the intermediate C is the desired product:

$$A_g + B \rightarrow^1 C$$
$$A_g + C \rightarrow^2 D$$

Both reactions are assumed first order to A and to B or C, and the ratio of rates determines the selectivity:

$$\frac{r_1}{r_2} = \frac{k_1 C_A C_B}{k_2 C_A C_C} = \left(\frac{k_1}{k_2} \right) \times \left(\frac{C_B}{C_C} \right)$$ (7.54)

The overall selectivity is the net amount of C produced per amount of B consumed, and the local or instantaneous selectivity is based on the rates of reaction:

$$S = \frac{r_1 - r_2}{r_1} = 1 - \frac{r_2}{r_1}$$ (7.55)

$$S = 1 - \left(\frac{k_2}{k_1} \right) \left(\frac{C_C}{C_B} \right)$$ (7.56)

For a semibatch reaction, C_B decreases with time and C_C increases, so the local selectivity is constantly decreasing. If there are no mass transfer effects, the equations for C_B and C_C can be integrated to get the yield of C, which goes through a maximum with time:

$$\frac{C_C}{C_{Bo}} = \frac{k_1}{k_2 - k_1}\left(e^{-k_1 t} - e^{-k_2 t}\right) \tag{7.57}$$

$$\frac{C_{C,\max}}{C_{Bo}} = \left(\frac{k_1}{k_2}\right)^{k_2/(k_2 - k_1)} \tag{7.58}$$

When mass transfer of A influences the rate, the gradient for A in the liquid film does not affect the yield of C if both reactions are first order to A. However, if the reaction is fast enough to make the gradient for B in the film important, the selectivity is decreased because of the lower value of C_B and the higher concentration of C_C in the film than in the bulk. Figure 7.9 shows a typical case.

The average value of B in the reaction zone can be estimated using Figure 7.7, provided the second reaction is still much slower than the first. The value of ϕ is assumed to depend mainly on $\sqrt{M}$ and ϕ_a for the first reaction. If $\phi \cong \sqrt{M}$, the gradient for B is negligible and there is no selectivity change. However, if ϕ is less than $\sqrt{M}$, the difference between ϕ and $\sqrt{M}$ can be used to estimate C_B. Since the rate varies with the square root of D_B in the pseudo-first-order region, the following equation holds:

$$\frac{\overline{C_B}}{C_{Bo}} \cong \left(\frac{\phi}{\sqrt{M}}\right)^2 \tag{7.59}$$

This approximation is not valid if ϕ is much less than $\sqrt{M}$, but then the selectivity would be too low to be practical. If the diffusivities of B and C are nearly the same,

$$\Delta C_B = C_{Bo} - \overline{C_B} \cong \Delta C_C = \overline{C_C} - C_{Co} \tag{7.60}$$

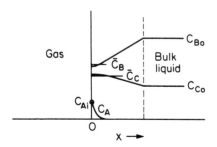

FIGURE 7.9   Concentration gradients for fast consecutive reactions.

Then the average values in the reaction zone can be used to estimate the local selectivity.

Example 7.4

Tests in a homogeneous system show that $k_2 = 0.09k_1$ for the reaction sequence

$$A + B \rightarrow^1 C$$
$$A + C \rightarrow^2 D$$

For $C_{Ai} = 0.02$ M, $C_{Bo} = 3$ M, $D_A = D_B = D_C = 10^{-5}$ cm^2/sec, and $k_1 = 10^4$ L/mol-sec, will mass transfer change the selectivity if $k_L^* = 0.015$ cm/sec?

Solution. For reaction 1,

$$\sqrt{M} = \frac{\sqrt{3 \times 10^4 \times 10^{-5}}}{0.015} = 36.5$$

At the start of the reaction, from Eq. (7.51),

$$\phi_a = 1 + \frac{3}{0.02} = 151$$

From Figure 7.7,

$$\phi = 33$$

From Eq. (7.59),

$$\frac{\overline{C_B}}{C_{Bo}} = \left(\frac{33}{36.5}\right)^2 = 0.82$$

From Eq. (7.60),

$$\Delta C_B = 0.18(3) = 0.54 \cong \Delta C_C$$
$$\overline{C_C} = 0 + 0.54$$

From Eq. (7.56),

$$S = 1 - 0.09\left(\frac{0.54}{2.46}\right) = 0.98$$

At 50% conversion,

$$C_B = 1.5, \quad C_C + C_D = 3 - 1.5 = 1.5$$

$$\sqrt{M} = \frac{\sqrt{1.5 \times 10^4 \times 10^{-5}}}{0.015} = 25.8$$

$$\phi_a = 1 + \frac{1.5}{0.02} = 76$$

$$\phi = 23$$

$$\frac{\overline{C_B}}{C_{Bo}} = \left(\frac{23}{25.8}\right)^2 = 0.79$$

$$\overline{C_B} = 0.79(1.5) = 1.19, \qquad \Delta C_B = 1.50 - 1.19 = 0.31 \cong \Delta C_C$$

If $C_D \cong 0.1$,

$$C_C \cong 1.4,$$

$$\overline{C_C} = 0.31 + 1.4 = 1.71$$

$$S = 1 - 0.09\left(\frac{1.71}{1.19}\right) = 0.87$$

In comparison, with no mass transfer effect,

$$S = 1 - 0.09\left(\frac{1.4}{1.5}\right) = 0.92$$

The local selectivity is lowered from 100% to 98% at zero conversion and from 92% to 87% at 50% conversion because of mass transfer limitations.

For parallel reactions, the selectivity may be altered by mass transfer effects if the reactions are of different order. For:

$$A + B \xrightarrow{1} C \qquad r_1 = k_1 C_A$$
$$A + 2B \xrightarrow{2} D \qquad r_2 = k_2 C_A C_B$$

$$\frac{r_1}{r_2} = \frac{k_1}{k_2 C_B}$$

When the reaction is rapid enough to deplete B in the film, reaction 2 will be affected more than reaction 1 and the selectivity to product C is enhanced.

SUMMARY OF POSSIBLE CONTROLLING STEPS

A gas–liquid reaction may be controlled by the rate of mass transfer or the rate of chemical reaction, or both steps may affect the overall rate. When interpreting laboratory data or designing a large reactor, it is important to know if any one step is controlling and whether the relative importance of the mass transfer and reaction steps is likely to change on scaleup. The transition from reaction control for very slow reactions to mass transfer control for very fast reactions is more complex than for solid-catalyzed gas reactions because the gas and liquid reactants are introduced in different phases and because there are several mass transfer processes to consider.

The different regimes for a gas–liquid reaction are shown in Table 7.1. For consecutive mass transfer of gas A through the liquid film and reaction in the bulk liquid, the process goes from reaction control to mass transfer control as the reaction rate constant goes from a very low value (relative to the mass transfer coefficient) to a quite high value. For a very fast reaction that takes place entirely in the liquid film, neither mass transfer nor reaction controls, but both influence the rate of reaction. The reaction may be pseudo first order to A because of a constant concentration of B in the reaction zone. When the reaction rate is increased further or when the concentration of B is lower, diffusion of A and B in the film and the reaction rate all influence the overall rate and the selectivity with consecutive reactions decreases. When the reaction is instantaneous, it may occur at a plane near the gas–liquid interface, and the rate may depend on diffusion rates of A and B. For moderately soluble gases and an instantaneous reaction, the limiting step may be diffusion of A through the gas film.

TABLE 7.1 Possible Controlling Steps for $A_g + B \rightarrow C$

Reaction rate	Controlling step	Equations	Notes
Very slow	Reaction in bulk	(7.10)–(7.12)	Reaction in film negligible
Moderate	Both steps affect rate	(7.10)–(7.12)	Reaction in film negligible
Fast	Mass transfer of A	(7.10)–(7.12)	Reaction in film negligible
Very fast	Simultaneous diffusion of A and reaction in film	(7.34)–(7.35)	$\bar{C}_{AL} = 0$ pseudo 1st order to A
Very fast	Diffusion of A and B and reaction	Fig. 7.7	$\bar{C}_B > C_{Bo}$ selectivity changes
Instantaneous	Mass transfer of A and B in liquid film	(7.45)–(7.48)	
Instantaneous	Mass transfer of A in gas film	(7.51)–(7.52)	

TYPES OF GAS–LIQUID REACTORS

When carrying out a gas–liquid reaction, the gas may be dispersed in the liquid, as in bubble-column reactors or stirred tanks, or the gas phase may be continuous, as in spray contactors or trickle-bed reactors. The fundamental kinetics are independent of the reactor type, but the reaction rate per unit volume and the selectivity may differ because of differences in surface area, mass transfer coefficient, and extent of mixing. In the following sections, gas holdup and mass transfer correlations and other performance data for gas–liquid reactors are reviewed and some problems of scaleup are discussed.

BUBBLE COLUMNS

A bubble-column reactor is a vertical cylindrical vessel with a height/diameter ratio that is usually at least 1.5 and may be as large as 20. Gas is introduced near the bottom of the column through a set of nozzles, or a sparger. Spargers may have an array of parallel pipes connected to a manifold or several radial arms in a spider pattern or concentric pipe circles, all with downward-facing holes every few inches. The holes are sized to give exit velocities of 100–300 ft/sec, and the gas enters the liquid as jets, which break up into bubbles after a short distance. As the bubbles rise, they tend to coalesce and then break up again; the average bubble size depends mainly on the gas flow rate and physical properties of the liquid rather than on the orifice size. For small laboratory reactors, a fritted-glass disc or a sintered-metal plate is sometimes used to disperse the gas. These give smaller initial bubble sizes, but they are generally not used for large reactors because of the cost and the greater chance of plugging. Also, any benefit from decreased initial bubble size becomes quite small for tall columns.

In the simplest type of bubble column, gas is dispersed at the bottom and bubbles are present throughout the reactor, as shown in Figure 7.10a. In a loop or air-lift reactor (Fig. 7.10b), gas is introduced beneath a central draft tube, and rising bubbles carry liquid upward. After most of the bubbles disengage, liquid flows downward in the annulus. The direction of flows could be reversed by feeding gas to the annulus. The columns may have internal coils for heat transfer and baffles to decrease axial mixing.

One of the main design variables for bubble columns is the superficial velocity of the gas, which affects the gas holdup, the interfacial area, and the mass transfer coefficient. The superficial velocity changes as gas passes up the column because of the decrease in hydrostatic head and changes in the total molar flow. When these changes are small, an average of the inlet and exit velocities can be used to predict the performance, though in some

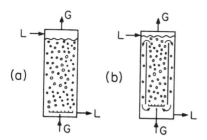

FIGURE 7.10 Bubble-column reactors: (a) simple type; (b) gas-lift type.

studies the velocity reported corresponds to the gas feed rate at the exit pressure. An important factor that is sometimes overlooked is the vapor pressure of the solution. The gas bubbles quickly become saturated with solvent vapor, which decreases the partial pressure of the reaction gas and increases the volumetric flow rate of gas in the reactor.

Gas Holdup

At very low gas velocities (less than 3 cm/sec or 0.1ft/sec), bubbles formed near the sparger rise at almost constant velocity, with little interaction. This is called the *quiescent regime* or the *bubbling regime*. An increase in gas flow rate gives a nearly proportional increase in gas holdup or void fraction ϵ. The holdup is related to the average bubble velocity v_b and the superficial velocity u_G by the equation

$$u_G = v_b \epsilon \tag{7.61}$$

In laboratory tests, the gas holdup is usually determined visually from the height of aerated liquid relative to the original height h_o:

$$h(1 - \epsilon) = h_o \tag{7.62}$$

$$\epsilon = 1 - \frac{h_0}{h} \tag{7.63}$$

At moderate gas velocities, when the holdup is about 0.1, some large bubbles are formed by coalescence, and they rise very rapidly, producing vigorous mixing or churning of the liquid. This has been called the *churn-turbulent regime* or the *heterogeneous regime*, but it will be referred to simply as the *turbulent regime*. In this regime, increases in gas flow produce a greater proportion of large, fast-moving bubbles, so the holdup does not increase as rapidly with flow rate as in the quiescent regime. Figure 7.11 shows holdup data from several sources taken at about 20°C in columns

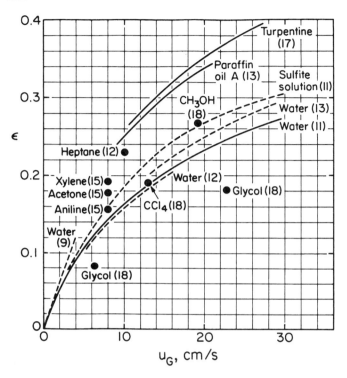

FIGURE 7.11   Void fractions for bubble columns.

10 cm or larger in diameter using single nozzles or multiple-orifice spargers. Studies [9–11] have shown that the holdup is slightly greater for columns smaller than 10 cm when porous plate distributors are used; correlations for these systems are not included here.

Data from a comprehensive study of the air–water system by Yoshida and Akita [11] are shown as a heavy solid line in Figure 7.11. They used columns 7.7, 15, 30, and 60 cm in diameter and found the holdup was the same for the three larger sizes. They found no effect of nozzle size when using a single injector and no effect of liquid depth. The gas holdup ϵ is nearly proportional to u_G at very low velocities but shows a smaller dependence on u_G in the turbulent regime. If the data are plotted on a log-log plot, the slope gradually changes from 1.0 to about 0.5 at high velocities.

Figure 7.11 also shows the air–water data of Fair et al. [9] for a 46-cm tank at low gas velocities, the data of Wilkinson et al. [12] for 15-cm and 23-cm columns in the turbulent regime, and the data of Krishna and

Ellenberger [13] for a 63-cm column in the turbulent regime. These three sets of data are in reasonably good agreement with those of Yoshida and Akita, as are the results from other studies of the air–water systems [14]. Yoshida and Akita also used air with 0.3 N Na_2SO_3 solution, since sulfite solutions provide an easy way to measure oxygen absorption rates. Although the density, viscosity, and surface tension of the sulfite solution are almost the same as for water, the gas holdup is appreciably greater, as shown by the upper dashed line in Figure 7.11. The average bubble size is smaller in sulfite solutions because coalescence is retarded by electrostatic effects. The effect of salts has been observed by many workers but cannot be accurately predicted. Similar changes in bubble size and gas holdup [15] occur with some mixtures of organic liquids, probably because coalescence is retarded by surface tension gradients or monolayers at the interface.

With organic liquids, the gas holdup is generally higher than for water, except for high-viscosity fluids. Most of the solvents represented in Figure 7.11 have lower density, lower surface tension, and lower viscosity than water, and there is no general agreement about the effect of each variable on the holdup. A few of the many published correlations are given in Table 7.2. The Akita–Yoshida correlation is based on three dimensionless groups, the Bond, Galileo, and Froude numbers. Each group includes the column diameter D_t as the length parameter; and since ϵ does not depend on D_t, only certain combinations of exponents are possible. This may prejudice the conclusions for other variables. The term $\epsilon/(1 - \epsilon)^4$ was used to account for the changing effect of u_G on ϵ. To show more directly the effects of key variables on ϵ, their correlation can be converted to the approximate form shown in Table 7.2. The three correlations differ significantly in the predicted effects of surface tension, with exponents of -0.07, $-3/8$, and -0.18, and the viscosity exponents are -0.08, $+0.25$, and -0.05. The difference in sign for the viscosity exponents is a result of studying different ranges of viscosity. For low-viscosity liquids, ϵ increases with μ_L because of the decrease in bubble velocity, but ϵ decreases with viscosity when the viscosity is large [16]. For viscous liquids, very large bubbles form and move rapidly through the column, giving low values of ϵ. The correlations in Table 7.2 indicate a slight increase in ϵ with ρ_L, but this may not be correct, since the range of densities was small, and the holdup for CCl_4, the most dense liquid studied, is less than predicted based on its surface tension and viscosity. Ignoring the density effect, the holdup data for most organic liquids varies with about $\sigma^{-0.4}$ and $\mu_L^{0.2}$ for low-viscosity liquids and with $\sigma^{-0.4}\mu_L^{-0.1}$ for viscous liquids.

The effects of gas properties are often ignored, but studies at high pressure and with gases of different molecular weight show that ϵ increases with about $\rho_G^{0.2}$ in the highly turbulent regime [13,15,17]. There is less effect

TABLE 7.2 Holdup Correlations for Pure Liquids in Bubble Columns

1.   Akita and Yoshida [18]

$$\frac{\epsilon}{(1-\epsilon)^4} = 0.2 \left(\frac{gD_T^2 \rho_L}{\sigma}\right)^{1/8} \left(\frac{gD_T^3 \rho_L^2}{\mu_L^2}\right)^{1/12} \frac{u_G}{(gD_T)^{1/2}}$$

or

$$\frac{\epsilon}{(1-\epsilon)^4} = 0.2 N_{BO}^{1/8} \, N_{Ga}^{1/12} \, N_{Fr}^{1.0}$$

ϵ varies with: $\rho_L^{0.15} \sigma^{-0.07} \mu_L^{-0.08} u_G^{0.5}$

2.   Van Dierendonck [19]

$$\epsilon = 1.2 \left(\frac{\mu_L \mu_G}{\sigma}\right)^{1/4} \left(\frac{u_G}{\left(\frac{\sigma g}{\rho_L}\right)^{1/4}}\right)^{1/2}$$

ϵ varies with: $\rho_L^{1/8} \sigma^{-3/8} \mu_L^{1/4} u_G^{3/4}$

3.   Hikita et al. [20]

$$\epsilon = 0.672 \left(\frac{\mu_L \mu_G}{\sigma}\right)^{0.578} \left(\frac{\mu_L^4 g}{\rho_L \sigma^3}\right)^{-0.131} \left(\frac{\rho_G}{\rho_L}\right)^{0.062} \left(\frac{\mu_G}{\mu_L}\right)^{0.107}$$

ϵ varies with: $\rho_L^{0.069} \sigma^{-0.185} \mu_L^{-0.053} u_G^{0.578} \rho_G^{0.062} \mu_G^{0.107}$

of gas density at low gas velocity. The effect of gas viscosity, if present at all, is small enough to be neglected.

The following procedure is recommended for predicting holdup for systems other than air–water. The average superficial velocity in the column is calculated allowing for the solvent vapor pressure and the average absolute pressure, and ϵ for the air–water systems is predicted from Figure 7.11. Then the ratios of physical properties are used to adjust the value of ϵ if the viscosity is about 1 cp or less (10^{-3} Pa-sec):

$$\frac{\epsilon}{\epsilon_{air-H_2O}} = \left(\frac{\sigma}{\sigma_{H_2O}}\right)^{-0.4} \left(\frac{\mu_L}{\mu_{H_2O}}\right)^{0.2} \left(\frac{\rho_G}{\rho_{air}}\right)^{0.2} \tag{7.64}$$

For viscous liquids, the exponent on the viscosity term should be changed to -0.1, but predictions for this regime are quite uncertain because of the limited amount of data.

In recent detailed studies of bubble columns, separate equations have been proposed for the small-bubble holdup and the holdup of large bubbles, which form above a critical gas velocity [13]. This is a more fundamental approach that may eventually lead to better prediction of column performance.

Mass Transfer Coefficient

The liquid-phase mass transfer coefficient $k_L a$ can be measured by following the rate of change of concentration in the liquid during a gas absorption or a stripping experiment. When a pure gas is used, there is no gas-phase resistance and no change in gas concentration, which simplifies the analysis, though for oxygen in air, the gas-phase resistance is negligible because of the low solubility. One problem with transient absorption tests is that for high values of $k_L a$, the liquid becomes nearly saturated in a short time, and the data have to be corrected for measurement lag. If a pure gas is used to keep the equilibrium concentration constant and if the liquid is well mixed, the absorption equation is:

$$V\left(\frac{dC}{dt}\right) = k_L a(C^* - C)V \tag{7.65}$$

$$\int \frac{dC}{C^* - C} = \ln\left(\frac{C^* - C_0}{C^* - C}\right) = k_L a t \tag{7.66}$$

For $k_L a = 0.1 \sec^{-1}$, a typical value for moderate gas velocity, the liquid would be 90% saturated after only 23 seconds. Taking data for longer times reduces the error due to measurement lag; but for a very close approach to equilibrium, the difference $(C^* - C)$ cannot be accurately determined.

For slightly soluble gases, such as oxygen, there is an advantage to using stripping tests rather than absorption tests to measure $k_L a$. The driving force for stripping is $(C - C^*)$, and C^* is generally much less than C even at the gas exit, so the test can be extended to longer times and quite low values of C with little error because of uncertainty in C^*. However, when stripping CO_2 or other moderately soluble gases, the exit gas may be close to saturation, and the average driving force $(C - C^*)$ is subject to greater error, particularly when the degree of gas-phase mixing is uncertain. Early studies of CO_2 desorption in bubble columns [7] gave anomalous results because of changes in the extent of gas mixing.

Although absorption plus chemical reaction sounds more complex than physical absorption or desorption, the absorption of oxygen in sulfite solutions is often used to characterize the performance of bubble columns or stirred reactors. With a 0.2–1.0 N solution of sodium sulfite and a small amount of copper sulfate as catalyst (10^{-4} M), the rate of oxidation is independent of sulfite concentration, and the oxygen absorption rate is constant until nearly all the sulfite has reacted:

$$SO_3^= + \frac{1}{2}O_2 \xrightarrow{Cu^{++}} SO_4^= \tag{7.67}$$

The time for complete oxidation is generally 20–100 minutes, but the reaction needs to be carried out only long enough to show that the rate is constant, by sampling either the liquid or the exit gas. The key to the steady-state measurement of $k_L a$ is that the Cu^{++}-catalyzed chemical reaction in the bulk liquid is fast enough to keep the dissolved oxygen concentration almost zero, yet not fast enough for significant enhancement of the mass transfer process due to reaction in the liquid film. The reaction rate is then proportional to the partial pressure of oxygen in the gas phase. The following equation holds whether or not the liquid is perfectly mixed:

$$-\frac{dC_{SO_3^=}}{dt} = \frac{2K_L a C_{O_2}^*}{1 - \epsilon} \tag{7.68}$$

The equilibrium oxygen concentration $C_{O_2}^*$ is the product of the solubility and the oxygen partial pressure. Most workers have used either the average oxygen pressure or the log-mean pressure, which assumes plug flow of the gas; but for vigorous mixing, the exit pressure would be more appropriate. The effect of back-mixing of the gas will be discussed later. The gas-film resistance is negligible, so $k_L a \cong K_L a$.

The Akita values of $k_L a$ for oxygen absorption in sulfite solution in a 60-cm column are shown in Figure 7.12. The coefficients were based on the

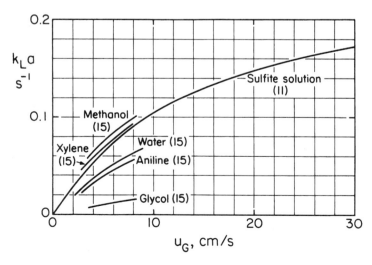

FIGURE 7.12 Mass transfer coefficients for oxygen absorption in bubble columns.

aerated volume of the reactor and the arithmetic mean partial pressure of oxygen, taking into account the average hydrostatic head. The coefficients were independent of the catalyst concentration in the range 10^{-5} M to 10^{-4} M, showing that oxygen transfer was not enhanced by the chemical reaction. They also found that $k_L a$ was about the same for different-diameter nozzles and different liquid depths. At low gas velocity, $k_L a$ is proportional to u_G; but in the turbulent regime, $k_L a$ increases less rapidly with velocity. The plots for ϵ and $k_L a$ are quite similar, indicating that the increase in $k_L a$ with u_G comes mainly from increased gas holdup and interfacial area. A surprising result was the increase in $k_L a$ with column diameter from 7.7 to 60 cm, which was fitted with a 0.17 exponent for the diameter. There was no obvious explanation for the effect, and the authors suggested that designs for large columns be based on the 60-cm data rather than extrapolating with $D_t^{0.17}$.

The effects of system properties on $k_L a$ are uncertain, because there are relatively few mass transfer studies for organic liquids, and most were transient tests, which are inherently less accurate than the steady-state oxygen/sulfite tests. Data for oxygen absorption in a few organic solvents and in water are shown in Figure 7.12. The coefficients for methanol and xylene are greater than for water because the holdup and diffusivities are greater. With aniline, the coefficients are less than for water because the much lower diffusivity offsets the slightly higher holdup. Some published correlations for $k_L a$ are given in Table 7.3; there are wide differences in the recommended exponents for density, viscosity, and surface tension. These correlations can be used to get approximate values of $k_L a$, but a simpler approach should be satisfactory for preliminary analysis. The mass transfer coefficient is estimated using the O_2–water data and correcting for differences in gas holdup and diffusion coefficient:

$$k_L a = k_L a_{O_2-H_2O}\left(\frac{D}{D_{O_2-H_2O}}\right)^{0.5}\left(\frac{\epsilon}{\epsilon_{O_2-H_2O}}\right) \tag{7.69}$$

The diffusivity for oxygen in water at 20°C is 2.4×10^{-5} cm^2/sec. If holdup data are not available, Eq. (7.64) or Figure 7.11 can be used to estimate ϵ.

The use of the volumetric coefficient $k_L a$ for predicting reactor performance is appropriate if the reaction takes place almost entirely in the bulk liquid, and the equations for consecutive mass transfer and reaction apply [Eqs. (7.10) and (7.12)]. For simultaneous diffusion and reaction in the liquid film, the coefficient per unit area, k_L, must be known to predict the enhancement factor. Unfortunately, the data for k_L show much more scatter than the data for $k_L a$ because of the difficulty in measuring a or calculating a from bubble size measurements. Also, k_L has been shown to vary appreci-

TABLE 7.3   Mass Transfer Correlations for Bubble Columns

1.   Akita and Yoshida [18]

$$k_L a = 0.6 \left(\frac{\mu_L}{\rho_L D}\right)^{0.5} \left(\frac{gD_T^2 \rho_L}{\sigma}\right)^{0.62} \left(\frac{gD_T^3 \rho_L^2}{\mu_L^2}\right)^{0.31} \epsilon^{1.1}$$

2.   Hikita et al. [21]

$$k_L a = \frac{14.9 g f}{u_G} \left(\frac{u_G u_L}{\sigma}\right)^{1.76} \left(\frac{\mu_L^4 g}{\rho_L \sigma^3}\right)^{-0.248}$$

$$\left(\frac{\mu_G}{\mu_L}\right)^{0.243} \left(\frac{\mu_L}{\rho_L D}\right)^{-0.604}$$

$f = 1.0$ for non electrolytes

$f = 10^{0.06811}$ for $I < 1 kg$-ion/m^3

3.   Ozturk et al. [15]

$$\frac{k_L a d_B^2}{D} = 0.62 \left(\frac{\mu_L}{\rho_L D}\right)^{0.5} \left(\frac{g \rho_L d_B^2}{\sigma}\right)^{0.33} \left(\frac{g \rho_L^2 d_B^2}{\mu_L^2}\right)^{0.243}$$

$$\left(\frac{u_G}{\sqrt{g d_B}}\right)^{0.68} \left(\frac{\rho_G}{\rho_L}\right)^{0.04}$$

ably with bubble size, and use of an average k_L introduces further uncertainty. Small bubbles tend to act as rigid spheres, and for 1- to 2-mm air bubbles in water, the oxygen absorption coefficient is about 0.01–0.015 cm/sec. For larger bubbles, internal circulation and oscillations in shape increase k_L to about 0.03–0.05 cm/sec. These approximate values can generally be used to show whether a reaction is in the regime of simultaneous diffusion and reaction. The value of k_L could be adjusted for differences in diffusivity following Eq. (7.69).

Effect of Liquid Depth

For some gas–liquid reactions, it is advantageous to use a very tall reactor rather than one that is shorter but larger in diameter. With a tall bubble column, the hydrostatic head increases the driving force for gas absorption at the bottom, and this effect plus the increase in gas residence time permits a greater fraction of the reactant gas to be absorbed. A tall reactor also requires less space for installation. Other factors to consider are the increased work of compression, though the work does not go up in proportion to the depth, and the effect of hydrostatic head on the volumetric gas rate. The holdup and mass transfer coefficient are lower at the bottom because of the lower superficial velocity.

Jackson and Shen [22] studied oxygen absorption from air into sulfite solutions in bubble columns up to 21 m deep. They reported that $k_L a$ based on the mean oxygen pressure decreased with the -0.45 or -0.55 power of the liquid depth, h. An alternate correlation of their results, Figure 7.13, shows that $k_L a$ varies nearly inversely with the average pressure expressed as $h/2 + 10.34$, where h is in meters. This inverse dependence arises because $k_L a$ is proportional to about the first power of the gas velocity at the low velocities used (0.28 cm/sec and 0.83 cm/sec). The effects of depth on the driving force and on $k_L a$ therefore cancel, and the oxygen absorption rate for a given molar flow rate is proportional to the oxygen mole fraction and nearly independent of depth. For the air–sulfite system at low flow rates, about 30% of the oxygen was used in 10 meters; the predicted utilizations are 50% for 20 meters and 75% for 40 meters. For absorption from richer gases than air, the decrease in molar flow rate would have to be allowed for.

Axial Dispersion

Rising gas bubbles cause strong circulation currents in the liquid, and most bubble columns have nearly uniform concentration and temperature in the

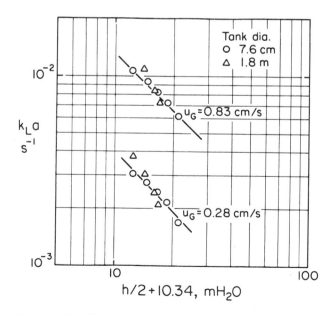

FIGURE 7.13 Oxygen absorption in tall bubble columns. (Data from Ref. 22.)

liquid phase. A good example of rapid liquid mixing was given by Jackson and Shen [22], who showed that temperature uniformity in a large (1.8-m × 12-m) stratified tank was achieved in less than 2 minutes at an air rate of only 0.1 m/min. Others have characterized liquid mixing by an axial dispersion coefficient E_L calculated from tracer tests for columns operating with continuous flow of gas and liquid. Dispersion increases with gas flow and with column diameter, and the following dimensional correlation was given for E_L in ft²/hr, D_t in ft, and u_G in ft/hr [23]:

$$E_L = 73.5D_t^{1.5}u_G^{0.5} \tag{7.70}$$

Equation (7.70) does not allow for the possible effect of column length, and the effect of diameter may level off for sizes greater than 4 feet. For $D_t = 4$ ft and $u_G = 100$ ft/hr, $E_L = 5900$ ft²/hr. With this value for E_L, the Peclet number for a reactor with $L = 20$ ft and a liquid velocity of 40 ft/hr would be $Pe_L = \frac{Lu_L}{E_L} = 0.14$. For Peclet numbers much less than 1.0, the conversion for a first-order reaction is almost the same as for a perfectly mixed reactor, so the assumption of perfect mixing is usually justified for large bubble-column reactors.

Gas mixing in bubble columns has not received much attention, and plug flow has often been assumed in analysis of laboratory data or design of reactors, although some workers assumed perfect gas mixing. Gas mixing can occur by the process of bubble coalescence and breakup as the bubbles move upward or by the entrainment of small bubbles in liquid flowing down near the walls. In the quiescent regime, the gas holdup is low, and most of the bubbles move upward at about the same velocity. In tests of sulfite oxidation with air at $u_G = 0.35$ or 0.67 cm/sec, gas samples showed a nearly linear change in oxygen concentration with depth, indicating very little gas mixing [22].

In the turbulent regime, the large bubbles promote rapid coalescence, and there is more entrainment of small bubbles due to the vigorous liquid circulation. Tracer tests showed a wide distribution of gas residence time from about $0.2\bar{t}$ to $3\bar{t}$ [23]. The distribution is similar to that for a well-mixed system except for the initial delay, but this distribution could be due largely to differences in bubble velocity rather than to actual gas mixing. Attempts to treat the residence time distribution with axial dispersion models have led to quite different correlations, though there is general agreement that the calculated dispersion coefficients increase with gas velocity and with column diameter. At present there is no reliable method to predict the effect of gas mixing on reactor performance. The plug-flow and back-mixed models can be used to show the range of possible reaction rates and to indicate whether more modeling studies or more experimental data are needed.

Example 7.5

The commercial production of acetic acid by the carbonylation of methanol is carried out continuously in bubble-column reactors at 170–190°C and 20–40 atmospheres using a soluble rhodium catalyst with methyl iodide as a promoter. Tests in small stirred reactors showed that the reaction rate is proportional to the product of the rhodium and iodide concentrations but is independent of the methanol concentration and the partial pressure of carbon monoxide for partial pressures greater than about 2 atmospheres [24,25]:

$$r = 158.8 \times 10^6 \exp\left(\frac{-8684}{T}\right)(Rh)(CH_3I)mol/L\text{-sec}$$

a. Use the general correlations for bubble columns to predict the maximum rate of CO absorption at 180°C and a total pressure of 20 atm if the reaction takes place entirely in the bulk liquid. Assume $(Rh) = 4 \times 10^{-3}$ M, $(CH_3I) = 1M$, $u_G = 10$ cm/sec at the bottom of the reactor, and 80% utilization of the CO fed. The solubility of CO at reaction conditions is 7×10^{-3} mol/L-atm.

b. Give the approximate dimensions for a bubble column reactor to produce acetic acid at the rate of 0.1 kmol/sec (about 400 MM-lb/year).

c. What would be the effect of using a lower gas velocity—say, 5 cm/sec—at the bottom of the reactor?

Solution.

a. The liquid in the reactor will be almost completely mixed with a high concentration of acetic acid and little methanol, so predictions are based on the properties of acetic acid. At 180°C (453 K),

$$P' = 5.0 \text{ atm}, \qquad P_{co} = 20 - 5 = 15 \text{ atm}$$

$$\mu = 0.19 \text{ cp}$$
$$\sigma = 28 \text{ dynes/cm at } 20°C \cong 20 \text{ at } 180°C$$

From the Wilke–Chang equation [Eq. (4.17)], with $M_B = 60.05$, $V_A = 30.7$,

$$D_{CO} = \frac{7.4 \times 10^{-8}(60.05)^{1/2}453}{0.19(30.7)^{0.6}} = 1.75 \times 10^{-4} \text{ cm}^2/\text{sec}$$

Since 80% of the CO reacts, the exit gas velocity is 2 cm/sec, and calculations will be based on the average of 6 cm/sec. For a gas with 75% CO, 25% CH_3CO_2H, $M = 0.75(28) + 0.25(60) = 36$.

Assume ideal gas:

$$\rho_G = \frac{36(20)}{(82.056 \times 10^{-3})(453)} = 19.4 \text{ kg/m}^3$$

At 20°C, 1 atm,

$$\rho_{air} = 1.21 \text{ kg/m}^3$$

From Figure 7.11,

$$\epsilon_{air-water} = 0.12 \text{ at } 6 \text{ cm/sec}$$

From Eq. (7.64),

$$\epsilon = 0.12 \left(\frac{72}{20}\right)^{0.4} \left(\frac{0.19}{1.0}\right)^{0.2} \left(\frac{19.4}{1.21}\right)^{0.2} = 0.25$$

From Figure 7.12 for $u_G = 6$ cm/sec,

$$k_L a_{air-water} = 0.051 \text{ sec}^{-1}$$

From Eq. (7.69),

$$\text{predicted } k_L a = 0.051 \left(\frac{17.5 \times 10^{-5}}{2.4 \times 10^{-5}}\right)^{0.5} \left(\frac{0.25}{0.12}\right) = 0.29 \text{ sec}^{-1}$$

At $P_{CO} = 15$ atm,

$$C^*_{CO} = 7 \times 10^{-3} \ (15) = 0.105 \text{ mol/L}$$

$$r_{max} = k_L a (C^*_{CO} - 0) = 0.29 \ (0.105) = 3.04 \times 10^{-2} \text{ mol/L-sec}$$

Compare with the kinetic rate at 180°C:

$$r = 158.8 \times 10^6 \exp\left(\frac{-8684}{453}\right)(4 \times 10^{-3})(1.0)$$

$$= 3.0 \times 10^{-3} \text{ mol/L-sec}$$

At 20 atm, the possible rate of CO absorption is 10 times the kinetic rate, so the solution would be 90% saturated with CO, and the rate should be independent of P or P_{CO}. A driving force of 1.5 atm would be needed in the bubble column to make the rate of CO absorption match the kinetic rate. If the true kinetics (no mass transfer effect) show a dependence on P_{CO} below 2 atm, then the bubble-column reactor would show a CO dependence below $P_{CO} = 2 + 1.5 = 3.5$ atm or at a total pressure below 8.5 atm.

b. To react 0.1 kmol/sec, feed $0.1/0.8 = 0.125$ kmol/sec. Gas phase is 75% CO, total flow $= 0.125/0.75 = 0.167$ kmol/sec. Neglecting hydrostatic head, at 20 atm, 180°C:

$$\text{volume flow} = \frac{0.167(82.056 \times 10^{-3})(453)}{20} = 0.310 \text{ m}^3/\text{sec}$$

For $u_G = 0.1$ m/sec, $S = 3.1$ m², $D_t = 1.99$, or 2.0 m.

$$\text{Liquid volume} = \frac{0.1 \text{ kmol/sec}}{3 \times 10^{-6} \text{kmol/L-sec}} \times 10^{-3} = 33.3 \text{ m}^3$$

$$h_o = \frac{33.3}{3.1} = 10.8 \text{ m clear liquid}$$

$$h = \frac{10.8}{1 - 0.25} = 14.4 \text{ m aerated liquid}$$

Reactor is 2.0 m in diameter and 15 m tall.

c. If u_G is 5 cm/sec at the bottom, the reactor cross section would be $0.310/0.05 = 6.2$ m², or $Dt = 2.8$ m. For the same volume, $h_o = 10.8/2 = 5.4$ m. At $u_G = 3$ cm/sec,

$$\epsilon = 0.07 \text{ for air–water}$$

$$\epsilon = \frac{0.07}{0.12}(0.25) = 0.15$$

$$h = \frac{5.4}{0.85} = 6.4 \text{ m, or 7 m}$$

A reactor 2.8 m × 7 m might be preferred to one 2 m × 15 m, even though the value of $k_L a$ would be about 40% lower than with $u_G = 10$ cm/sec. The maximum mass transfer rate for CO would still be six times the kinetic rate, but the rate would start to be affected by CO pressure at a higher total pressure.

Heat Transfer

The temperature in a bubble-column reactor can be controlled by removing or adding heat through the column wall or a cooling coil. Based on studies with an electrically heated wall, the heat transfer coefficient varies with about the 0.2–0.25 power of the gas velocity and is greater for water than for viscous solutions or organic liquids [9,26,27]. Dimensionless correlations have been developed using a modified Stanton number, where the gas velocity is combined with the physical properties of the liquid. The following equation, proposed by Hart, was based on tests in a 9.9-cm column, but it is probably applicable to larger columns as well (26):

$$\left(\frac{h_w}{\rho_L c_p u_G}\right)\left(\frac{c_p \mu_L}{k_L}\right)^{0.6} = 0.125\left(\frac{u_G^3 \rho_L}{g \mu_L}\right)^{-0.25} \tag{7.71}$$

The dimensionless group on the right-hand side is the product of a Reynolds number and the Froude number and thus takes account of inertial, viscous, and gravitational forces. The terms in Eq. (7.71) can be combined to show the net effect of each variable:

$$h_w \propto u_G^{0.25} g^{0.25} \rho_L^{0.75} k_L^{0.6} \mu_L^{-0.35} c_p^{0.4} \tag{7.72}$$

Note that the column dimensions, the gas properties, and the bubble size do not appear in the correlation. However, some effect of tube diameter is expected for tubes or cooling coils placed in the bubble column, since coefficients for coils in stirred tanks are larger for small-diameter tubes. Also, Eq. (7.71) could be modified by adding the term $(\mu/\mu_w)^{0.14}$ if there is a large difference between the bulk and wall viscosities.

Several sets of heat transfer data were reviewed by Hikita et al. [27], and they presented additional results for different liquids over a wide range of conditions. They showed that h_w varied with a decreasing power of u_G as the gas velocity increased, and the exponent was about 0.15 for $u_G = 5$–35 cm/sec. However, their values of h_w are 10–50% above those for other studies at comparable conditions, perhaps because they used quite short (2 or 5 cm) heated sections in their columns. For short sections, the thermal boundary layer is not fully established, and the coefficient is higher than for a long heated section. For design calculations, Eq. (7.71) is recommended for gas velocities up to 5 cm/sec; for extrapolation to higher velocities, an exponent of 0.15 should be used.

STIRRED-TANK REACTORS

When a high-speed agitator is used to disperse the gas for a reaction, the interfacial area per unit volume and the mass transfer coefficient are usually much larger than for a bubble column. The more vigorous agitation also leads to better heat transfer to a cooling coil or jacket and more uniform suspension of any solids that are present. The main disadvantages of the stirred reactor are the extra cost of the agitator system and the greater degree of gas back-mixing, which lowers the driving force for gas absorption and may offset the higher mass transfer coefficient.

A typical stirred reactor is shown in Figure 7.14. Gas is introduced under the impeller through a single pipe or a sparger ring. The impeller is typically a flat-blade or curved-blade turbine with a diameter about one-third the tank diameter. Four vertical baffles with a width one-tenth the tank diameter are installed close to the wall to prevent vortex formation. For tall vessels, multiple stirrers may be used, spaced about one tank diameter apart on the vertical shaft. When radial-flow turbines are used for the upper agitators as well as for the gas dispenser, liquid circulation loops develop near each impeller, giving good local mixing, but there is limited mixing between zones. Sometimes axial-flow impellers are used for the upper stirrers to promote end-to-end mixing when uniform liquid composition is desired.

Power Consumption

The power consumption of the stirrer, expressed as kW/m^3 or $HP/1000$ gal, and the superficial velocity of the gas are the main adjustable parameters that determine the mass transfer coefficient. The agitator power can be estimated from the power number N_p and the standard equation for liquids modified by a factor (P_g/P_o) to allow for the effect of aeration:

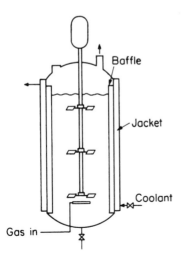

Gas in

FIGURE 7.14 Stirred-tank reactor with multiple impellers.

$$P_g = \left(\frac{P_g}{P_o}\right) \rho N_p N^3 D_a^5 \tag{7.73}$$

The ratio (P_g/P_o) depends mainly on the gas velocity and to some extent on the stirrer type and speed, tank size, and liquid properties. The simple correlation in Figure 7.15 shows this ratio as a function of the aeration number, $N_{Ae} = Q/ND_a^3$. For a standard turbine the ratio drops rapidly to about 0.5 at $N_{Ae} = 0.05$, and then it slowly approaches 0.4 at high values of N_{Ae}. More detailed studies show a family of (P_g/P_o) curves for different stirrer speeds, gas velocities, or Froude members, but there is no generally accepted correlation [28]. The steep drop in P_g/P_o occurs as u_G is increased to about 1 cm/sec; at velocities greater than 5 cm/sec, the impeller may not be able to disperse the gas effectively.

The decrease in power consumption is due to the formation of gas pockets behind the turbine blades and not to the change in density of the fluid, since the suspension density is decreased only about 10% because of gas holdup. With the curved-blade turbine, which has concave blades facing forward, the convex rear profile leads to smaller gas pockets and a smaller change in power on aeration.

Mass Transfer

Mass transfer coefficients for gases in stirred tanks have been measured using transient absorption or desorption tests with pure liquids or by following the psuedo-steady-state absorption and reaction of oxygen in sulfite solutions. Unlike the case with bubble columns, there are large differences in the coefficients found with these two approaches, as shown in Figure 7.16. The upper two lines are based on the study of Rushton, Gallagher, and Oldshue [29] for oxygen absorption at room temperature in sulfite solutions

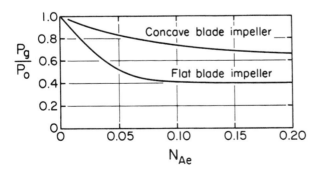

FIGURE 7.15 Relative power consumption for gassed impellers.

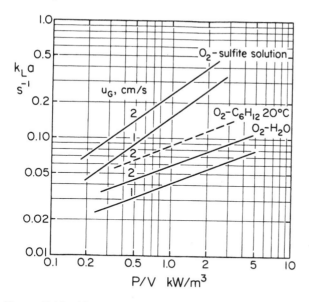

FIGURE 7.16   Mass transfer coefficients for oxygen absorption in stirred tanks.

using 6- or 12-inch tanks with standard six-blade turbines. Their values of $K_g a$ were converted to $k_L a$ using the estimated solubility of oxygen in sulfite solution and assuming no gas-film resistance. The following equation gives $k_L a$ in $\sec^{-1}$ as a function of P_g/V in kW/m^3 and u_G in cm/sec for oxygen in sulfite solutions:

$$k_L a = 0.146\left(\frac{P_g}{V}\right)^{0.74} u_G^{0.7} \tag{7.74}$$

Several other studies of sulfite oxidation have given similar results, though the exponents for gas velocity and power dissipation vary. A detailed comparison of these results is difficult, because most authors don't tell whether $k_L a$ is based on the clear liquid or the aerated volume or on the exit, arithmetic mean, or log-mean oxygen pressure. The classic paper by Cooper, Fernstrom, and Miller [30] is worth reading even though their data for vaned-disk stirrers and flat paddles are not as useful as data for turbine agitators. They showed that at constant stirrer speed $K_g a$ first increased and then leveled off or went through a maximum as gas velocity increased and that using actual power per unit volume gave a simple correlation that was valid for scaleup.

The lower lines in Figure 7.16 are for oxygen transfer in pure water, where bubble coalescence makes the interfacial area and the values of $k_L a$ much less than for sulfite solution. The lines are based on the correlation of van't Riet [31], which gives coefficients in reasonable agreement with a number of other studies. For oxygen in water,

$$k_L a = 0.04 \left(\frac{P_g}{V}\right)^{0.4} u_G^{0.5} \tag{7.75}$$

The effect of dissolved salts was clarified by Robinson and Wilke [32], who showed that adding any electrolyte to water inhibited bubble coalescence and that the agitation exponent in the equation for $k_L a$ gradually increased from 0.4–0.9 as the ionic strength changed from 0 to 0.40. Since surfactants can also affect the bubble size, it is difficult to predict a or $k_L a$ for multi-component solutions.

Example 7.6

An aerobic fermentation is to be carried out in a 200-m^3 reactor (4-m dia. × 16 m) with a normal liquid depth of 12 m and atmospheric pressure at the top. A flat-blade turbine will be used to disperse the air, and two axial-flow impellers will be installed on the same shaft to promote end-to-end mixing. Air will be supplied below the turbine at a superficial velocity of 3 cm/sec (based on 30°C and 1 atm). The Henry's law constant for oxygen is 5.2×10^4 atm/m.f. (10% greater than for pure water), and the peak oxygen demand is estimated to be 45 mmol/L-hr. Tests in a small unit show that $k_L a$ for this solution is 70% of the value for oxygen absorption in sodium sulfite solution. The solution viscosity is about 1.5 cp.

a. At the peak demand, what fraction of the oxygen supplied is used?
b. How much power is required for the agitators before and after the air is turned on?
c. Assuming that the oxygen mass transfer coefficient is determined by the average velocity and power dissipation, estimate $k_L a$ and the average dissolved oxygen concentration.
d. If the mass transfer coefficient in the upper region of the reactor is 40% less than the average because of the reduced rate of energy dissipation, will the dissolved oxygen concentration still be greater than the critical value of 1 mg/L?

Solution.

a.
$$S = \frac{\pi}{4}(4)^2 = 12.6\text{-m}^2 \text{ cross section}$$

$$V = 12.6 \times 12 = 151\text{-m}^3 \text{ solution volume}$$

At the top of the reactor, if there's no change in molar flow ($\Delta CO_2 = -\Delta O_2$):

$$u_G = 3 \text{ cm/sec}$$

$$F_{air} = \frac{12.6(0.03)(3600)10^3}{(0.08206)(303)} = 5.47 \times 10^4 \text{ mol/hr}$$

$$F_{O_2} = 0.21 F_{air} = 1.15 \times 10^4 \text{ mol/hr}$$

At peak demand,

$$O_2 \text{ used} = (45 \times 10^{-3})(151 \times 10^3)$$
$$= 6.8 \times 10^3 \text{ mol/hr}$$
$$O_2 \text{ left} = 1.15 \times 10^4 - 6.8 \times 10^3$$
$$= 4.7 \times 10^3 \text{ mol/hr}$$
$$\% \, O_2 \text{ in exit gas} = \frac{4.7 \times 10^3}{5.47 \times 10^4} \times 100 = 8.59\%$$
$$\text{Fraction oxygen used} = \frac{21 - 8.59}{21} = 0.591$$

b.   Assume:

$$\frac{D_a}{D_t} = \frac{1}{3},$$

$$D_a = 1.333 \text{ m}$$

$$N = 120 \text{ rpm} = 2 \text{ sec}^{-1}$$

$$\rho = 1000 \text{ kg/m}^3$$

$$\text{Re} = \frac{2(1.333)^2 1000}{1.5 \times 10^{-3}} = 2.37 \times 10^5$$

For a standard turbine, $N_p = 6.0$ [28]:

$$P_o = \frac{6.0(1000)(2)^3(1.333)^5}{1000 \text{ W/kW}} = 202 \text{ kW}$$

If the turbine is 2 m from the bottom, or 10 m below the surface, the pressure is about 2 atm:

$$u_G = \frac{3}{2} = 1.5 \text{ cm/sec} = 0.015 \text{ m/sec}$$

$$N_{Ae} = \frac{Q}{ND_a^3} = \frac{0.015(12.6)}{2(1.333)^3} = 0.0399$$

From Figure 7.15, $P_g/P_o \cong 0.55$:

$$P_g = 0.55(202) = 111 \text{ kW}$$

For a pitched-blade turbine, $N_p = 1.7$ [28, p 253]:

$$P_o = \frac{1.7}{6.0} \times 202 = 57 \text{ kW}$$

Solution reaching the upper stirrers is already aerated, so assume

$P_g/P_o \cong 0.8$:
$$P_g = 0.8(57) = 46 \text{ kW}$$

Total power $= 111 + 2(46) = 203 \text{ kW}$ aerated (272 HP)
$$= 202 + 2(57) = 316 \text{ kW}$$ no air (423 HP)

c.
$$\text{ave}\left(\frac{P}{V}\right) = \frac{203}{151} = 1.34 \text{ kW/m}^3$$

$$\text{ave } u_G = \frac{3 + 1.5}{2} = 2.25 \text{ cm/sec}$$

Using Figure 7.16,

$$k_L a = 0.7(0.32) = 0.22 \text{ sec}^{-1}$$

If gas is back-mixed,

$$y_{O_2} = 0.086$$
$$\text{Average depth} = 6 \text{ m}$$
$$\text{Average pressure} = 1 + \frac{6}{10.3} = 1.58 \text{ atm}$$

$$C_{O_2}^* = \frac{1.58(0.086)}{5.2 \times 10^4} \times 55.5 = 1.45 \times 10^{-4} \text{ mol/L}$$

Since

$$r = \frac{45 \text{ mol}}{\text{L-hr}} = \frac{45 \times 10^{-3}}{3600} = 1.25 \times 10^{-5} \frac{\text{mol}}{\text{L-sec}}$$

and

$$r = k_L a(C^* - C)$$

$$(C^* - C) = \frac{1.25 \times 10^{-5}}{0.22} = 5.68 \times 10^{-5} \text{ mol/L}$$

$$C = 14.5 \times 10^{-5} - 5.68 \times 10^{-5} = 8.82 \times 10^{-5} \text{ mol/L}$$

$$C = 8.82 \times 10^{-5} \times 32 = 2.82 \ 10^{-3} \text{g/L} = 2.8 \text{ mg/L}$$

d. If $k_L a = 0.22(1 - 0.4) = 0.13$ in the top part of the reactor and oxygen demand is unchanged, then

$$C^* - C = \frac{5.68 \times 10^{-5}}{0.6} = 9.47 \times 10^{-5} \text{ mol/L}$$

Using the same average C^*,

$$C = 1.45 \times 10^{-4} - 9.47 \times 10^{-5} = 5.03 \times 10^{-5} \text{ mol/L}$$

$$C \times 32 \times 1000 = 1.6 \text{ mg/L}$$

This is close to the critical value of 1 mg/L. If C^* is based on 1 atm, the pressure at the top of the tank,

$$C^* = \frac{1.45 \times 10^{-4}}{1.58} = 9.2 \times 10^{-5} \text{ mol/L}$$

Then at the peak demand, $(C - C^*) > C^*$, or C would drop to zero.

A higher air rate could be used or the reactor operated under a slight pressure to make sure all regions have more than 1 mg O_2/L.

The limited data for mass transfer of gases to pure organic liquids show trends similar to those for oxygen in water. Sridhar and Potter [33] followed the transient absorption of oxygen in cyclohexane in a 2-L stirred vessel and reported that $k_L a$ varied with $(P_g/V)^{0.4} uG^{0.75} D^{0.5}$. As shown in Figure 7.16, the mass transfer coefficients are about 1.5 times the values for oxygen in water. Other studies [34] have shown that the interfacial area for gas dispersed in pure liquids depends on $(P_g/V)^{0.4} \sigma^{-0.6} \rho_L^{0.2}$. Comparing cyclohexane with water, the lower surface tension and liquid density and the higher diffusivity should give about a 2.4-fold higher $k_L a$. The difference between the predicted factor of 2.4 and the 1.5 shown in Figure 7.16 could be due to changes in k_L with bubble size or may just illustrate the uncertainty in comparing coefficients obtained by different methods in different equipment. For predicting coefficients for other liquids, either the water or the cyclohexane data for a given power input can be used with corrections for changes in physical properties:

$$k_L a = k_L a_{\text{ref}} \left(\frac{\sigma_{\text{ref}}}{\sigma}\right)^{0.6} \left(\frac{\rho}{\rho_{\text{ref}}}\right)^{0.2} \left(\frac{D}{D_{\text{ref}}}\right)^{0.5} \tag{7.76}$$

From a fundamental standpoint, it might seem better to develop separate correlations for a and k_L, and several authors have taken this approach. However, k_L is usually estimated from $k_L a$ and values of a calculated from photographs; there is more scatter in the published k_L values than in the $k_L a$ data. Furthermore, theory and some supporting data indicate that k_L is several-fold greater for large deformable bubbles than for small bubbles, which act as rigid spheres [34], and the effect of using a simple average k_L is not clear. The main purpose of the correlation presented here is to help in the interpretation and scaleup of lab or plant data rather than for use as the main basis for final reactor designs, and the use of simpler correlations based on $k_L a$ should be satisfactory.

When mass transfer has a significant effect on the overall reaction rate, it is important to make laboratory or pilot-plant tests at agitations condition close to those anticipated for a large reactor. Often, tests in small stirred reactors are made at very high power consumption, which cannot be matched on large equipment, and the reaction rate and effects of system properties may differ appreciably from what will be found in a large reactor. For example, in a study of the hydrogenation of ethyl anthroquinone [35] in a 2-L vessel, the stirrer speed was 3000 rpm, corresponding to a power input of about 100 kW/m^3, and the value of $k_L a$ at 55°C was 1.65 sec^{-1}. While intense agitation may make it easier to study the fundamental kinetics because of small or negligible mass transfer resistance, mass transfer would be much more important and the overall rate lower for practical power consumption in a large reactor. When the probable reaction conditions for a large reactor have been determined, a scaledown approach should be used and some tests made in a lab or pilot-plant reactor at conditions that will give comparable mass transfer rates.

Example 7.7

The partial oxidation of cyclohexane to cyclohexanol and cyclohexanone was carried out continuously in a pilot plant with three 35-liter reactors operating in series at 155°C and 8.2 atm [36]. The total conversion was only 10–20%, and the selectivity was 70–80%, depending on the conversion and agitation conditions. Low conversions were necessary because the products are readily oxidized to a variety of byproducts. For a typical run, the feed rates were 100

L/hr cyclohexane, 9.9 nm³/hr air, the conversion and selectivity were 9.5% and 73%, and the oxygen in the vent gas was about 0.2%.

a. Assuming that oxygen mass transfer is controlling, calculate the apparent value of $k_L a$, and compare it with predicted values of $k_L a$ for stirred reactors. What does this comparison show about the regime in which the reactor operates?

b. Would the selectivity be expected to depend on variables such as agitator speed, gas flow rate, and degree of gas mixing?

Solution.

a. Get moles O_2 consumed and check with moles C_6H_{12} reacted:

$$F_{O_2} = \frac{9.9 \times 10^6}{3600} \text{ cm}^3/\text{sec} \times \frac{0.21}{22{,}400 \text{ cm}^3/\text{mol}}$$
$$= 2.58 \times 10^{-2} \text{ mol/sec}$$

Since $y_{O_2} = 0.002$ in vent gas (assume solvent-free basis), 99% of the oxygen is consumed:

$$\Delta N_{O_2} = 0.99(2.58 \times 10^{-2}) = 2.55 \times 10^{-2} \text{ mol/sec}$$

For C_6H_{12},

$$\rho_{20°C} = 0.779 \text{g/cm}^3$$

$$F_{C_6} = \left(\frac{100{,}000}{3600}\right)\left(\frac{0.779}{84.16}\right) = 0.257 \text{ mol/sec}$$

$$\text{products} = F_{C_6} \times 0.095 \times 0.73 = 1.78 \times 10^{-2} \text{ mol/sec}$$

Main reactions:

$$C_6H_{12} + 0.5O_2 \xrightarrow{1} C_6H_{11}OH$$
$$C_6H_{12} + O_2 \xrightarrow{2} C_6H_{10}O + H_2O$$

With the main products formed in a 3/2 ratio, the oxygen used for these products is $1.78 \times 10^{-2} [0.6(0.5) + 0.4(1)] = 1.25 \times 10^{-2}$ mol/sec, or 0.7 mol O_2/mole product
The remaining O_2 used,

$$(2.55 - 1.25)10^{-2} = 1.3 \times 10^{-2} \text{ mol/sec}$$

corresponds to

$$\frac{1.3 \times 10^{-2}}{0.257(0.095)(1 - 0.73)} = \frac{1.97\text{mol moL } O_2/\text{mol } C_6 \text{ to}}{\text{byproducts}}$$

Use the total O_2 consumption to evaluate $k_L a$. O_2 solubility from Wild et al. [37]:

$$\log X_{O2} = 0.366 \log T - 3.8385$$

At $T = 155 + 273 = 428$ K,

$$X = 1.33 \times 10^{-3} \text{ m.f./atm}$$

Assume vapor pressure of solution is that of C_6H_{12} at 155°C, $P' = 5.8$ atm:

$$P_{O_2} + P_{N_2} = 8.2 - 5.8 = 2.4 \text{ atm} \quad \text{(neglect any CO,CO}_2, \text{ etc.)}$$

If gas is back-mixed,

$$P_{O_2} = 0.002(2.4) = 0.0048 \text{ atm}$$
$$x = 0.0048(1.33 \times 10^{-3}) = 6.38 \times 10^{-6} \text{ m.f. } O_2$$

For C_6H_{12},

$$\rho_M \cong \frac{650}{84.16} = 7.72 \text{ mol/L at } 155°C$$

$$C_{O_2}^* = 6.38 \times 10^{-6}(7.72) = 4.93 \times 10^{-5} \text{ mol/L}$$

Assume each reactor has 30 L solution:

$$\text{apparent } k_L a = \frac{2.55 \times 10^{-2}}{3 \times 30} \times \frac{1}{4.93 \times 10^{-5}} = 5.7 \text{ sec}^{-1}$$

To predict $k_L a$ from Figure 7.16, P/V and u_G should be known and D estimated for O_2 at 155°C. Allowing for solvent vapor, the total vapor flow is 8.2/2.4 times the air flow:

$$F_{\text{total}} \cong \left(\frac{9.9 \times 10^{-6}}{3600}\right)\left(\frac{428}{273}\right)\left(\frac{1}{8.2}\right)\left(\frac{8.2}{2.4}\right) = 1.8 \times 10^3 \text{ cm}^3 \text{ sec}$$

If the reactors are about 30-cm dia × 50 cm tall,

$$u_G = \frac{1.8 \times 10^{-3}}{3\pi \, 30^2/4} = 0.85 \text{ cm/sec}$$

The power consumption was not measured, but it may have been as high as 5 kW/m³ (25 HP/1000 gal). From Figure 7.16, for O_2–C_6H_{12} at 20°C, 2 cm/sec, 5 kW/m³:

$$k_L a \cong 0.16 \text{ sec}^{-1}$$

At 155°C,

$$\mu \cong 0.2 \text{ cp}, \qquad \mu_{20} = 0.98 \text{ cp}$$

Correct for diffusivity using T/μ ratio:

$$\frac{D_{155}}{D_{20}} = \left(\frac{428}{293}\right)\left(\frac{0.98}{0.2}\right) = 7.16$$

Correct for lower gas velocity using $u_G{}^{0.5}$:

$$\text{Predicted } k_L a = 0.16(7.16)^{0.5}\left(\frac{.85}{2}\right)^{0.5} = 0.28 \text{ sec}^{-1}$$

The apparent value of $k_L a = 5.7 \text{ sec}^{-1}$ is about 20 times the predicted mass transfer coefficient, so the absorption of oxygen is greatly enhanced by chemical reactions in the liquid film. The products are formed by a chain reaction with complex kinetics. But if the kinetics can be approximated by a first-order expression, the reaction would fall in the pseudo-first-order regime, where the rate varies with the square root of the oxygen diffusivity and the rate constant. Errors in the estimated physical properties or the power consumption will not affect this conclusion. The greatest uncertainty may be in the gas mixing. If the log-mean oxygen pressure was used, the apparent $k_L a$ would be much smaller, but still greater than the predicted value:

$$P_{O_2 lm} = 0.107 \text{ atm}$$

By ratio,

$$k'_L a = \frac{0.0048}{0.107} \times 57 = 2.6 \text{ sec}^{-1}$$

b. The change in selectivity with agitation condition at the same conversion suggests concentration gradients for cyclohexane and products in the liquid film or different reaction orders for the main and side reactions.
    If the gas is perfectly mixed and Eq. (7.51) is used, then

$$\phi_a = 1 + \frac{C_{Bo}}{n C_{Ai}}\left(\frac{D_B}{D_A}\right)^{0.5}$$

$$C_{Bo} = (1 - 0.095)(7.72) = 6.99 \text{ mol/L}$$

$$C_{Ai} = 4.93 \times 10^{-5}$$

Assume $D_B = 0.5 \, D_A \; n = 1/0.7 = 1.4$:

$$\phi_a = 7.2 \times 10^4$$

No significant gradient for C_6H_{12} or products would be expected, since $\phi_a/\phi \cong 3500$. However, for newly formed gas bubbles that have not yet mixed with other bubbles, the much higher oxygen concentration could lead to very fast local oxygen absorption and gradients of cyclohexane and cyclohexanol near the gas–liquid interface. A stirrer that promotes rapid gas mixing should give better selectivity.

PACKED-BED REACTORS

Packed-bed absorbers are often used to remove small amounts of pollutant from an air stream by absorption plus reaction in a liquid. They can also be used to recover and recycle a valuable reactant. With packed beds, the gas pressure drop is much lower than in a bubble column or a stirred tank. Another advantage is that the gas in a tall packed column is essentially in plug flow, which makes it easier to get a high percent removal. With counterflow of gas and liquid, the operating range is limited by flooding. But when absorption is accompanied by an irreversible reaction, counterflow has no advantage, and the system can be operated with downflow of both phases at very high rates.

The equation for consecutive mass transfer and reaction is the same as for a tank [Eq. (7.10)], except that the liquid holdup h is used instead of the fraction liquid, $(1 - \epsilon)$:

$$\frac{P_A}{r} = \frac{1}{k_g a} + \frac{H}{k_L a} + \frac{H}{k_2 C_B h} = \frac{1}{K_O} \quad \text{or} \quad \frac{1}{K_g a} \tag{7.77}$$

The mass transfer coefficients $k_g a$ and $k_L a$ can be predicted from empirical equations or by using data for simple gases, such as NH_3, O_2, and CO_2, and correcting for differences in diffusivity and packing characteristics. Correlation for $k_L a$ and $k_g a$ and examples of applications are given in mass transfer texts [28].

NOMENCLATURE

Symbols

a	interfacial area per unit reactor volume
a'	interfacial area per unit liquid volume
Bo	Bond number, $gD_t^2 \rho_L/\sigma$
C	molar concentration
C^*	equilibrium concentration
c_p	heat capacity

D	diffusivity, diameter
D_a	agitator diameter
D_t	tank diameter
E_L	axial-mixing coefficient
F	feed rate, flow rate
Fr	Froude number, $u_G/(gD_t)^{1/2}$
G	gas flow rate
Ga	Galileo number, $gD_t^3\rho_L^2/\mu_L^2$
g	gravitational constant
H	Henry's law constant
h	height of aerated liquid
h_o	height of clear liquid
h_w	wall heat transfer coefficient
$K_g a$	overall rate constant, gas-phase driving force
$K_L a$	overall rate constant, liquid-phase driving force
k	reaction rate constant, thermal conductivity
k_2	second-order rate constant
k_g	gas mass transfer coefficient per unit area
$k_g a$	gas-phase volumetric mass transfer coefficient
k_L	liquid mass transfer coefficient per unit area
k_L^*	normal value of k_L
$k_L a$	liquid-phase volumetric mass transfer coefficient
L	length, liquid flow rate
M	molecular weight, molar concentration
$\sqrt{M}$	square-root modulus, Eq. (7.27)
N	stirrer speed
N_A	flux of A or rate per unit area
N_{Ae}	aeration number $= Q/ND_a^3$
N_P	power number $= P/\rho N^3 D_a^5$
n	ratio of reactants
P	partial pressure, power consumption
Pe_L	Peclet number $= Lu_L/E_L$
P_g	power consumption with gas and liquid present
P_o	power consumption with liquid only
P'	vapor pressure
Q	volumetric gas flow rate
Re	Reynolds number for stirrer, $ND_a^2\rho/\mu$
r	reaction rate per unit reactor volume
r'	reaction rate per unit volume of liquid
S	selectivity, cross-sectional area
t	time
u_G	superficial gas velocity

V	volume
v_b	bubble velocity
x	distance from interface
x_L	film thickness
y	mole fraction in gas

Greek Letters

a	distance to reaction plane, proportionality factor
ϵ	void fraction, gas holdup
μ	viscosity
ρ	density
ρ_G	density of gas
ρ_L	density of liquid
σ	surface tension
τ	time constant
ϕ	enhancement factor
ϕ_a	asymptotic value
π	pi

PROBLEMS

7.1   A liquid-phase oxidation is carried out in aqueous solution in an agitated, sparged reactor. Batch tests show that the absorption rate decreases as the conversion increases, judging from the change in exit gas composition. The dissolved reactant and the product are nonvolatile, and the vapor pressure of the solution is about 80% that of water. The solution density is 60 lb/ft^3.

Test conditions:

$T = 70°C$

$P = 1$ atm above the liquid

$h = 8$-ft liquid depth (before aeration)

$u_G = 0.08$ ft/sec (feed gas at 1 atm, 70°C)

$y_{in} = 0.21$ (fraction oxygen, dry basis)

$\dfrac{P}{V} = 4$ HP/1000 gal unaerated liquid

at start: $x = 0$,	$y_{out} = 0.017$
at: $x = 0.5$,	$y_{out} = 0.023$
at: $x = 0.9$,	$y_{out} = 0.046$

a. Calculate the oxygen absorption rate in lb-mol/hr-ft^2 (of reactor cross section), allowing for the change in molar flow rate of the gas.
b. Calculate $K_g a$ for the three conversion levels, assuming perfect mixing of the gas phase and instantaneous saturation of gas bubbles with water vapor. Compare the coefficients with those reported for oxidation of sulfite solution.
c. Show how $K_g a$ varies with the concentration of the liquid reactant. What does this suggest as the rate-limiting step?
d. What would be the advantages and disadvantages of carrying out the oxidation continuously in a packed-bed reactor rather than in a stirred-tank reactor?

7.2   The chlorination of *p*-cresol was carried out at atmospheric pressure in a small stirred reactor operated with continuous addition of chlorine gas. For an initial concentration of 4.5 M, the maximum yield of monochlorocresol was 3.5 M. From separate tests in a homogeneous reactor, the rate constants for the chlorination of cresol and monochlorocresol were determined to be 20 L/mole-sec. and 1.2 L/mole-sec.

a. What is the maximum yield of monochlorocresol and the selectivity in the absence of diffusion effects?
b. Could the selectivity be increased by changing the initial cresol concentration or by changing the chlorine concentration in the gas phase?

7.3   A partial oxidation is carried out by bubbling oxygen through a solution of reactant B in a stirred reactor. Batch tests show the oxygen absorption rate decreases as the conversion of B increases (Table 7.4). The initial reaction rate divided by the oxygen solubility is less than the value predicted from correlations for $k_L a$, though the predicted rate is uncertain by ±30%.

TABLE 7.4   Data for Problem 7.3

Conversion %	C_B, mole/L	r, mole/L-hr
0	2	3.08
50	1	2.21
70	0.6	1.63
85	0.3	0.97

    a.   Correlate these results and predict the initial reaction rate for C_B = 4 M.

    b.   For C_B = 2 M, how close is the bulk liquid to saturation with oxygen?

7.4   The partial oxidation of a hydrocarbon (M = 90) is carried out in the liquid phase with a soluble catalyst. Air is passed through a bubble column (1.6-m diameter × 8 m) at u_G = 100 m/hr. Shortly after the start of a semibatch run, the exit gas had 2% O_2, 92% N_2, and 6% solvent vapor. The oxygen solubility is 1.5×10^{-3} mol/L, atm, about twice that for sulfite solution, but the density and viscosity are comparable to that of aqueous sulfite solution.

    a.   Calculate mass transfer coefficients $K_g a$ and $K_L a$.

    b.   In which of the several regimes for mass transfer plus reaction does this system fall?

7.5   A gas A (M = 30) is removed from an air stream by absorption and reaction with a solute B (M = 60) dissolved in aqueous solution. Tests at 1 atm and 25°C in a column packed with 0.5-inch Raschig rings gave these results:

$$G = 300 \text{ lb/hr-ft}^2$$

$$L = 1500 \text{ lb/hr-ft}^2$$

$$K_g a = 0.25 \text{ moles/hr-ft}^3\text{-atm}$$

The calculation for $K_g a$ was based on the log-mean partial pressure of A in the gas, since the equilibrium pressure was zero. Correlations from the literature give the following values of the gas-film and liquid-film coefficients [28]:

    for $NH_{3\text{-air}}$: $k_g a = 9.4$ moles/hr-ft^3-atm

    for $O_{2\text{-water}}$: $k_L a = 33 \text{ hr}^{-1}$

The Henry's law coefficient for the gas is 1.5 atm/mole fraction.

    a.   Make a quantitative analysis of these results to see if the rate-limiting step can be determined.

    b.   If another type of reactor is to be tested, what type would you recommend, and why?

7.6   The absorption of H_2S in basic solutions is sometimes said to be "gas-film controlled," even though H_2S has a very low solubility in water

($H = 4.78 \times 10^2$ atm/mole fraction at 20°C). Under what condition is the foregoing statement likely to be correct? Use calculations and diagrams to support your reasoning.

7.7 The following reaction is to be carried out in a bubble-column reactor at about 1–2 atmospheres:

$$A_g + B \xrightarrow{\text{Cat}} C + D_g$$

Gases A and D have about the same solubility; for A, $H = 200$ atm/mol/L at reaction conditions.

$$r = k_2 C_A C_B, \quad k_2 = k(\text{cat}) = 160 \text{ L/mol/sec for (cat)} = 0.01$$
$$D_A = 4 \times 10^{-5} \text{cm}^2/\text{sec}$$
$$D_B = 2 \times 10^{-5} \text{ cm}^2/\text{sec}$$

Preliminary tests of physical absorption of A gave

$$k_L^* a = 0.1 \text{ sec}^{-1}$$

Photographs showed an average bubble size of 1 mm, and the liquid level rose 25% when the gas was turned on. The bubble column will be designed for counterflow operation with $C_{Bo} = 4$ M and 80% conversion of B. Pure A is fed, and 95% conversion of A is expected.

    a.   If $P = 2$ atm at the bottom of the column and 1 atm at the top, predict the rate of absorption of A at the top and at the bottom. The solution vapor pressure is negligible.

    b.   How much change in rate would be produced by doubling the catalyst concentration?

REFERENCES

1. DW Van Krevelen, PJ Hoftyzer. Chem Eng Sci 2:145, 1953.
2. S Hatta. Tohoku Imperial U Tech Rep 8:1, 1928; 10:119, 1932.
3. HE Benson, JH Field, RM Jimeson. Chem Eng Progr 50:366, 1954.
4. R Higbie. Trans AIChE 31:365, 1935.
5. PLT Brian, JF Hurley, EH Hasseltine. AIChE J 7:226, 1961.
6. H Inoue, T Kobayashi. Proc IV European Symposium Chem Reaction Eng, Brussels, 1968.
7. HL Shulman, MC Molstad. Ind Eng Chem 42:1058, 1950.
8. P Harriott. Can J Chem Eng 48:109, 1970.
9. JR Fair, AJ Lambright, JW Anderson. Ing Eng Chem Proc Des Dev 1:33, 1962.
10. YT Shaw, BG Kelkar, SP Godbole, WD Deckwer. AIChE J 28:353, 1982.
11. F Yoshida, K Akita. AIChE J 11:9, 1965.
12. PM Wilkinson, AP Spek, L van Dierendonck. AIChE J 38:544, 1992.

13. R Krishna, J Ellenberger. AIChE J 42:2627, 1996.
14. WD Deckwer, A Schumpe. Chem Eng Sci 48:889, 1993.
15. SS Ozturk, A Schumpe, WD Deckwer. AIChE J 33:1473, 1987.
16. SH Eissa, K Schugerl. Chem Eng Sci, 30:1251, 1975.
17. DJ Vermeer, R Krishna. Ind Eng Chem Process Des Dev 20:475, 1981.
18. K Akita, F Yoshida. Ind Eng Chem Proc Des Dev 12:76, 1973.
19. LL Van Dierendonck. PhD dissertation, Twente University, Netherlands, 1970.
20. HS Hikita, S Asai, K Tanigawa, K Segawa, M Kitao. Chem Eng J 20:59, 1980.
21. HS Hikita, S Asai, K Tanigawa, K Segawa, M Kitao. Chem Eng J 22:61, 1981.
22. ML Jackson, CC Shen. AIChE J 24:63, 1978.
23. GD Towell, GH Ackerman. Proc 2nd Inter Symp Chem Reaction Eng, Amsterdam, 1972.
24. SB Dake, R Jaganathan, RV Chaudhari. Ind Eng Chem Res. 28:1107, 1989.
25. LS Nowicki, S Ledakowicz, R Zarzycki. Ind Eng Chem Res 31:2472, 1992.
26. WF Hart. Ind Eng Chem Proc Des Dev 15:109, 1976.
27. HS Hikita, S Asai, H Kikukawa, T Zaike, M Ohue. Ind Eng Chem Proc Des Dev 20:540, 1981.
28. WL McCabe, JC Smith, P Harriott. Unit Operations of Chemical Engineering. 6th ed. New York: McGraw-Hill, 2000, p 275, 581–585.
29. JH Rushton, JB Gallagher, JY Oldshue. Chem Eng Progr 52:319, 1956.
30. CM Cooper, GA Fernstrom, SA Miller. Ind Eng Chem 36:504, 1944.
31. K Van't Riet. Ind Eng Chem Proc Des Dev 18:357, 1979.
32. CW Robinson, CR Wilke. Ferment Technol Today 73, 1972.
33. T Sridhar, OE Potter. ACS Symp Ser 196 p 499, 1981.
34. PH Calderbank: In: VW Uhl, JB Gray, eds. Mixing: Theory and Practice. Vol II. New York: Academic Press, 1967, p 23.
35. T Bergelin, NH Schoon. Ind Eng Chem Process Des Dev 20:615, 1981.
36. JWM Steeman, S Kaarsenaker, PJ Hoftyzer. Chem Eng Sci 14:139, 1961.
37. JD Wild, T Sridhar, OE Potter. Chem Eng Journal 16:209, 1978.

8

Multiphase Reactors

The solid-catalyzed reaction of a gas with a liquid can be carried out in a slurry reactor, where fine catalyst particles are suspended in the liquid, or in a fixed bed of catalyst pellets, where gas and liquid flow continuously through the bed. For both types, there are several mass transfer steps to consider in modeling the reactor, since the gas dissolves and diffuses into the liquid and then both reactants diffuse to the catalyst and into the pores. The models are therefore more complex than those given in Chapter 7 for gas absorption plus reaction in a liquid.

A major difference between the two types of multiphase reactors is that the amount of catalyst in the slurry reactor is only 0.01–1% of the total volume, whereas it is 50–60% of the volume of the packed bed. In this chapter the slurry reactor is considered first, because it is the more common type and because having the catalyst concentration as a variable makes it easier to evaluate the kinetic models.

SLURRY REACTORS

Slurry reactors are widely used for metal-catalyzed hydrogenations of alkenes and alkynes, including the important process of converting vegetable oils to margarine and other fats. Hydrogenations of ketones, aldehydes, and nitro compounds are also performed in slurry reactors, usually with small amounts of metal catalysts. Catalytic reactions involving carbon monoxide, oxygen, or other slightly soluble gases can be carried out in slurry reactors, though there are not as many examples as for hydrogenation. The model presented here is written for the hydrogenation of olefins rather than for the general reaction $A + B \longrightarrow C$. This makes it easier to keep track of the concentration gradients, and the model is readily adapted for other types of reactions.

The reaction of hydrogen with olefin B is exothermic and essentially irreversible at typical reaction conditions:

$$H_2 + B \longrightarrow C \tag{8.1}$$

A semibatch reaction is considered with continuous addition of H_2 to a charge of B or B plus solvent in a stirred tank. The reactor is well mixed and operates at constant temperature and pressure, with small quantities of catalyst completely suspended. The sequence of steps at any stage of the conversion is:

1. H_2 dissolves and diffuses into the bulk liquid.
2. H_2 diffuses to the catalyst particles.
3. Reactant B diffuses to the catalyst particles.
4. H_2 and B diffuse into the catalyst pores, adsorb, and react.
5. Product C desorbs and diffuses out of the catalyst and into the bulk liquid.

If the hydrogen is diluted with solvent vapor or product gases, mass transfer of hydrogen from the gas phase to the gas–liquid interface is an additional step in the overall process. However, the high diffusivity of hydrogen and its low solubility usually make the gas-film resistance negligible, as is assumed here. For reaction with diluted gases of moderate solubility, the resistance in the gas film can be significant, as in the absorption of SO_2 from flue gas in lime or limestone slurries. In such cases, a gas-phase diffusion step is added to the foregoing sequence.

First-Order Model

The model for slurry hydrogenation is based on the three steps involving hydrogen mass transfer and reaction. The surface reaction is assumed first order to hydrogen, so the rate constant can be combined with the mass transfer coefficients in a simple equation. Step 3 is omitted, since the bulk concentration of B is normally much greater than the hydrogen concentration in the liquid. Step 5 is not included in the model but will be considered later, in the discussion of selectivity with consecutive reactions. Because of the low solubility of hydrogen in most liquids, the accumulation of hydrogen in the solution is much less than the hydrogen consumed in the reaction. Therefore, the accumulation of hydrogen is neglected, and all steps in the series are assumed to take place at the same rate.

The gas absorption rate per unit volume of reactor is

$$r_1 = k_b a_b (C_i - C_b) \tag{8.2}$$

where

$$k_b a_b = \text{volumetric coefficient for gas bubbles, sec}^{-1}$$
$$\text{(equivalent to } k_L a \text{ of Chapter 7)}$$
$$C_i, C_b = \text{concentration of } H_2 \text{ at the gas-liquid interface}$$
$$\text{or in the bulk liquid}$$

The rate of hydrogen diffusion to the catalyst is proportional to the external mass transfer coefficient, k_c, the external area per gram, a_c, and the catalyst concentration, m, in g/cm^3:

$$r_2 = k_c a_c m (C_b - C_s) \tag{8.3}$$

The units of $k_c a_c m$ are sec^{-1}, the same as for $k_b a_b$. The external area is based on the surface/volume ratio for a sphere, modified with a shape factor λ. For crushed solids, λ is typically 1.2–1.4:

$$a_c = \frac{6\lambda}{d_p \rho_{\text{cat}}} \tag{8.4}$$

The rate of diffusion plus reaction in the porous catalyst is proportional to m, the effectiveness factor η, and the surface concentration C_s:

$$r_4 = \eta k m C_s \tag{8.5}$$

Since these three first-order steps are in series, Eqs. (8.2), (8.3), and (8.5) can be combined to give an overall rate equation:

$$r = KC_i$$

$$\frac{C_i}{r} = \frac{1}{K} = \frac{1}{k_b a_b} + \frac{1}{m} \left(\frac{1}{k_c a_c} + \frac{1}{\eta k} \right) \tag{8.6}$$

Separating the Resistances

Equation (8.6) shows that the overall resistance $1/K$ consists of a gas absorption resistance, $1/k_b a_b$, and a catalyst resistance,

$$\frac{1}{m} \left(\frac{1}{k_c a_c} + \frac{1}{\eta k} \right)$$

These two terms can be determined by making a series of runs with different catalyst concentrations and plotting C_i/r versus $1/m$. The plot should be linear, with a positive slope, and $1/k_b a_b$ can be obtained by extrapolating to $1/m = 0$. The value of C_i is calculated from the hydrogen pressure and the solubility, but to test the model, $1/r$ can be plotted instead of C_i/r if all the runs are at the same hydrogen pressure.

This method of plotting the rate data was suggested by Davis et al. in 1932 [1], and it has been widely used in analyzing hydrogenation data. Typical reciprocal plots and the corresponding concentration gradients are sketched in Figure 8.1. The intercept is the gas absorption resistance divided by C_i. For the second point of Figure 8.1a, the gas absorption resistance is 75% of the total resistance, which means that C_b is 25% of the saturation value. Under these conditions, the reaction rate increases only slightly with increased catalyst concentration but is strongly affected by agitator speed and superficial gas velocity, which affect $k_b a_b$.

When the reciprocal plot has an intercept close to the origin, as in Figure 8.1b, the catalyst resistance is controlling, and the rate increases almost in proportion to the catalyst charge. The solution is nearly saturated with hydrogen, so changes in the gas absorption coefficient have little effect on the reaction rate.

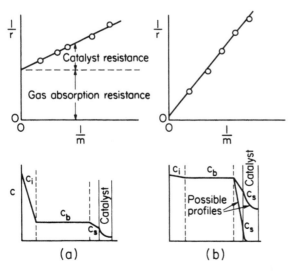

FIGURE 8.1 Reciprocal plots and concentration profiles for catalytic hydrogenation: (a) Gas absorption resistance large but not controlling; (b) catalyst resistance controlling.

If the reaction rate is directly proportional to the catalyst concentration, there is no need to make a reciprocal plot, since the catalyst resistance is obviously controlling, and the solution is saturated with hydrogen. In some laboratory studies, the stirrer speed was increased until there was no further effect on reaction rate, and under these conditions the catalyst resistance was probably controlling. Such tests are useful in studying the reaction kinetics, but the results may not apply on scaleup, since the energy dissipated per unit volume could be an order of magnitude greater than can be obtained in a large reactor, and the gas-absorption resistance may become significant or even controlling.

In some cases, the hydrogenation rate increases with catalyst concentration, but the reciprocal plot is not linear, making the separation of the resistances difficult. One possible explanation is catalyst poisoning. When a small amount of catalyst is used, impurities in the solution can poison a significant fraction of the catalyst, but the poison will have only a small effect at moderate catalyst concentrations. The reciprocal plot would then have the shape shown in Figure 8.2a, and the extrapolation to $1/m = 0$ could be made by ignoring the points at high $1/m$. On the other hand, very high catalyst concentrations may lead to agglomeration or incomplete suspension of the catalyst and a nonlinear plot, as shown in Figure 8.2b.

Deviations from the model could also be caused by complex kinetics. Some hydrogenations appear half order or zero order to hydrogen at high pressures, and Eq. (8.6) is then not strictly correct. However, if the resistances for gas absorption and mass transfer to the catalyst are relatively large, the reciprocal plot will be almost linear, and it can be extrapolated to get the gas absorption resistance. The kinetics for B could be zero

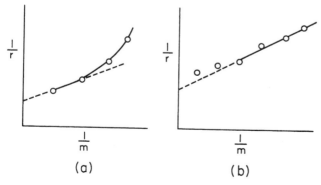

FIGURE 8.2 Reciprocal plots for catalytic hydrogenation: (a) Catalyst poisoning at low m; (b) agglomeration or incomplete catalyst suspension at high m.

order, first order, or of the Langmuir–Hinshelwood type, but this will not affect the analysis if the hydrogenation rates are measured at the same concentration of B. If the observed rate decreases with increasing conversion, plots based on the initial rate and the rate at high conversion might show a change in gas-absorption resistance due to changes in physical properties of the solution.

A key assumption for the model is that of *consecutive* mass transfer and reaction, which means negligible reaction in the liquid around the gas bubbles. When the bulk concentration of hydrogen is very low, significant reaction may take place in catalyst particles inside the liquid film, where the average hydrogen concentration is much greater than the bulk value. This involves *simultaneous* mass transfer and reaction, and a model similar to that in Chapter 7 could be developed [see Eq. (7.37)]. The reaction rate might then vary with the square root of the catalyst concentration. Reaction in the liquid film is usually unimportant for slurry hydrogenations, but it could be a factor in fermentations due to cells in the liquid films around air bubbles.

An alternate method proposed for separating the absorption and catalyst resistances is to vary the gas flow rate and to plot $1/r$ versus $(1/u_o)^n$, assuming $k_b a_b$ varies with u_0^n [2]. However, reported values for the exponent n cover a wide range, and there is no consensus on the best value. The correct value for n for a series of runs cannot be determined by trial, since different assumed values give apparently good linear plots. Whatever method of analysis is used to separate the resistances, the values of the gas-absorption resistance and the catalyst resistance should be checked for reasonableness using common sense and published correlations.

Gas-Absorption Coefficient

The gas solubility is needed to calculate $k_b a_b$ from the rate data, since the intercept on the $1/r$ plot is $1/k_b a_b C_i$. The solubility of hydrogen in organic liquids is quite low, but it increases with temperature, unlike the trend for most gases. A correlation for hydrogen in nonpolar solvents is given by Schaffer and Prausnitz [3], and data for several polar solvents are given by Brunner [4]. Some of these results are plotted in Figure 8.3 [5–7] along with data for vegetable oils. For many of the solvents, the hydrogen solubility increases slightly more rapidly than the increase in absolute temperature, so if the data are available at only one temperature, the solubility could be assumed proportional to T. For most systems, the solubility is directly proportional to hydrogen pressure up to at least 10 atmospheres and then increases somewhat less rapidly for higher pressures [5].

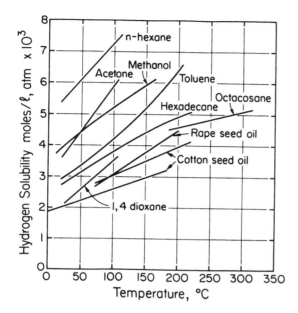

FIGURE 8.3 Hydrogen solubility in organic liquids.

There are two approaches to estimating $k_b a_b$ for a particular system. One is to use a general correlation for gas absorption in stirred reactors, and the other is to start with published data for a hydrogenation reaction and correct for differences in physical properties and agitation conditions. There are many correlations for $k_L a$, most of them based on the absorption of O_2 or CO_2 in water or in aqueous solutions, and, as was discussed in Chapter 7, there are wide differences in the $k_L a$ values and in the exponents assigned to the dimensionless groups. The following correlation of Van't Riet [8] for gas absorption in low-viscosity liquids was recommended in Eq. (7.75), and it should be applicable to slurry reactors if the catalyst concentration is very low and does not affect the bubble size. For O_2/H_2O,

$$k_L a = k_b a_b = 0.04 \left(\frac{P}{V}\right)^{0.4} u_0^{0.5} \qquad \text{in sec}^{-1} \tag{8.7}$$

where

$$\frac{P}{V} = \text{kW/m}^3$$

$$u_0 = \text{cm/sec}$$

For hydrogen absorption, Eq. (8.7) must be adjusted for the effect of diffusivity. Limited data show that the diffusion coefficient of hydrogen in low-viscosity organic liquids is 3–10 times the diffusivity of oxygen in water, or 2- to 5-fold greater than predicted by the widely used Wilke–Chang equation [9] [see Eq. (4.17)]. The effects of viscosity and solvent molecular weight on D_{H_2} are uncertain, but pending further data, the Wilke–Chang equation is recommended with a correction factor of 3 for H_2. With a measured or a predicted diffusivity, Eq. (8.7) can be corrected for hydrogen absorption using the penetration theory: For H_2,

$$k_b a_b = 0.04 \left(\frac{P}{V}\right)^{0.4} (u_0)^{0.5} \left(\frac{D_{H_2-\text{oil}}}{D_{O_2-H_2O}}\right)^{0.5} \tag{8.8}$$

At 25°C,

$$D_{O_2-H_2O} = 2.4 \times 10^{-5} \text{ cm}^2/\text{sec}$$

The predicted absorption coefficients for hydrogen at $u_0 = 1$ cm/sec, assuming $D_{H_2} = 4 \times D_{O_2}$ are plotted as a dashed line in Figure 8.4.

The data points in Figure 8.4 are hydrogen absorption coefficients obtained by three different methods. In the study by Cordova and Harriott [10], methyl linoleate was hydrogenated in a 5-liter reactor, and the gas absorption resistance was determined by extrapolation to $1/m = 0$ on a reciprocal rate plot. The hydrogenation of rapeseed oil was carried out in 30-, 500-, and 24,000-liter reactors by Bern et al. [7]

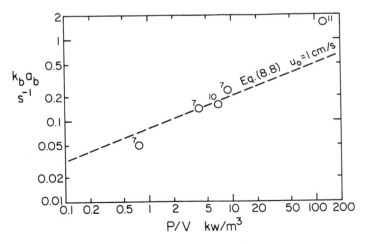

FIGURE 8.4 Gas absorption coefficients for hydrogen in stirred reactors.

using so much catalyst that the gas absorption resistance was controlling. The hydrogenation of ethyl anthroquinone in xylene–octanol solvent was studied by Bergelin and Schöön [11], who made separate desorption tests to measure $k_b a_b$ for their 2-liter reactor. For each reactor, the power per unit volume of liquid was estimated from correlations for turbine agitators with a correction factor for aeration. The correction factor was 0.5 for a single impeller or the sparged impellers on a common shaft. If only the lowest impeller was sparged, a factor of 0.7 was used for the others. The values of P/V in Figure 8.4 are approximate, but they cover such a wide range that accurate values are not needed to show the trend.

Judging from Figure 8.4, the effect of P/V on $k_b a_b$ appears similar to that predicted by Eq. (8.7), though perhaps the exponent on P/V should be higher than 0.4. The true effect of P/V cannot be determined without accounting for differences in diffusivity and gas velocity. The data in Figure 8.4 were not converted to a standard velocity because the inlet and exit velocities were quite different for some tests. When a large fraction of the incoming gas is consumed, as is often the case in hydrogenations, it is not clear whether the exit velocity or some type of average velocity should be used in correlating the data.

One of the main conclusions from Figure 8.4 is that laboratory and pilot-plant reactors generally have gas-absorption coefficients that are much greater than what can be achieved in a large industrial reactor. A power input of 2 kW/m^3, or 10 HP/1000 gal, is very vigorous agitation for a large reactor, but lab reactors can be operated at 50–500 HP/1000 gal. Most laboratory studies have been made with stirrer speeds of 1000–3000 rpm, giving more than 100 HP/1000 gal. The reaction rate is then often independent of stirrer speed and gas velocity, and the solution is saturated with hydrogen. This permits a study of reaction kinetics, if the catalyst mass transfer resistance can be accounted for. However, unless some tests are made under conditions closer to those predicted for a large reactor, the selectivity as well as the overall reaction rate may change on scaleup, since, in the large reactor, the solution will probably have a much lower concentration of hydrogen.

Example 8.1

Soybean oil was hydrogenated at 204°C and 45 psig using nickel catalyst in a stirred reactor. With 0.005% Ni, the iodine value (IV) decreased from 130 to 80 in 26 minutes, and with 0.0125% Ni, it took 17 minutes to reach IV = 80 [12]. Estimate the gas absorption coefficient and the fraction of the overall resistance due to gas absorption.

Solution. The relationship between iodine value and hydrogen consumption is [7]:

$$r = \frac{d(IV)}{dt}(0.039)\rho$$

Assume an oil density of 0.8 g/cm^3:

For run 1, $r_{ave} = \dfrac{130 - 80}{26}(0.039)(0.8)\dfrac{1}{60} = 0.0010\,\dfrac{\text{mole } H_2}{\text{L-sec}}$

For run 2, $r_{ave} = \dfrac{130 - 80}{17}(0.039)(0.8)\dfrac{1}{60} = 0.00153\,\dfrac{\text{mole } H_2}{\text{L-sec}}$

From Figure 8.3, the estimated H_2 solubility is 4×10^{-3} mol/L-atm:

$$P_{H_2} = \frac{45}{14.7} + 1 = 4.06 \text{ atm}$$

$$C_{H_2 i} = 4 \times 10^{-3} \times 4.06 = 0.0162 \text{ mol/L}$$

Since there are only two data points, a plot is not needed, and the weight percent of Ni can be used instead of *m*:

Run 1: $\dfrac{C_i}{r} = \dfrac{0.0162}{0.001} = 16.2 = \dfrac{1}{k_b a_b} + \dfrac{R_{cat}}{0.005}$

Run 2: $\dfrac{C_i}{r} = \dfrac{0.0162}{0.00153} = 10.6 = \dfrac{1}{k_b a_b} + \dfrac{R_{cat}}{0.0125}$

Solving simultaneously,

$$\frac{1}{k_b a_b} = 6.86 = \text{gas-absorption resistance}$$

Relative gas-absorption resistance $= \dfrac{6.86}{16.2} = 42\%$ for run 1

$$= \dfrac{6.86}{10.6} = 65\% \text{ for run 2}$$

$$k_b a_b = \frac{1}{6.86} = 0.146, \text{ or } 0.15 \text{ sec}^{-1}$$

The reactor size and conditions were not given, but 0.15 sec^{-1} is a typical absorption coefficient for a vigorously agitated pilot-plant reactor.

External Mass Transfer

The reciprocal plot separates the catalyst resistance from the overall resistance, but it does not show the relative importance of external mass transfer and internal diffusion plus reaction. If the average particle size is known, a_c

can be calculated from Eq. (8.4), and k_c can be predicted from studies of mass transfer to suspended particles. The slip velocity theory of Harriott [13] states that k_c, the mass transfer coefficient for a suspended particle, is at least somewhat greater than k_c^*, the coefficient for the same particle falling at its terminal velocity in the liquid. A particle suspended in a stirred reactor is subject to frequent acceleration and deceleration, and the ratio k_c/k_c^* depends on the frequency and magnitude of the velocity fluctuations, which in turn depend on the agitator type and the energy dissipation rate.

The minimum coefficient, k_c^*, is obtained from the correlation for single spheres [see Eq. (5.37)] using the terminal velocity to calculate the Reynolds number.

$$Sh^* = 2 + 0.6 \ Re^{1/2} Sc^{1/3} \tag{8.9}$$

where

$$Re = \frac{d_p v_t \rho}{\mu}$$

$$Sc = \frac{\mu}{\rho D}$$

$$Sh^* = \frac{k_c^* d_p}{D}$$

As shown in Figure 8.5, the minimum coefficient is nearly independent of particle size in the range 100–1000 microns. The terminal velocity is approximately proportional to $d_p^{1.0}$ in this range, so Re varies with d_p^2. The Sherwood number increases with about $d_p^{1.0}$, leading to almost no effect of d_p on k_c^*. Many experimental studies have found little or no effect of particle size in this range. However, for very small particles, $Sh^* \cong 2.0$ and k_c^* varies inversely with d_p. Since k_c cannot be less than k_c^*, small particles should show a similar increase in k_c as d_p decreases.

Some mass transfer data for spheres in water and viscous solutions are shown in Figure 8.6. For $d_p = 10 - 100$ microns, k_c varies with $d_p^{-0.7}$, and the ratio k_c/k_c^* is about 2.6 [13]. The mass transfer coefficient is independent of tank size if P/V and D_a/D_t are constant. The coefficients for small particles increase with the 0.1 power of the energy input, and for large particles the exponent is about 1/6. For the tests shown in Figure 8.6, $P/V \cong 1.0$ HP/1000 gal, or 0.2 kW/m^3; for very vigorous agitation in a large tank—say, 10 HP/1000 gal—k_c would be 25–45% greater than at 1 HP/1000 gal.

Density difference has no effect on k_c for $\Delta\rho < 0.4$ g/cm^3, in contrast to the effect on k_c^*. However, at higher values of $\Delta\rho$, k_c increases with about $\Delta\rho^{0.3-0.4}$, matching the change in k_c^*. In applying slip velocity theory to get

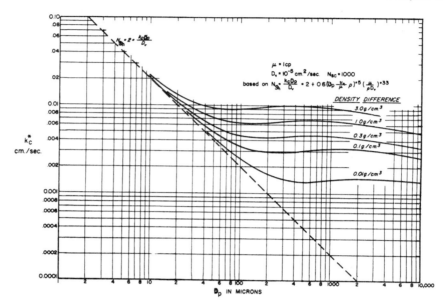

FIGURE 8.5 Mass transfer coefficients for particles falling in water. (From Ref. 13. Reproduced with permission of the American Institute of Chemical Engineers. Copyright 1962 AIChE. All rights reserved.)

k_c^*, a value of 0.3 g/cm^3 should be used for $\Delta\rho$ if the true value is less than this. Other correlations for k_c have been presented based on the energy dissipation rate. Levins and Glastonbury [14] propose the following equation for particles close to neutral buoyancy:

$$\text{Sh} = 2 + 0.47\left(\frac{d_p^{4/3}\varepsilon^{1/3}}{\mu/\rho}\right)^{0.62}\left(\frac{D_a}{D_t}\right)^{0.17}Sc^{0.36} \tag{8.10}$$

The increase in coefficient with increase in agitator diameter at the same power input was also observed by Harriott [13], and it is probably due to the more uniform dissipation of energy with the larger impeller.

Another way to predict k_c utilizes the relationship derived by Batchelor based on turbulence theory [15]. His correlation gives a fairly good fit to two sets of data [13,14], but it does not include the effects of density difference or D_a/D_t ratio. Whatever method is used to predict k_c, the uncertainty in k_c and in a_c should be kept in mind. The product $k_c a_c$ may be known only to $\pm 50\%$, but even a rough estimate may be useful in evaluating kinetic data.

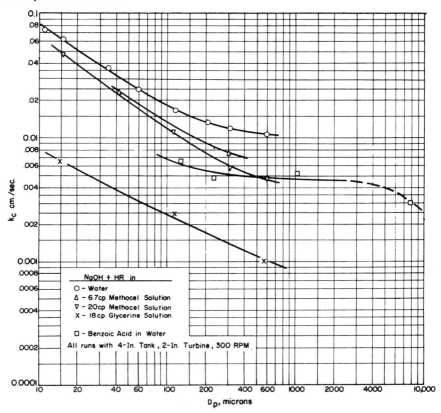

FIGURE 8.6 Effect of particle size on the mass transfer coefficient for particles in a stirred tank. (From Ref. 13. Reproduced with permission of the American Instsitutre of Chemical Engineers. Copyright 1962 AIChE. All rights reserved.)

Changing particle size can provide experimental evidence about the internal and external catalyst resistances. For small particles, $k_c a_c$ varies with $d_p^{-1.7}$, whereas ηk may vary with d_p^0 to d_p^{-1}, depending on the value of the Thiele modulus. Figures 8.7 and 8.8 show data for the hydrogenation of methyl linoleate using Pd-carbon as a catalyst [10]. The two steps in the reaction are the conversion of doubly unsaturated linoleate, L, to oleate, O, and the reaction of oleate to saturated methyl stearate, S:

$$L + H_2 \xrightarrow{1} O \qquad\qquad (8.11)$$

$$O + H_2 \xrightarrow{2} S \qquad\qquad (8.12)$$

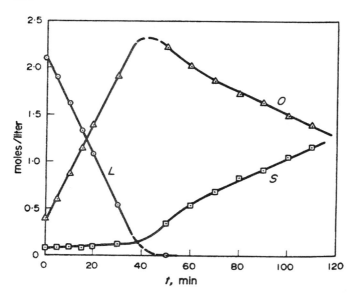

FIGURE 8.7 Hydrogenation of methyl linoleate with catalyst size 1 at 121°C. (From Ref. 10 with permission from Elsevier Science.)

Tests were made with the catalyst as received and with fractions having average sizes from 12 to 93 microns. A run with the smallest size is shown in Figure 8.7. The fact that L decreases at a constant rate does not prove anything about the intrinsic kinetics, since zero-order behavior could be caused by hydrogen mass transfer control as well as by strong chemisorption of linoleate. Runs with other catalyst sizes and concentrations led to the reciprocal plot in Figure 8.8. The data for all sizes converge to the same intercept, as expected, since the gas-absorption resistance should not depend on the particle size. The slopes of the lines increase with particle size, because of increases in both internal and external resistance. Calculations using slip velocity theory showed that the external resistance was about 10% of the catalyst resistance for the smallest size and 50% for the largest size.

Note that the runs with Pd/carbon catalyst were made with less than 1.0 grams of catalyst per liter, and this is typical of slurry hydrogenations. The initial reaction rate from Figure 8.7 is 7.5 moles/hr, g cat, which is orders of magnitude greater than for most reactions carried out in fixed beds of catalyst.

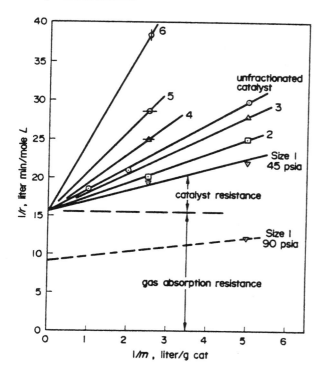

FIGURE 8.8 Determination of bubble and catalyst resistances for methyl linoleate hydrogenation at 121°C. (From Ref. 10 with permission from Elsevier Science.)

Example 8.2

The palladium-catalyzed hydrogenation of nitrobenzene in methanol at 1 atm and 30°C was studied by Acres and Cooper [16]. The rate was proportional to the amount of catalyst used and independent of the nitrobenzene concentration in the range 0–20% nitrobenzene. The reaction rate was 2.4 L H_2/min, g cat, and the average particle size of the Pd/C catalyst was estimated to be 10 μm. Estimate the effect of external mass transfer for these conditions.

Solution. Assume D_{H_2} is three times the value given by the Wilke–Chang equation:

$$D_{H_2} = 3 \times \frac{7.4 \times 10^{-8}(\chi M_B)^{0.5}T}{\mu V_A^{0.6}}$$

for methanol: $\chi = 1.9,$ $M_B = 32,$ $\rho = 0.79$ g/cm^3, $\mu = 0.52$ cP

for hydrogen: $V_A = 14.3$

$$D_{H_2} = 3 \times \frac{7.4 \times 10^{-8}(1.9 \times 32)^{0.5} 303}{0.52 \times (14.3)^{0.6}} = 2.04 \times 10^{-4} \text{ cm}^2/\text{sec}$$

$$Sc = \frac{\mu}{\rho D} = \frac{0.0052}{0.79(2.04 \times 10^{-4})} = 32$$

Assuming dry $\rho_{cat} = 1.2$ g/cm^3 (40% porosity with $\rho_s = 2$), in methanol

$$\rho_{cat} = 0.6(2) + 0.4(.79) = 1.516 \text{ g/cm}^3 \qquad \text{(pores full of methanol)}$$

$$\Delta\rho = \rho_{cat} - \rho = 1.516 - 0.79 = 0.726 \text{ g/cm}^3$$

From Stokes' law [see Eq. (9.4)],

$$v_t = \frac{980(10^{-3})^2(0.726)}{18(0.0052)} = 7.6 \times 10^{-3} \text{ cm/sec}$$

$$Re = \frac{10^{-3}(7.6 \times 10^{-3})(0.79)}{0.0052} = 1.15 \times 10^{-3}$$

From Eq. (8.9),

$$Sh^* = 2 + 0.6(1.15 \times 10^{-3})^{0.5}(32)^{1/3} = 2.06$$

$$k_c^* = \frac{2.06(2.04 \times 10^{-4})}{10^{-3}} = 0.42 \text{ cm/sec}$$

With vigorous agitation,

$$k_c \cong 2k_c^* = 0.84 \text{ cm/sec}$$

From Eq. (8.4) with $\lambda \cong 1.3$,

$$a_c = \frac{6(1.3)}{10^{-3}(1.2)} = 6500 \text{ cm}^2/\text{g}$$

$$k_c a_c = 0.84(6500) = 5460 \frac{\text{cm}^3}{\text{sec-g}} \quad \text{or} \quad \frac{\text{mol}}{\text{sec-g mol/cm}^3}$$

For a measured rate of 2.4L H_2/min-gcat

$$r = \frac{2.4}{22.4} \times \frac{1}{60} = 1.79 \times 10^{-3} \frac{\text{mol}}{\text{sec-gcat}}$$

$$r = k_c a_c (\Delta C_{\text{ext}})$$

$$\Delta C_{\text{ext}} = \frac{1.79 \times 10^{-3}}{5460} = 3.3 \times 10^{-7} \text{ mol/cm}^3 = 3.3 \times 10^{-4} \text{ mol/L}$$

From Figure 8.3, the $C_{H_2 i} \cong 4.1 \times 10^{-3}$ mol/L

$$\frac{\Delta C_{\text{ext}}}{C_{H_2 i}} = 0.08$$

The external mass transfer resistance is about 8% of the overall resistance, which is barely significant. However, for 20-μm particles, a_c and k_c would each be half as great, and external mass transfer would be important.

Selectivity

With consecutive reactions that are affected by mass transfer rates, the selectivity depends on reaction conditions and may change on scaleup. In hydrogenation of vegetable oils, it is desirable to convert all multiple double bonds in the triglycerides to monoenes without forming many saturated species. For the methyl linoleate system [Eqs. (8.11), (8.12)], the selectivity is defined as the relative rates of reactions 1 and 2 corrected to the same concentration:

$$S = \frac{r_1/L}{r_2/O} = \frac{r_1}{r_2} \times \frac{O}{L} \tag{8.13}$$

With the palladium catalyst, the selectivity was quite high, as Figure 8.7 shows, and little stearate was formed until the linoleate was completely reacted. The selectivity ranged from 30 to 60, changing with reaction conditions.

A more detailed model was developed to show the effects of reaction conditions on selectivity [17]. The model was based on simultaneous diffusion and reaction of H_2, L, and O inside the particles and on competitive adsorption of L and O on the surface. Allowing for hindered diffusion in small pores, the effective diffusivities for L and O were estimated to be 100-fold lower than D_e for H_2. This made the internal gradients for L and O appreciable, even though the solution concentrations were very much greater than the hydrogen concentration. Typical gradients are sketched in Figure 8.9. The gradient for L has little effect on the reaction rate, because L is strongly chemisorbed on

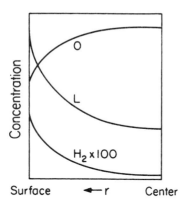

Surface   ◄─ r   Center

FIGURE **8.9** Internal concentration profiles for catalytic hydrogenation of methyl linoleate.

the catalyst, and the reaction is almost zero order to L. However, the average L/O ratio in the catalyst is about half the ratio at the surface, which reduces the selectivity by 50%.

With the Pd/carbon catalyst, the selectivity increased with increasing catalyst charge, which was attributed mainly to the change in hydrogen concentration. When both gas-absorption and catalyst resistances are important, an increase in the catalyst concentration raises the overall rate somewhat but lowers the dissolved hydrogen level. Each catalyst particle then has a lower reaction rate and smaller gradients for L and O, which leads to a higher selectivity. Another effect of lower reaction rate per gram of catalyst is that the surface is closer to equilibrium coverage by reacting species. At steady state, the rate of chemisorption of reactant L equals the rate of desorption plus the rate of reaction, so the surface coverage is less than the equilibrium value for the reaction. The opposite is true for the product O, which has a higher coverage than the equilibrium value. Allowing for the departure from equilibrium coverage and the gradients due to internal diffusion led to a good fit for the selectivities observed with the palladium catalyst. For hydrogenation of edible oils, nickel catalysts are generally used, but the same trend of increased selectivity with increased catalyst loading or decreased hydrogen concentration is observed [6]. Decreased internal gradients of reactants seems the likely explanation, since the effective diffusivities of the triglycerides in small pores are even lower than those of methyl linoleate.

Industrial Oil Hydrogenation

Hydrogenation of edible oils and fats is normally carried out batchwise in a tank equipped with cooling coils. There may be two or three agitators on a single shaft, as shown in Figure 7.14. Hydrogen is added through a sparger ring below the bottom agitator or below the two lower agitators, which are flat-blade turbines. The upper agitator is a pitched-blade turbine or an axial-flow mixer, to promote vertical circulation and entrain hydrogen from the gas space. Some reactors are operated dead-end, with hydrogen added only to maintain pressure. However, a higher rate of gas absorption is obtained by feeding excess hydrogen and recycling the vent gas.

The pressure for oil hydrogenation is usually 2–5 atmospheres and the temperature 120–200°C. Finely divided Ni/SiO_2 or Raney nickel is the catalyst, and the concentration is 0.01–0.5 wt%. The amount of catalyst used is often large enough to make gas absorption the limiting step. The nickel catalyst is not very expensive, and it is removed by filtration and discarded.

The progress of the reaction is followed by monitoring the hydrogen consumption or the iodine number, a measure of the concentration of double bounds. The hydrogenation is rarely carried to completion, since the goal is to make a product with a certain melting point range, a moderate degree of unsaturation, and a desirable distribution of the cis-trans isomers. The reaction order, based on the change in iodine number, has been reported as 0, $1/2$, 1, or greater than 1, and the range of values is due in part to different operating conditions. When hydrogen absorption is controlling, the reaction will appear zero order. When the catalyst resistance controls, external and internal diffusion effects make it difficult to determine the intrinsic kinetics. Also, the triply and doubly unsaturated branches of the triglycerides in the oil react faster than the monoenes, so the reaction order based on iodine value is not the same as that for individual species.

FIXED-BED REACTORS

Fixed-bed reactors are selected for solid-catalyzed gas–liquid reactions that are relatively slow and require a large amount of catalyst compared to that used in slurry reactors. One advantage of fixed-bed operation is that the catalyst particles are held in the reactors and don't need to be separated from the liquid by filtration. This makes it easier to operate continuously. Another advantage is the absence of an agitator, which permits operation at very high pressures in reactors with large L/D ratios. Finally, with a fixed bed of catalyst, high conversion of one or both reactants is easier to achieve, because the flows of the gas and liquid through the reactor are close to ideal plug flow.

Disadvantages of fixed beds include lower effectiveness factors because of the larger particle size and lower coefficients for gas–liquid and liquid–solid mass transfer. Fixed beds with two fluid phases are also difficult to scale up or scale down because of incomplete wetting and changes in gas and liquid flow distribution. Finally, there is a risk of temperature excursions with exothermic reactions in packed beds, since radial heat transfer is poor. Large reactors are often operated adiabatically, but hot spots may occur because of uneven flow distribution.

There are three ways of operating a multiphase fixed-bed reactor. The two fluid phases can flow downward in parallel or upward in parallel, or the flows can be countercurrent. The most common method is parallel down-flow, as shown in Figure 8.10a, and reactors of this type are called *trickle beds*. The name originated from laboratory studies using small reactors and very low flow rates. The liquid was observed to trickle in rivulets over pieces of partially wetted packing, and portions of the liquid seemed stagnant. At low flows, the liquid holdup is small and the gas phase is continuous. Tall, large-diameter reactors operate at high liquid and gas velocities with flows that are more turbulent, but the reactors are still classified as trickle beds if both phases flow downward.

When gas and liquid are passed upward through the bed, as in Figure 8.10b, the bed is flooded and gas bubbles are dispersed in the upflow-ing liquid. This method of operation is not often used because of the higher pressure drop for gas flow, even though it may give a higher degree of wetting of the catalyst surface. With a high ratio of gas flow to liquid flow, the gas phase may become continuous, with liquid entrained as small droplets of spray.

The third method of operation, counterflow of gas and liquid, as in Figure 8.10c, is hardly ever used for catalytic reactions. For any of the three methods of operation, the catalyst particles must be quite small, often a few millimeters in diameter, to have reasonable effectiveness factors. However, with counterflow operation, the flow rates are limited by flooding, and the flooding rates are much lower than for gas absorp-tion or stripping columns, where packings several centimeters in diameter are common. Since hydrogenations and most other reactions carried out in trickle beds are irreversible, the driving force for mass transfer and reaction is the same for parallel or countercurrent flow, and there is no inherent advantage of counterflow operation. Therefore, parallel-flow operation is usually selected to permit operation over a wide range of flow rates.

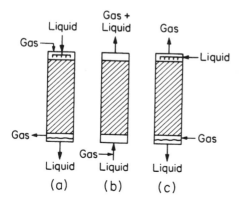

FIGURE 8.10   Fixed-bed reactors for gas–liquid–solid systems: (a) Trickle bed; (b) upflow flooded; (c) counterflow.

Trickle-Bed Reactors

Trickle-bed reactors are widely used for hydrogenations in the petroleum industry, including hydrodesulfurization (HDS) of heavy oils and gasoline, hydrodenitrogenation (HDN), hydrocracking, and hydrofinishing of lubricating oil [18]. A great deal of the published work on trickle beds has been directed at understanding and improving the operation of these processes. Trickle beds are also used for some other chemical processes, including the hydrogenation of glucose to make sorbitol, the hydrogenation of methyl styrene to cumene, the selective removal of acetylenic compounds from olefins, and the hydrogenation of ethyl anthraquinone, a step in the synthesis of hydrogen peroxide.

There have been many studies of the hydrodynamics of trickle beds that describe the different flow regimes and give empirical correlations for the pressure drop, liquid holdup, and the partial wetting of the catalyst. Only a few of these studies are discussed here, since extensive reviews are available [18–21]. A recent review [20] includes over 170 references.

Flow Regimes

When gas and liquid flow downward through a bed of solids, the flow regime may be trickle flow, pulsing flow, bubble flow, or spray flow, depending on the flow rates and properties of the fluids and solid. At low gas and liquid rates, the gas phase is continuous and liquid flows in a thin laminar film over wetted portions of the particle surface. This is called the *gas-continuous region* or the *trickle-flow region*. As either the gas or liquid flow is increased, ripples form on the liquid surface, and occasional pulses of liquid may be observed. In early work [22] this was called the *transition region*, but in recent studies this brief transition is ignored and the next regime is called *pulsing flow*. At high flow rates, liquid passes through the column as a series of pulses, which are zones of high liquid holdup extending across the column diameter in small columns. The pulse frequency ranges from 2 to 6 sec^{-1}, depending on liquid flow rate and particle size. At very high liquid flow and low gas flow, the liquid may become the continuous phase, with gas dispersed as fine bubbles in the liquid. At high gas and low liquid flow, shear at the interface may form small drops of liquid that are carried along in the gas in the *spray-flow* regime.

Different flow regimes were described in early work by Weekman and Myers [22], who passed air and water downward through beds of glass beads or catalyst spheres. They presented the results as a flow map, an arithmetic plot of gas mass velocity versus liquid mass velocity, with lines marking regime boundaries. Results from a similar study by Tosun [23] are shown in Figure 8.11, where a log-log plot is used. Other workers have used liquid and gas superficial velocities or Reynolds numbers as the coordinates on the flow maps.

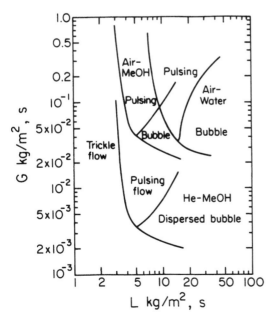

FIGURE 8.11 Flow map for trickle-bed reactors. (After Ref. 26.)

In Tosun's work, the gas was air or helium, and water, methanol, or glycerine solutions were used to show the effect of liquid properties. The main interest is in the transition from trickle flow to pulsing flow, which occurs as L increases or as both L and G increase at a constant ratio. With methanol, the transition occurs at a liquid rate about half that for water, which can be attributed to the threefold difference in surface tension (26 vs 72 dyne/cm). Liquids with low surface tension will wet more of the surface and have a greater holdup, leaving less space for gas flow. Viscous liquids will also have greater holdup and an earlier transition to pulsing flow (data not included in Figure 8.11).

The flow map for helium–methanol is similar to that for air–methanol but displaced to 10-fold-lower gas rates. This shows that the velocity of the gas is more important than the mass flow rate. The plot for helium would be close to that for air if based on linear velocity. In many high-pressure reactors, such as HDS or HDN reactors, the gas density is several times that of air at STP; although some data are available, the effect of high gas density on the flow transitions is still uncertain [20,23].

The combined effects of liquid and gas properties, flow rates, and particle size were included in generalized flow maps by Talmor [24] and Charpentier and Fauvier [25]. Separate correlations were made for foaming and nonfoaming liquids. However, the data show considerable scatter, and no single approach to predicting the transition flows can be recommended. Many of the studies were made using glass beads or catalyst spheres a few millimeters in diameter; more data are needed for finer crushed catalyst or for 1/32- or 1/16-in-diameter catalyst extrudates.

Pressure Drop

The pressure drop for concurrent downflow of gas and liquid in a packed bed can be predicted using correlations of the Lockhart–Martinelli type [22]. The pressure drop for each phase flowing separately through the bed is calculated using the Ergun equation [Eq. (3.64)], and these values define a parameter χ:

$$\chi = \frac{(\Delta P/\Delta L)_L}{(\Delta P/\Delta L)_G}^{1/2} \tag{8.14}$$

The pressure drop for gas–liquid flow relative to that for liquid alone is then found from an empirical relationship:

$$\phi_L = [(\Delta P/\Delta L)_{LG}/(\Delta P/\Delta L)_L]^{1/2} = f(\chi) \tag{8.15}$$

Several empirical correlations show ϕ_L decreasing with increasing χ, and a regression analysis by Tosun [26] led to the following equation of the Midoux [27] type:

$$\phi_L = 1 + \frac{1}{\chi} + \frac{1.424}{\chi^{0.576}} \tag{8.16}$$

The data for Eq. (8.16) cover a range of ϕ_L values from 1.1 to 5 or pressure drops up to 25 times greater than for liquid flow alone. The mean error in predicted ϕ_L was 18%, but the mean error in ΔP was larger. More accurate correlations might be developed based on equations for the liquid holdup, since this determines the available space for gas flow.

Liquid Holdup

Liquid holdup, which is expressed as the volume of liquid per unit volume of bed, affects the pressure drop, the catalyst wetting efficiency, and the transition from trickle flow to pulsing flow. It can also have a major effect on the reaction rate and selectivity, as will be explained later. The total holdup, h_t, consists of static holdup, h_s, liquid that remains in the bed after flow is stopped, and dynamic holdup, h_d, which is liquid flowing in thin films over part of the surface. The static holdup includes liquid in the pores of the catalyst and stagnant packets of liquid held in crevices between adjacent particles. With most catalysts, the pores are full of liquid because of capillary action, and the internal holdup is the particle porosity times the volume fraction particles in the bed. Thus the internal holdup is typically $(0.3 - 0.5)(0.6)$, or about 0.2–0.3. The external static holdup is about 0.03–0.09 for 0.3-cm particles and up to about 0.10 for particles 0.05–0.10 cm in size.

The dynamic holdup depends mainly on the particle size and the flow rate and physical properties of the liquid. For laminar flow, the average film thickness is predicted to vary with $L^{1/3}$, as in flow down a wetted-wall column or an inclined plane. In experiments with water in a string-of-spheres column, where the entire surface was wetted, the holdup did agree with theory [28]. For randomly packed beds, the dynamic holdup usually varies with a fractional power of the flow rate, but the reported exponents range from 0.3 to 0.8, and occasionally agreement with the 1/3 power predicted by theory may be fortuitous.

Dynamic holdup data from a few sources are shown in Figure 8.12. The lowest two lines are for water in beds of glass beads 0.48 cm in diameter [29] or 0.41 cm in diameter [30]. The holdup increases with $L^{0.7-0.8}$, because raising the flow rate increases the fraction of the packing wetted as well as the film thickness. The middle line shows the data of Ross [31] for 0.48-cm

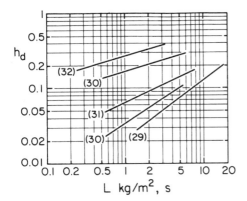

F_{IGURE} **8.12** Dynamic holdup in trickle beds.

catalyst particles, and h_d varies with $L^{0.5}$. The greater holdup than for glass beads of the same size is probably due to better wetting of the porous particles. The upper two lines are for quite small catalyst particles, 0.054 cm in diameter [30] or 0.071 cm in diameter [32]. The holdup is two to four times greater than for the larger particles, and it increases with $L^{1/3}$. The apparent agreement with laminar flow theory is a coincidence, since the surface is not completely wetted. The data of Schwartz et al. [32] are for hexane and Al_2O_3, and those of Goto and Smith [30] are for water and CuO/ZnO. The higher holdup for hexane is probably due to the lower surface tension, since the lower viscosity would tend to decrease the dynamic holdup.

The gas properties have no effect on liquid holdup at low pressure and low gas rates, when the liquid flow is affected only by gravity forces. At high gas velocity the holdup decreases because of shear at the gas–liquid interface. Several correlations have been proposed to account for the effects of liquid and gas properties on holdup, but these correlations are complex and quite different in form [20], which makes comparisons difficult. Furthermore, most of the data are from studies at ambient conditions using water or low-molecular-weight solvents. More data are needed from reactors operating at industrial conditions.

Wetted Area

In the trickle-flow regime, only part of the catalyst surface is covered by a film of liquid, and it is sometimes assumed that only this part of the catalyst is effective. Correlations for the fraction wetted area, a_w/a_i, have been used to interpret trickle-bed reaction data and to predict trends. However, correlations for a_w/a_i for large particles of packing may not be valid for small catalyst pellets, and it is also incorrect to assume that the reaction rate is directly proportional to the wetted surface. Consider possible concentration profiles for a partially wetted particle in a hydrogenation reactor, as shown in Figure 8.13a. The pores are full of liquid, so reactant B can diffuse through the particles to the "dry" side. If the concentration gradients are moderate, as shown by the solid lines in this example, and the reaction is first order to B, then the average reaction rate in the dry half of the pellet would be somewhat less than in the wetted half. Reaction would take place throughout the pellet, and the effectiveness factor would be slightly less than for complete wetting. However, if the reaction was approximately zero order to B because of strong chemisorption, the average rate could be nearly equal to or even greater that for complete wetting, because the absence of an external liquid film leads to a higher hydrogen concentration at the dry surface.

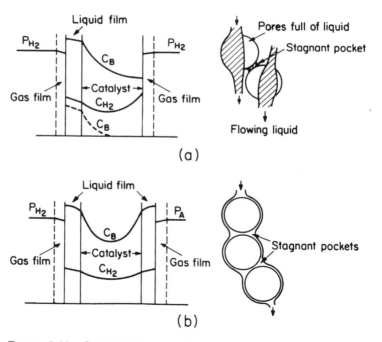

FIGURE 8.13 Concentration profiles in catalyst particles: (a) a partially wetted particle; (b) a fully wetted particle (solid line: medium reaction rate; dashed line: very fast reaction).

The dotted line in Figure 8.13a shows the gradient for B when the reaction rate is very fast and C_B falls to zero inside the pellet. There is no reaction near the dry side, and the observed rate would increase with the fraction wetted, perhaps with L^n. However, because the internal diffusion of B is in all directions, and not just normal to the surface, B will spread over a broader region than just below the wetted area. This makes the fraction of catalyst utilized greater than the wetted fraction and adds to the difficulty of developing a detailed reactor model.

When the catalyst is covered by flowing liquid, as in Figure 8.13b, the concentration profiles across the particle might be symmetrical, and the local reaction rate could be estimated from the intrinsic rate, the external mass transfer coefficient, and the Thiele modulus. However, a lower rate is expected where the particles almost touch because the thicker liquid layer increases the mass transfer resistance.

External Mass Transfer

The external mass transfer coefficients can be predicted using stagnant-film theory, penetration theory, or complex correlations with several dimensionless groups [20]. Since the external resistances are usually small for industrial reactors, simple film theory is probably satisfactory. Mass transfer of the gas might involve two steps in series, as in slurry reactors, where gas diffuses into bulk liquid and then diffuses to the catalyst surface. However, the liquid film in trickle-bed reactors is very thin, and a single gradient is shown in Figure 8.13. The average film thickness, z, can be calculated from the dynamic holdup and the wetted area, and the coefficient for hydrogen is then

$$k_{gl} = \frac{D_{H_2}}{z} \qquad (8.17)$$

Since B is already in the film, the effective diffusion distance is half the film thickness, so the coefficient for B is

$$k_{ls} = \frac{2D_B}{z} \qquad (8.18)$$

Because D_{H_2} is several times D_B, the coefficient for B is much less than for H_2.

Reactor Models

In kinetic models for trickle beds, the reaction is often assumed to be first order to both reactants, but attention is focused on the liquid reactant, since the gas concentration doesn't change very much in the reactor. For HDS, HDN, and other high-pressure purification processes, the hydrogen concentration in the liquid is often as great as or greater than the hydrocarbon concentration, in contrast to slurry hydrogenations, where the dissolved hydrogen concentration is much lower than that of the other reactant.

For the ideal case of plug flow and completely wetted catalyst, the conversion for a first-order reaction is given by the same equation used for gas–solid reactions in Chapter 3:

$$FC_0 \, dx = k\eta C_0(1 - x) \, dW \tag{8.19}$$

$$\ln\left(\frac{1}{1 - x}\right) = \frac{k\eta W}{F} = \frac{k\eta V \rho_b}{F} \tag{8.20}$$

In petroleum processing, the conversion may be given as a function of the liquid hourly space velocity (LHSV), and the apparent rate constant, k_{app}, includes the effect of partial wetting as well as the effect of internal concentration gradients:

$$\ln\left(\frac{1}{1 - x}\right) = \frac{k_{app}\rho_b}{\text{LHSV}} \tag{8.21}$$

where

$$\text{LHSV} = \frac{F}{V} = \frac{L/\rho}{l} \tag{8.22}$$

The reciprocal of LHSV has the units of time, but it is not the average residence time, since the liquid occupies only a fraction of the bed volume.

When conversion data for HDS or HDN processes are analyzed, the semilog plot of $1/(1 - x)$ versus $1/\text{LHSV}$ suggested by Eq. (8.21) usually results in an upward-curving line. Figure 8.14a shows some data of Ross [31] for sulfur removal in a pilot-plant reactor. The weight hourly space velocity used (WHSV) here differs from LHSV by the factor ρ/ρ_{bed}. The curvature could be due in part to a reaction order higher than 1.0 or to a distribution of reactivities in the sulfur compounds. The main reason is probably a change in wetted area with flow rate, since the fraction of surface wetted was probably small at the low flow rates that were used. If the rate constant k_{app} in Eq. (8.21) is assumed to depend on a fractional power of the liquid flow rate L, as in $k_{app} = k'L^n$, Eq. (8.22) can be modified to show a fractional dependence on LHSV and on the bed length l:

$$\ln\left(\frac{1}{1 - x}\right) = \frac{k'l^n}{(\text{LHSV})^{1-n}} \tag{8.23}$$

Choosing $n = 0.4$ gives a reasonable fit to the data, as shown in Figure 8.14b, though the fit is about as good for $n = 0.3$ or 0.5.

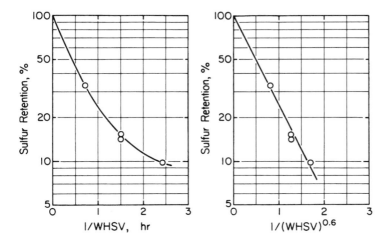

FIGURE 8.14 Sulfur removal in a pilot-plant reactor. (Data from Ref. 31.)

Similar results were found by Henry and Gilbert [33], who studied sulfur removal, nitrogen removal, and hydrocracking in small reactors. Most of the first-order plots were curved upward when LHSV^{-1} was used, but straight lines were obtained with LHSV$^{-2/3}$. The explanation proposed was that the reaction rate was directly proportional to the dynamic holdup, which was predicted to increase with $L^{1/3}$, following laminar-flow theory. Their recommended correlation in simplified form is

$$\ln\left(\frac{1}{1-x}\right) = \alpha(\text{LHSV})^{-2/3} l^{1/3} d_p^{-2/3} v^{1/3} \tag{8.24}$$

There is no fundamental basis for Eq. (8.24), since reaction occurs on the catalyst surface and not in the liquid phase. Also, the laminar-flow holdup theory is not generally valid for packed beds, as was shown by Figure 8.12. Equation (8.24) may be satisfactory for correlating some sets of laboratory data, but it is likely to have considerable error if used for other systems or for large changes in l, d_p, or LHSV.

A different model and alternate explanation for the (LHSV)$^{-2/3}$ term was given by Mears [34], who said the rate should be proportional to the wetted area and used the empirical correlation of Puranik and Vogelpohl for a_w/a_t [35]. This correlation was based on data for gas-absorption packings, and the fraction wetted area was given as a function of Reynolds number, the Weber number, and a surface tension ratio:

$$\frac{a_w}{a_t} = 1.05 \text{Re}^{0.047} \text{We}^{0.135} \left(\frac{\sigma_c}{\sigma}\right)^{0.206} \tag{8.25}$$

Combining terms and rounding off exponents gives the following result:

$$\frac{a_w}{a_t} \propto L^{0.32} d_p^{0.18} \mu^{-0.05} \sigma^{-0.34} \rho^{-0.13} \tag{8.26}$$

Then, since a_t varies inversely with d_p and k is assumed proportional to a_w, the predicted effects of major variables are

$$\ln\left(\frac{1}{1-x}\right) = \alpha(\text{LHSV})^{-0.68} l^{0.32} d_p^{-0.82} \tag{8.27}$$

The exponents for LHSV and l are almost the same as in Eq. (8.24), but basing the derivation on wetted area is more logical than using liquid holdup. However, as discussed earlier, the fraction of catalyst utilized can be greater than or less than the fraction wetted, depending on the relative rates of diffusion and reaction. Furthermore, Eq. (8.25) is based on data for packings 1 cm and larger and is probably not valid for small (0.1–0.3 cm) catalyst particles. Some data on holdup and wetted area for small particles are available, and empirical correlations for liquid–solid contacting efficiency have been presented [36,20]. The value of these correlation for predicting the performance of large trickle-bed reactors has not been demonstrated.

Scaleup

A major problem with Eqs. (8.24) and (8.27) is the predicted dependence of $\ln(1/1 - x)$ on $l^{1/3}$. There is no intrinsic dependence on l, and the effect actually comes from a change in liquid rate. Laboratory reactors are relatively short, and they are operated at low liquid velocities to get the desired high conversion. If the plant reactor is much taller and L is increased to get the same LHSV, the wetted area will increase, and the conversion should also increase, but perhaps not as much as predicted by Eq. (8.24). For example, consider a 1-m lab reactor where 90% conversion is obtained at low values of L and G. If an industrial reactor 20 m tall is planned and LHSV is constant, using Eq. (8.24) indicates an increase in effective rate constant of $20^{1/3}$, or 2.71, corresponding to a plant conversion of 99.8%. However, if a_w/a_t is 0.5 or higher, the wetted area can't increase by a factor of 2.7, and the actual change in effective rate constant is uncertain. The ratio a_w/a_t and the ratio k_{app}/k^* (where $k^* = \eta k$, the rate constant for complete wetting) both generally increase with L, but at decreasing rate as complete wetting is approached. If the change in k_{app} could be accurately predicted, the large reactor could be designed for a higher LHSV, by either decreasing the height or reducing the diameter. However, to be safe, the reactor might be designed for the same LHSV and any higher conversion accepted as a bonus.

To reduce the uncertainty, scaleup might be done in stages. Perhaps a 1-m lab reactor would be followed by a 4-m pilot plant before designing a 20-m reactor. The possible changes in reaction conditions are sketched in Figure 8.15. The increase in L and G move the operation from the trickle-flow regime closer to the pulsing regime. Many industrial reactors are reported to operate near the transition or in the pulsing regime, but performance data for large reactors are rarely published. A paper by Ross [31] does give some results for sulfur removal in a large reactor and a pilot-plant unit. Surprisingly, the conversion was significantly lower in the 6.5-ft × 63-ft commercial reactor than in the 0.18-ft × 13-ft pilot unit. Residence time measurements showed abnormally low liquid holdup in the large reactor, which was attributed to poor liquid distribution. It is difficult to get uniform distribution of liquid in a large reactor, and the distribution may change along the reactor length, as is known to happen in gas-absorption columns.

Pilot-plant tests are very expensive, and other approaches to the scaleup problem should be considered. One method is to determine the apparent rate constant for completely wetted catalyst by using an upflow flooded reactor with no gas phase present, as suggested by Baker [37]. The liquid is presaturated with hydrogen in a separate contactor, and if only

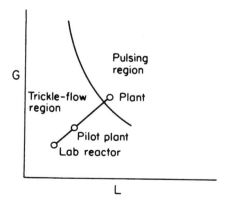

Figure 8.15 Transition to pulsing flow in a trickle-bed reactor.

part of the hydrogen is used, the drop in hydrogen concentration can be corrected for. Assuming a pseudo-first-order reaction (which should be checked), the value of k^* is calculated using Eq. (8.21) and compared to k_{app}, the value from trickle-bed tests. If the ratio k_{app}/k^* is quite low, a large increase in catalyst effectiveness should be possible on increasing the liquid rate. However, if k_{app}/k^* is 0.8 or higher, there may be no improvement in performance on scaleup to higher liquid rates, particularly when the difficulty of getting good liquid distribution in a large reactor is considered. This scaleup method is based on the assumption that the gas-absorption resistance is negligible, which is often true for high-pressure hydrotreating reactions.

A second approach to more accurate scaleup is to dilute the catalyst in the lab reactor with small, inert particles to get better wetting. This also decreases the effect of axial dispersion by spreading out the catalyst over a greater length of reactor, though this is not a major problem for reactors 1 meter long. Baker [37] found that the apparent rate constants for the diluted trickle bed were the same as or only slightly less than the rate constants from the flooded bed. The use of a bed diluted with fines is also recommended by Al-Dahhan et al. [20] to improve wetting efficiency for small reactors at low flow rates. Care must be taken in packing the bed to get reproducible performance [38,39].

A third approach is to operate the short laboratory reactor at very high liquid and gas rates to match expected conditions in a large plant reactor. The conversion will be quite low, but the kinetics can be studied by making runs at several feed concentrations to simulate conditions at various points in the large reactor. However, with this approach it might be difficult to match the effects of product inhibition (H_2S has a retarding effect on HDS reactions) and gradual aging or poisoning of the catalyst.

Effect of Particle Size

Some early studies of trickle-bed reactors used 3/16-in. or 1/8-in. catalyst pellets, but the trend has been to smaller sizes because of higher reaction rates. Extrudates with diameters of 1/16-in. or 1/32-in. are often used. Shah and Paraskos [40] studied desulfurization and demetallization of crude oil in a small reactor at 400°C and 200 psia. They used 1/32-in. extrudates (790 μm × 2–3 mm) or the same catalyst crushed and sieved to give an average size of 550 μm. The apparent rate constants for the crushed catalyst were greater than for extrudates by a factor of about 1.3 for sulfur removal and 2–3 for vanadium and nickel removal. The higher factor for metal removal indicates a strong pore diffusion limitation. The metals are present in the higher molecular fractions of the crude oil and have lower diffusivities than the sulfur compounds [41]. Further studies of desulfurization showed a large decrease in apparent rate constant for particles larger than 1 mm, but not much change from 1 mm to 500 μm [42]. As expected, the effect of particle size was more pronounced at 420°C than at 400°C.

Some of the observed increase in rate with decreasing particle size may come from increased wetting. The model of Mears, Eq. (8.27), predicts that the reaction rate varies with $d_p^{-0.82}$, but this is without any consideration of pore diffusion limitations. If the Thiele modulus is large and there are strong internal gradients, the effectiveness factor would vary with d_p^{-1}, and if Eq. (8.27) held, the overall rate would vary with $d_p^{-1.82}$. No such strong dependence on particles size has been reported. For smaller particles and moderate rates, the effectiveness factor should be close to 1.0, and it is unlikely that the rate would vary with $d_p^{-0.82}$, as suggested by Eq. (8.27), or with any other constant power of d_p. Because of the unsymmetrical concentration gradients in a partially wetted pellet, it is difficult to predict the exact effect of particle size, but probably the benefit of smaller sizes is due primarily to decreased internal gradients and only partly to an increase in wetted area. Pore diffusion calculations are useful in showing the relative importance of hydrogen and hydrocarbon internal gradients.

Other Trickle-Bed Reactions

Although trickle beds are used primarily for petroleum processing at high temperatures and pressures, many of the published research studies deal with simple reactions at moderate temperatures and pressures. For hydrogenation of pure liquids at 1–3 atm, the concentration of dissolved hydrogen may be orders of magnitude smaller than the hydrocarbon concentration, and mass transfer becomes more important for hydrogen than for the hydrocarbon. A partially wetted catalyst may have a higher rate of reaction than a fully wetted particle because of better mass transfer of hydrogen to the dry surface. The reaction is then said to be gas limited. Also, if the hydrocarbon reactant is quite volatile, mass transfer of hydrocarbon through the gas phase can further increase the reaction rate at dry surfaces.

The hydrogenation of α-methyl styrene to cumene is a popular reaction for laboratory studies [43–45], and it can be either gas limited or liquid limited, depending on reaction conditions. With a 2.5% Pd/Al$_2$O$_3$ catalyst, the reaction was very rapid at 1 atm and 40°C, and the effectiveness factor was less than 0.1 for 0.3-cm particles [43]. Under there conditions, the reaction rate decreased by 25% as the liquid rate was increased about 10-fold. The decrease occurred because the effect of lower hydrogen concentration at the wetted surface was more important than the effect of slightly higher-concentration of α-methyl styrene. However, with a less active catalyst (0.5% Pd), the reaction was no longer gas limited, and the rate increased with increasing liquid flow rate, because the external mass transfer resistance was a smaller part of the overall resistance.

Another example of a gas-limited reaction is the hydrogenation of benzene to cyclohexane over Pt/Al_2O_3 catalyst [46]. At 76°C and 1 atm, the reaction rate decreased 25% as the liquid rate was increased fourfold. In this case, the higher rate on nonwetted surface was due to a combination of higher hydrogen concentration and diffusion of benzene in the vapor phase. Whether a reaction is gas limited or liquid limited in a trickle bed depends on the relative concentrations of gaseous and liquid reactants, the reaction orders, the diffusion coefficients, and the fraction of the surface that is wetted. A reaction that is gas limited at inlet conditions may become liquid limited at high conversion.

Example 8.3

A trickle-bed reactor 2.5 cm in diameter and 60 cm long was used by Baker to study sulfur and nitrogen removal from a heavy oil [37]. The catalyst in the form of 1.5-mm × 4.5-mm extrudates was diluted with an equal volume of 1.0-mm silicon carbide particles. Characteristics of the oil are given in Table 8.1. Test results are given in Table 8.2.

TABLE 8.1 Data for Example 8.3: Oil Properties

Specific gravity	0.8868
% C	86.19
% S	2.04
% N	0.13
MW	374
50% distilled	450°C

TABLE 8.2 Test Results

Run	T, °C	P, atm	LHSV^{-1}	% S removal
1	365	65	0.75 hr	77
2	365	65	1.39 hr	83

a. Assuming a first-order reaction, determine the apparent rate constant for the two runs. Is the difference consistent with Eq. (8.23) or Eq. (8.24)?

b. For the higher rate constant, estimate the internal effectiveness factor based first on the diffusion of sulfur compounds and then on the diffusion of hydrogen. Assume 1.5 moles H_2 consumed per mole S reacted.

Solution. Use Eq. (8.21).

a.   Run 1: $k_{app}\rho_b = \ln\left(\dfrac{1}{0.23}\right) \times \dfrac{1}{0.75}$

$$= 1.96 \dfrac{\text{moles S}}{\text{hr}, l \text{ bed}, \text{ mole S/L}}, \text{ or hr}^{-1}$$

Run 2: $k_{app}\rho_b = \ln\left(\dfrac{1}{0.17}\right) \times \dfrac{1}{1.39} = 1.27 \text{ hr}^{-1}$

Run 1 is at a liquid rate 1.39/0.75, or 1.85-fold, greater than Run 2. If k_{app} varies with L^n, then

$$\dfrac{1.96}{1.27} = 1.54 = 1.85^n$$

$$n = 0.70$$

This is a greater dependence on liquid rate than the 1/3 exponent reported by Henry and Gilbert [33] and others. Some of the apparent effect of flow rate may be due to an error in assuming a first-order reaction. The sulfur compounds vary in reactivity, and the data may be better fitted by a second-order equation.

b. The effective diffusivities are needed to calculate the Thiele modulus, and the following approach gives only rough estimates because experimental data are not available.

Estimate based on diffusion of sulfur compounds: At 365°C, $\rho \cong 0.64 \text{ g/cm}^3$, $\mu/\rho \cong 0.75$, and $\mu \cong 0.5$ cP (From TEMA standards + ESSO databook). Use the Wilke–Chang equation [Eq. (4.17)] to estimate the diffusivity at $T = 638$ K. For CHS compounds, $M \cong 374$, $V_A \cong 374/0.6 = 623$:

$$D_{CHS} = \dfrac{7.4 \times 10^{-8}(374)^{1/2}(638)}{0.5(623)^{0.6}} = 3.84 \times 10^{-5} \text{ cm}^2/\text{sec}$$

In the catalyst, assume $\epsilon/\tau = 0.1$ and $D_{pore}/D_{bulk} = 0.5$ (hindrance factor for large molecules). So

$$D_e = 0.1(0.5)(3.84 \times 10^{-5}) = 1.92 \times 10^{-6} \text{ cm}^2/\text{sec}$$

If the bed is 60% catalyst, 40% voids:

$$k_{app} = \frac{1.96 \text{ mole}}{3600 \text{ sec}}, l \text{ bed, mole S/L} \times$$

$$\frac{1\text{-Lbed}}{0.6\text{-Lcatalyst}} = 9.07 \times 10^{-4} \text{ sec}^{-1}$$

$$\phi_{app} = R\left(\frac{k_{app}}{D_e}\right)^{1/2} = 0.095\left(\frac{9.07 \times 10^{-4}}{1.92 \times 10^{-6}}\right)^{1/2} = 2.06$$

From Figure 4.8, $\eta = 0.74$.

Estimate based on diffusion of hydrogen: For a similar oil at 367°C and 56 atm, the hydrogen solubility was estimated to be 0.48 mol/L [47]. Correcting to 65 atm:

$$C_{H_2} = 0.48\left(\frac{65}{56}\right) = 0.56 \text{ mol/L}$$

$$\text{Initial S conc.} = 640 \text{ g/L} \times 0.0204/32 = 0.41 \text{ mol/L}$$
$$\text{Initial rate} = 1.96 \times 0.41 = 0.80 \text{ mol/hr}, l \text{ bed}$$

For H_2,

$$k_{app} = 1.5 \times \frac{0.80}{3600} \times \frac{1\text{-}l \text{ bed}}{0.6\text{-}l \text{ cat}} \times \frac{1}{0.56} = 9.9 \times 10^{-4} \text{ sec}^{-1}$$

Use Wilke–Chang equation with a factor of 3.0 for H_2:

$$D_{H_2} = \frac{3 \times 7.4 \times 10^{-8}(374)^{1/2}(638)}{0.5(14.3)^{0.6}} = 1.11 \times 10^{-3} \text{ cm}^2/\text{sec}$$

$$D_{eH_2} = 0.1D_{H_2} = 1.11 \times 10^{-4}$$

$$\phi_{app} = 0.095\left(\frac{9.9 \times 10^{-4}}{1.11 \times 10^{-4}}\right)^{1/2} = 0.28$$

The effect of hydrogen gradients is negligible.

NOMENCLATURE

Symbols

a	area per unit volume
a_b	area of bubbles
a_w	wetted area
a_t	total area
a_c	external area per gram of catalyst
C	molar concentration or H_2 concentration
C_i	molar concentration at interface
C_b	molar concentration in bulk liquid
C_o	molar concentration in feed
C_s	molar concentration at catalyst surface
D	diffusivity
D	diameter
D_a	diameter of agitator
D_t	diameter of tank
d_p	particle diameter
F	feed rate
G	gas mass velocity
H	Henry's law constant
h	liquid holdup
h_d	dynamic holdup
h_s	static holdup
h_t	total holdup
IV	iodine value, g per 100 g oil
K	overall coefficient for mass transfer plus reaction
k	reaction rate constant
k_{app}	apparent reaction rate constant
k^*	reaction rate constant for full wetting
k_c	external mass transfer coefficient
k_c^*	external mass transfer coefficient for particle at its terminal velocity
$k_b a_b$	volumetric gas-absorption coefficient
$k_L a$	volumetric coefficient
k_{gl}	mass transfer coefficient for gas to liquid
k_{ls}	mass transfer coefficient for liquid to solid
L	liquid mass velocity, reactor length
LHSV	liquid hourly space velocity, often hr^{-1}
l	reactor length
m	catalyst charge, g/cm^3
N	stirrer speed

n	exponent for velocity effect
P	agitator power, pressure
R	gas constant, resistance
Re	Reynolds number for particle, $d_p u_o \rho/\mu$, or agitator, $ND_a^2\rho/\mu$
r	reaction rate, absorption rate, mass transfer rate
S	selectivity, cross-sectional area
Sc	Schmidt number, $\mu/\rho D$
Sh	Sherwood number, $k_c d_p/D$
Sh*	Sherwood number based on k_c^*
T	absolute temperature
u_o	superficial velocity
V	reactor volume, volume of solution
v_t	terminal velocity of a particle
W	mass of catalyst
We	Weber number, $G^2 d_p/\sigma\rho$
x	conversion
z	film thickness

Greek Letters

η	effectiveness factor
λ	shape factor
υ	kinematic viscosity
μ	viscosity, μw at wall
ε	energy dissipation rate
ρ	density of fluid
ρ_b	density of catalyst bed
ρ_c	density of catalyst
σ	surface tension
ϕ	Thiele modulus
ϕ_L	pressure drop parameter, Eq. (8.15)
χ	parameter in Eq. (8.14)

PROBLEMS

8.1 The nickel-catalyzed hydrogenation of cottonseed oil was studied in a small batch reactor at 115–160°C and 60 psig [48]. Some data for 0.07% Ni catalyst and 1175 rpm are given in Table 8.3.

TABLE 8.3 Data for Problem 8.1

115°C		130°C		145°C		160°C	
IV	t	IV	t	IV	t	IV	t
103.5	0	103.5	0	103.5	0	103.5	0
99	31	100	16	95	7	85	7
80	87	79	47	68	22	68	16
57	150	53	76	52	35	57	25
41	210	37	105	38	55	44	35

IV = idodine value; t = time, in minutes.

a. Plot the data to test for first-order or zero-order kinetics, and discuss the results.
b. What is the apparent activation energy?

8.2 A nickel-catalyzed hydrogenation gives a curved plot when $1/r$ is plotted versus $1/m$. Data are given in Table 8.4.

a. Ignoring the curvature at high values of $1/m$, extrapolate to get the reaction rate when gas absorption is controlling.
b. Assume there is a small amount of poison, m_p, that inactivates the same amount of catalyst for each run. Find a value for m_p that gives a reasonable straight line for a plot of $1/r$ vs $1/(m - m_p)$. What is the estimated reaction rate for gas-absorption control?

8.3 The hydrogenation of soybean oil was studied at 204°C and 45 psig using 0.005% Ni and 0.0125% Ni [12].

TABLE 8.4 Data for Problem 8.2

m, g/L	r
.004	.025
.005	.033
.010	.055
.015	.067
.020	.077

TABLE **8.5** Data for
Problem 8.3

	t, min	
IV	0.005% Ni	0.0125% Ni
130	0	0
120	4	3
100	13	9
80	26	17
60	42	26

a. Plot the data in Table 8.5 to test for a first-order reaction.
b. For the run with 0.0125% Ni catalyst, use the initial rate data
    and then the rate at 50% conversion to determine the fraction of
    the overall resistance due to gas absorption. Why do these results
    differ?

8.4   In their tests of nitrobenzene hydrogenation, Acres and Cooper
[16] found the catalyst resistance was controlling, and the reaction rate was
higher for catalysts with more palladium surface area (measured by CO
chemisorption) (see Table 8.6).
     How would the reaction rate be expected to vary with metal area if the
controlling step was (1) external mass transfer, (2) pore diffusion, and (3)
surface reaction? What do you think is the controlling step?
     8.5   The hydrogenation of α-methylstyrene was studied at 60°C and
50 psia in a 2-inch-diameter reactor packed with 1/8-inch catalyst pellets.
The reactor was operated with downflow of liquid and gas using 3-ft and
6-ft beds.

TABLE **8.6**   Data for Problem 8.4

Metal area, m^2/g	Rate, L H_2/min, g cat
3.1	0.3
9.5	1.2
15	1.3
32	2.9

TABLE 8.7 Data for Problem 8.5

Run	l, ft	LHSV	χ	u_{oL}, ft/hr
1	3	1	0.75	3
2	3	2	0.60	6
3	6	1	0.83	6
4	6	2	0.68	12
5	6	0.5	0.94	3

a. Plot the results in Table 8.7 to test for an assumed first-order reaction, and discuss the trends shown.
b. What LHSV might be used for 95% conversion in a reactor 3 ft in diameter and 30 ft long?

8.6 Hydrogenation is used to remove sulfur from oil with 4% S, and 60% of the sulfur is in aliphatic compounds with a moderate reactivity. The other 40% is in aromatic compounds that are about one-fifth as reactive.

a. If both types follow first-order kinetics, plot the expected conversion as $\ln[1/(1 - x)]$ against $(LHSV)^{-1}$ assuming complete wetting.
b. Would a second-order plot give a better fit?
c. Would using $(LHSV)^{1-n}$ give a good fit?

8.7 The catalytic hydrogenation of nitrobenzene to aniline in a solvent was studied using a stirred semibatch reactor with 25-μm particles of catalyst. The reaction was first order to hydrogen and zero order to nitrobenzene. Tests at 70°C and 90 psia gave the results listed in Table 8.8.

TABLE 8.8 Data for Problem 8.7

W, g/L	r, mol/hr-L
0.5	0.81
1	1.32
3	2.34

a. Predict the rate for $W = 5$ g/L.
b. Calculate the concentration of hydrogen in the bulk liquid for W $= 5$ g/L. The hydrogen solubility is 0.0046 M at 1 atm, and the diffusivity is 4×10^{-5} cm²/sec.
c. Is external mass transfer a significant part of the catalyst resistance?

REFERENCES

1. HS Davis, G Thomson, GS Crandall. J Am Chem Soc 54:2340, 1932.
2. DL Johnson, H Saito, JD Polejes, OA Hougen. AIChEJ 3:411, 1957.
3. SK Shaffer, JM Prausnitz. AIChE J 27:844, 1981.
4 E Brunner. J Chem Eng Data 30:269, 1985.
5. J Wisniak, LF Albright. Ind Eng Chem 53:375, 1961.
6. M Bockisch. Fats and Oils Handbook. Champaign, IL: AOCS Press, 1997, p 579.
7. L Bern, M Hell, NH Schöön. JAOCS 52:391, 1975.
8. K Van't Riet. Ind Eng Chem Proc Des Dev 18:357, 1979.
9. CR Wilke, P Chang. AIChE J 1:264, 1955.
10. WA Cordova, P Harriott. Chem Eng Sci 30:1201, 1975.
11. T Bergelin, , NH Schöön. Ind Eng Chem Proc Des Dev 20:615, 1981.
12. RC Hastert. In: YH Hui, ed. Bailey's Industrial Oil and Fat Products. 5th ed. Vol 4. New York: Wiley, 1996, p 230.
13. P Harriott. AIChE J 8:93, 1962.
14. DM Levins, JR Glastonbury. Trans Instn Chem Engrs 50:132, 1972.
15. G Batchelor. J Fluid Mech 98(Part 3):609, 1980.
16. GJK Acres, BJ Cooper. J Appl Chem Biotechnol 22:769, 1972.
17. K Tsuto, P Harriott, KB Bischoff. Ind Eng Chem Fund 17:19, 1978.
18. CN Satterfield. AIChE J 21:209, 1975.
19. A Gianetto, G Baldi, V Specchia, S Sicardi. AIChE J 24:1087, 1978.
20. MH Al-Dahhan, F Larachi, MP Dudokovic, A Laurent. Ind Eng Chem Res 36:3292, 1997.
21. YT Shah. Gas–Liquid–Solid Reactor Design. New York: McGraw-Hill, 1979.
22. VW Weekman Jr, JE Myers. AIChE J 10:951, 1964.
23. G Tosun. Ind Eng Chem Process Des Dev 23:29, 1984.
24. E Talmor. AIChE J 23:868, 1977.
25. JC Charpentier, M Favier. AIChE J 21:1213, 1975.
26. G Tosun. Ind Eng Chem Process Des Dev 23:35, 1984.
27. N Midoux, M Favier, JC Charpentier. J Chem Eng Jpn 9:350, 1976.
28. CN Satterfield, AA Pelossof, TK Sherwood. AIChE J 15:226, 1969.
29. JM Hochman, E Effron. Ind Eng Chem Fund 8:63, 1969.
30. S Goto, JM Smith. AIChE J 21:706, 1975.
31. LD Ross. Chem Eng Progr 61(10):77, October 1965.
32. JG Schwartz, E Weger, MP Dudokovic. AIChE J 22:894, 1976.

33. HC Henry, JB Gilbert. Ind Eng Chem Proc Des Dev 12:328, 1973.
34. DE Mears. Advances Chem 133:218, 1974.
35. SS Puranik, A Vogelpohl. Chem Eng Sci 29:501, 1974.
36. PL Mills, MP Dudokovic. AIChE J 27:893, 1981.
37. B Baker III. ACS Symp Ser 65:425, 1978.
38. MH Al-Dahhan, Y Wu, MP Dudokovic. Ind Eng Chem Res 34:741, 1995.
39. Y Wu, MR Khadilkar, MH Al-Dahhan, MP Dudokovic. Ind Eng Chem Res 35:397, 1996.
40. YT Shah, JA Paraskos. Ind Eng Chem Proc Des Dev 14:368, 1975.
41. E Newson. Ind Eng Chem Proc Des Dev 14:27, 1975.
42. AA Montagna, YT Shah, JA Paraskos. Ind Eng Chem Proc Des Dev 16:152, 1977.
43. M Herskowitz, RG Carbonell, JM Smith. AIChE J 25:272, 1979.
44. AH Germain, AG Lefebvre, GA L' Homme. Adv Chem Ser 133:164, 1974.
45. S Morita, JM Smith. Ind Eng Chem Fund 17:113, 1978.
46. CN Satterfield, F Özel. AIChE J 19:1259, 1973.
47. CN Satterfield. Mass Transfer in Heterogeneous Catalysis. MIT Press, 1970, p 96.
48. IA Eldib, LF Albright. Ind Eng Chem 49:825, 1957.

9

Fluidized-Bed Reactors

A fluidized bed is a bed of solid particles that are supported by the drag of upward-flowing gas or liquid. The particles are in continuous motion, and the suspension behaves like a dense fluid, which can be drained from the bed through pipes and valves. If the bed is tilted, the top surface remains horizontal, and large objects will either sink or float on the bed, depending on the relative densities. A naturally occurring example of fluidization is quicksand, where fine sand particles are suspended by upflowing water. Most industrial applications of fluidization involve solids and gases, which is the focus of this chapter.

Fluidized beds of fine solids are used for catalytic reactions in the petroleum and chemical industries, where the main advantages are nearly uniform temperature, good heat transfer to the wall or immersed surfaces, high effectiveness factors (because of the small particle size), and easy transfer of solids from one vessel to another. Fluidized beds are also used for combustion of coal, reduction of ores, and other solid–gas reactions, and these processes often use moderately large particles.

MINIMUM FLUIDIZATION VELOCITY

Consider what happens as gas is passed upward at slowly increasing velocity through a bed of fine solids resting on a porous distributor plate, as shown in Figure 9.1. At low flows, the pressure drop is proportional to the superficial velocity, since only the laminar-flow term of the Ergun equation [Eq. (3.64)] is significant. The particles remain in close contact, and no movement is observed. When the pressure drop becomes equal to the weight of the bed per unit area, point A in Figure 9.1, any further increase in velocity results in unbalanced forces on the bed. Either the particles behave as a cohesive mass and are forced upward and out of the bed, or the particles move slightly apart, and the bed becomes fluidized. With continued increases in gas velocity, the pressure drop remains constant, but the bed height increases.

When the flow to a fluidized bed is slowly reduced, the bed height decreases and may eventually stop at slightly above the original height, as shown by point B. Solids in a fluidized bed that is allowed to settle slowly often have a lower bed density than solids poured into the bed. Repeating the experiment after the fluidized bed has been allowed to settle should give reproducible results with no hysteresis. The pressure drop for the fixed bed would follow the lower line in Figure 9.1, and the minimum fluidization velocity is the velocity at which ΔP becomes constant and h starts to increase. When the data points are somewhat scattered, the minimum fluidization velocity, u_{mf}, is sometimes defined by the point where the pressure drop lines or the bed height lines intersect. If these values do not agree, an average value can be used, since the exact value of u_{mf} is not important.

Equating the weight of the bed per unit area to the pressure drop given by the Ergun equation results in a quadratic equation for u_{mf}. The term ϵ_m is the external void fraction at the minimum fluidization point:

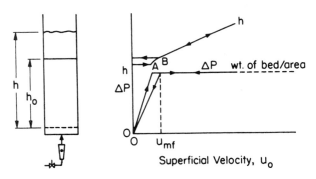

FIGURE **9.1**   Tests to determine minimum fluidization velocity.

$$\frac{\Delta P}{L} = g(1 - \epsilon_m)(\rho_p - \rho) = \frac{150\mu u_{mf}}{(\phi_s d_p)^2} \frac{(1 - \epsilon_m)^2}{\epsilon_m^3} + \frac{1.75\rho u_{mf}^2}{\phi_s d_p} \frac{(1 - \epsilon_m)}{\epsilon_m^3}$$

(9.1)

For small particles, only the first term of the Ergun equation is important, and the equation for minimum fluidization velocity is simplified:

$$u_{mf} = \frac{g(\rho_p - \rho)(\phi_s d_p)^2}{150\mu}\left(\frac{\epsilon_m^3}{1 - \epsilon_m}\right)$$

(9.2)

Uncertainty in the use of these equations arises because ϵ_m can range from 0.4 to 0.5, and it cannot be accurately predicted. A change from 0.4 to 0.45 means a 55% change in u_{mf}. For irregular particles, the shape factor ϕ_s also introduces uncertainty. Shape factors are 0.6–0.8 for crushed solids and 0.85–1.0 for rounded particles. The predicted dependence of u_{mf} on d_p^2 has been verified by many experimental studies, though some empirical equations for u_{mf} have a slightly lower exponent than 2.0, perhaps because ϵ_m changed with d_p.

Minimum fluidization velocities for spherical particles in air are shown in Figure 9.2. Equation (9.2) applies for particles up to about 300 microns in size, which includes most fluidized catalysts. For fluidized-bed combustion or metallurgical processes, the particles are much larger, and Eq. (9.1) must be used. For very large sizes, the laminar-flow term in Eq. (9.1) becomes unimportant, and u_{mf} varies with the square root of d_p:

$$u_{mf} = \left(\frac{\phi_s d_p(\rho_p - \rho)g\epsilon_m^3}{1.75\rho}\right)^{1/2}$$

(9.3)

The equations for minimum fluidization velocity are similar to those for the terminal velocity of a single particle, and it is instructive to examine the ratio of these velocities. For small spheres, the equation for Stokes' law divided by Eq. (9.2) gives

$$\frac{v_t}{u_{mf}} = \frac{gd_p^2(\Delta\rho)}{18\mu} \frac{150\mu}{g(\Delta\rho)d_p^2} \frac{(1 - \epsilon_m)}{\epsilon_m^3} = \frac{8.33(1 - \epsilon_m)}{\epsilon_m^3}$$

(9.4)

If $\epsilon_m = 0.45$, the terminal velocity is 50 times the minimum fluidization velocity, which indicates a wide range of possible operating conditions. For example, if $\bar{d}_p = 60$ µm and $u_{mf} = 0.2$ cm/sec, the bed could be operated at up to 10 cm/sec with no entrainment of the average-size particles. Most beds have a moderately wide distribution of particle sizes, and some entrainment of the fines is expected. Most of the entrained solids can be recovered via cyclone separators and returned to the bed. Some beds operate at $100 \times u_{mf}$

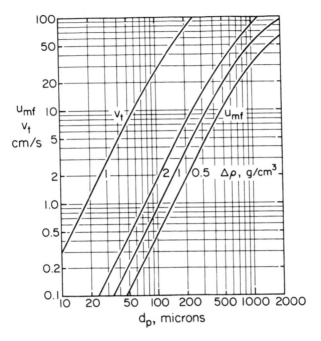

FIGURE 9.2 Minimum fluidization velocity and terminal velocity for spheres in air at 20°C with $\epsilon_m = 0.50$.

with high rates of entrainment; but with two cyclones in series, nearly complete recovery of the entrained solids can be achieved.

For large particles, the terminal velocity depends on $d_p^{1/2}$, as does u_{mf}, and the velocity ratio is

$$\frac{v_t}{u_{mf}} = \frac{2.32}{\epsilon_m^{1.5}} \tag{9.5}$$

For $\epsilon_m = 0.45$, $v_t/u_{mf} = 7.7$, a much lower ratio than for fine solids. Fluidized beds of large particles are usually operated at only 2–10 times the minimum fluidization velocity.

TYPES OF FLUIDIZATION

As the gas velocity is increased above u_{mf}, different types of behavior are observed, depending on the nature of the solid and the dimensions of the bed. For small porous particles such as FCC (fluid catalytic cracking) catalyst, the particles move further apart and the bed expands considerably for a small increase in velocity. The total pressure drop is constant, but the pressure drop per unit length decreases. For small changes in velocity, $\epsilon^3/(1 - \epsilon)$ is proportional to u_o, in accordance with Eq. (9.2). However, at a critical velocity, u_{mb}, small bubbles form, and the bed collapses with further increase in velocity, as shown by curve A in Figure 9.3. The bed height reaches a minimum and then increases again, as most of the additional gas passes through the bed as bubbles.

The behavior of the catalyst bed between u_{mf} and u_{mb} is called *particulate fluidization*, because the bed expands uniformly and is to an extent predictable from the drag on individual particles. Particulate fluidization is often found with water when using solids such as sand or ion-exchange beads, and the expanded bed may reach several times the original height. With gases, particulate fluidization is found only for some fine solids over a narrow range of velocities between u_{mf} and u_{mb}. For FCC catalyst, u_{mb} increases with $d_p^{0.6}$[1], and the difference between u_{mb} and u_{mf} decreases as d_p increases, as shown in Figure 9.4. For catalyst particles larger than about 180 microns, $u_{mf} \cong u_{mb}$, and there is no region of particulate fluidization.

For coarse solids and for fine catalyst at velocities greater than u_{mb}, most of the gas passes through the bed as bubbles, which may be several inches or a few feet in diameter. This regime is called *bubbling fluidization*, though in the older literature it is called *aggregative fluidization*. In bubbling fluidization, the bed has a *bubble* phase, which is almost free of particles, and a *dense* phase, where the particles are supported by upflowing gas at about the minimum fluidization velocity.

Solids that exhibit some region of particulate fluidization, are called type A (aeratable) in Geldart's classification system [2], while coarser solids

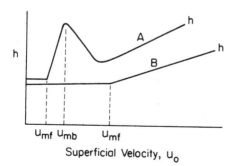

FIGURE 9.3 Bed expansion for particulate and bubbling fluidization: A, type A solids; B, type B solids.

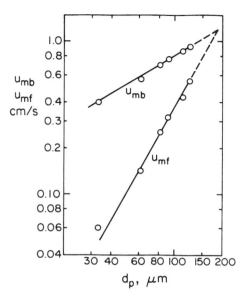

FIGURE 9.4   Effect of particle size on minimum bubbling velocity and minimum fluidization velocity for FCC catalyst.

showing only bubbling fluidization are called type B. Solids with $u_{mf} < 1$ cm/sec, which includes FCC catalysts and catalysts for several chemical processes, usually show type A behavior. Sand, glass beads, coal, and ores are generally type B solids, and the bed expansion is as shown in Figure 9.3. Although type A catalysts may be operated in the bubbling regime at velocities much greater than u_{mb}, they have better performance than type B catalysts at the same gas velocity. This is related to differences in bubble size, bed expansion, and flow patterns in the beds, as will be discussed later. Geldart's classification system also includes type C solids, which are very fine and difficult to fluidize because of cohesive forces, and type D solids, which have very large particles and may form spouted beds. This chapter deals only with type A and type B solids.

In bubbling fluidization, gas bubbles grow by coalescence as they pass up through the bed. In small-diameter units, when the bubble size approaches the column diameter, *slugging* is observed. The slugs may be shaped like round-nose bullets traveling in the center of the column, as shown in Figure 9.5, or they may be flat at the top and occupy the entire cross section. The bed height fluctuates sharply as the slugs reach the top of the bed, and severe vibrations may result. In small beds, slugging may start at velocities of only 5–10 cm/sec, and many laboratory studies of fluidized-

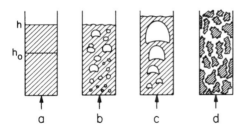

FIGURE 9.5 Types of fluidization: (a) Particulate fluidization; (b) bubbling bed; (c) slugging; (d) turbulent fluidization.

bed reactors have been carried out in the slugging regime, though large beds would be in the bubbling regime at comparable velocities.

At velocities much greater than u_{mf} or u_{mb}, the bed may exhibit *turbulent fluidization*, where there are local regions of low and high density but no distinct bubble phase. The solids are in erratic motion, flowing down in some areas and being carried upward in others. The average bed density is less than for bubbling fluidization, but the average bed height is hard to determine because of fluctuations in bed level and high entrainment of solids into the freeboard region above the bed. For catalytic reactions, operation in the turbulent regime is often desirable because of better contact between the gas and the solid, even though the bed density is relatively low and the rate of entrainment is high. Another advantage of turbulent fluidization is that pressure fluctuations in the bed are greatly reduced compared to bubbling fluidization.

The critical velocity for the transition to turbulent fluidization, u_c, depends on the gas and solid properties, and for porous catalysts it is generally between 0.2 and 0.5 m/sec. Complex empirical correlations for u_c have been published, but the transition can be understood by considering the major factor, which is the average bubble rise velocity, v_b, compared to the superficial velocity, u_o. In a bubbling bed, α is the fraction of bed volume occupied by bubbles, and $(1 - \alpha)$ is the fraction of dense bed, where the velocity is close to u_{mf}. The bed expansion and α are related as follows:

$$\alpha = \frac{\text{total bubble volume}}{\text{bed volume}}$$

$$u_o = \alpha v_b + (1 - \alpha)u_{mf} \tag{9.6}$$

$$\alpha = \frac{u_o - u_{mf}}{v_b - u_{mf}}$$

when

$$u_o \gg u_{\text{mf}}, \qquad \alpha \cong \frac{u_o}{v_b} \qquad\qquad (9.7)$$

If there are no solids in the bubbles, and the dense bed still has the void fraction ϵ_m, the bed expansion depends only on α:

$$\rho_s(1 - \epsilon_m)h_o = h\rho_s(1 - \epsilon_m)(1 - \alpha) \qquad\qquad (9.8)$$

$$\frac{h}{h_o} = \frac{1}{1 - \alpha} \qquad\qquad (9.9)$$

If the bed has expanded by 50%, $\alpha = 1/3$, and the bubbles would be almost touching. For higher values of α, the bubbles would be as close as particles in a packed bed or droplets in a concentrated emulsion; but since bubbles in a fluid bed have no skin or surface tension, high values of α are unlikely. As h/h_o approaches 2 or α approaches 0.5, frequent coalescence and breakup of bubbles will cause a transition to turbulent fluidization. The velocity of individual bubbles varies with the square root of the size. The predicted coefficient β is 0.71 [3], but data show values of 0.5–0.7 [4]:

$$v_b = \beta\sqrt{gD_b} \qquad\qquad (9.10)$$
$$\beta = 0.5\text{–}0.7$$

The average bubble size is hard to predict, since it varies with the gas velocity, the type of solids, and the bed height. Taking $D_b = 5$ cm as typical for type A solids, $v_b \cong 50$ cm/sec. Then if $u_o = 15$ cm/sec and $u_{\text{mf}} = 0.3$ cm/sec, $\alpha = 14.7/49.7 = 0.30$ and $h/h_o = 1.43$. Increasing u_o to 25 cm/sec would give $\alpha = 0.5$, if the bubbles still formed a dispersed phase. However, a transition to turbulent fluidization would probably occur before 25 cm/sec was reached. For larger bubbles, v_b is greater and α is smaller, so the transition to turbulent fluidization would take place at a higher superficial velocity.

At very high velocities, all particles fed to the reactor are carried up with the gas, and this mode of operation is sometimes called *fast fluidization*. However, since there is no definite bed level, a better term for this system is a *transport-line reactor* or, as it is called in catalytic cracking, a *riser reactor*. These are discussed in Chapter 10, on novel types of reactors.

REACTOR MODELS

Early studies of catalytic reactions in small fluidized beds showed considerably lower conversions than those measured for the same conditions in fixed beds [5–7]. Lower conversions were expected, because the uniform temperature of the fluid bed plus visual observations indicated vigorous back-mixing. However, in some cases, the fluid-bed conversion was even less than predicted for a completely back-mixed reactor! These low conversions were attributed to the two-phase nature of the bubbling bed. Most of the gas passes through the reactor in the bubble phase, where there is little or no reaction. A small amount of gas, just enough for fluidization, flows between the catalyst particles in the dense phase, where almost all the reaction occurs. (In early papers, the dense phase was called the *emulsion phase*.) Interchange of gas between the phases takes place as the bubbles rise through the bed, giving concentration profiles like those in Figure 9.6. The outlet concentration is a weighted average of C_D and C_B that is quite close to C_B because of the small flow in the dense phase.

Evidence for the two-phase model came from measurements of the gas concentration profile in a commercial catalytic cracking regenerator 40 ft in diameter with a 15-ft bed [8]. The exit gas had 1% O_2, but samples drawn from different bed depths had only 0.1–0.4% O_2. The bed samples also showed 12–14% CO_2, compared to 10% CO_2 in the exit gas. Although most of the gas flow was in the bubbles, the probe saw mainly dense-phase gas, where the conversion was higher than in the bubbles. Samples taken very rapidly showed wide fluctuations in oxygen content, since the probe was sometimes in a bubble and sometimes in the dense bed.

A greater understanding of bubbles in fluidized beds has come from theoretical studies and pictures of bubbles in two-dimensional and three-dimensional beds [3,9]. If the bubble rise velocity is greater than the superficial velocity, gas leaving the top of the bubble is carried back to the bottom

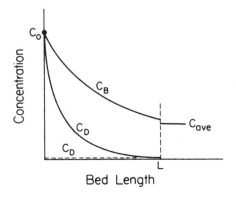

Figure 9.6 Concentration profiles in a bubbling fluidized bed. (solid line: plug flow in both phases; dashed line: mixed dense phase.)

by solids flowing down outside the bubble. The region where gas is recirculating is called the *bubble cloud*, which has the shape shown in Figure 9.7. The volume of the cloud is about 2–5% of the bubble volume for type A solids, but it can be a large fraction of the bubble volume for coarse solids. The wake includes solids carried upward by the rising bubble, and the volume of the wake is typically about one-fourth of the bubble volume. The movement of the wake is responsible for the mixing of solids and the nearly uniform temperature that is characteristic of fluid-bed reactors.

Pictures of bubbles and clouds have inspired some workers to develop reactor models based on the predicted behavior of individual bubbles [3,10]. In these models, the equations for gas interchange include a term for flow out of the bubble and a second term for mass transfer by molecular diffusion to the dense phase. In some models, the cloud is included as part of the bubble; in others, diffusion from bubble to cloud and cloud to dense phase are treated as mass transfer steps in series. In these models, the mass transfer coefficient is assumed to vary with $D_{AB}^{1/2}$, following the penetration theory, and the diffusion contribution is the major part of the predicted gas interchange rate.

Although the individual bubble models are included in many texts and research papers, they are not reliable for predicting reactor performance under practical conditions. One problem is that the bubble size must be assumed to use the model, and it is hard to tell what size to choose. A further problem is that the interchange rate does not show the predicted dependence on diffusivity. Fontaine and Harriott [11] used frequency response tests to compare bubble–dense bed interchange rates for different tracers. At 0.11 and 0.18 m/sec, there was no difference between the results for He and CO_2, in spite of the fourfold difference in diffusivity, and there was only a slight difference at 0.03 m/sec. DeVries and coworkers [12]

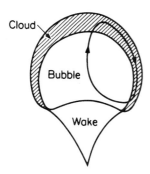

FIGURE 9.7   Gas circulation through the bubble and cloud.

reported no difference in residence time distributions with He and Ar as tracers in a bed fluidized at 0.1 m/sec. These two studies indicate that in a vigorously bubbling bed, the major contributions to gas interchange between bubbles and the dense phase must come from fluid dynamics effects such as bubble coalescence, bubble splitting, and wake shedding. This conclusion is supported by the work of Chiba and Kobayashi [13], who found that exchange coefficients for single bubbles injected into a bed at incipient fluidization were only one-third the values obtained with freely bubbling beds. At present it seems best to consider the gas interchange coefficient as a parameter to be determined by experiment or predicted by empirical correlations.

THE TWO-PHASE MODEL

There are many two-phase reactor models that treat gas interchange using a volumetric mass transfer coefficient. In this text, the coefficient is K, with units of $\sec^{-1}$ or ft^3 exchanged per second per ft^3 bed. (In some reports, this coefficient is q_b or F.) The simplest models assume plug flow in the bubble phase and either no mixing or perfect mixing in the dense phase. The concentration profiles have the shapes shown in Figure 9.6, and the basis for the model is given in Figure 9.8. Although the bubbles are dispersed in the dense phase, the conversion can be calculated as if there were separate parallel channels extending through the bed, as shown in the diagram. The total flow per unit cross-sectional area is u_o, the sum of u_B, the flow carried by bubbles, and u_D, the dense-phase flow. The actual bubble velocity, v_b, is much higher than u_B, but the value of v_b is not needed here.

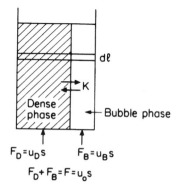

FIGURE 9.8 Two-phase model for fluid-bed reactor.

Model I: Plug Flow in Both Phases

The bubbles are assumed to contain no catalyst, so the concentration change comes only from transfer of reactant to the dense phase:

$$-u_B \, dC_B = K(C_B - C_D)dl \tag{9.11}$$

In terms of the fraction conversion in each phase,

$$C_B = C_0(1 - x_B)$$
$$C_D = C_0(1 - x_D)$$

So Eq (9.11) becomes

$$\frac{dx_B}{dl} = \frac{K}{u_B}(x_D - x_B) \tag{9.12}$$

The equation for the dense phase includes the gas interchange term and a reaction term, where ρ_b is the average density of the expanded bed. A first-order irreversible reaction is assumed:

$$-u_D \, dC_D = k\rho_b C_D \, dl - K(C_B - C_D)dl \tag{9.13}$$

or

$$\frac{dx_D}{dl} = \frac{k\rho_b(1 - x_D)}{u_D} - \frac{K}{u_D}(x_D - x_B) \tag{9.14}$$

Solutions for Eqs. (9.11) and (9.13) have been presented [5,14], but they are rarely used. For catalytic reactors operating with $u_o \gg u_{mf}$, the flow through the dense phase is small enough to be neglected; in Model II, u_B is assumed equal to u_o.

Model II: Plug Flow in Bubble Phase, No Flow or Mixing in Dense Phase

The equation for the bubble phase is similar to Eq. (9.12), but u_o replaces u_B:

$$\frac{dx_B}{dl} = \frac{K}{u_o}(x_D - x_B) \tag{9.15}$$

For the dense phase with no mixing, the interchange rate is equal to the reaction rate:

$$K(x_D - x_B) = k\rho_b(1 - x_D) \tag{9.16}$$

$$x_D = \frac{k\rho_b + Kx_B}{K + k\rho_b} \tag{9.17}$$

Using this value of x_D in Eq. (9.15) gives

$$\frac{dx_B}{dl} = \frac{K}{u_o}\frac{k\rho_b(1 - x_B)}{K + k\rho_b} \tag{9.18}$$

The terms K and $k\rho_b$ are for mass transfer and reaction in series, and they can be combined to give an overall rate coefficient K_o:

$$\frac{1}{K} + \frac{1}{k\rho_b} = \frac{K + k\rho_b}{Kk\rho_b} = \frac{1}{K_o} \tag{9.19}$$

Equation (9.18) can then be simplified for integration:

$$\frac{dx_B}{dl} = \frac{K_o}{u_o}(1 - x_B) \tag{9.20}$$

$$\ln\frac{1}{1 - x_B} = \ln\frac{1}{1 - x} = \frac{K_o L}{u_o} \tag{9.21}$$

or

$$x = 1 - e^{-K_o L/u_o}$$

Equation (9.21) is for plug flow in both phases but with negligible flow in the dense phase, so $x \cong x_B$.

Another way of combining the terms for mass transfer and reaction in series is to use N_r, the number of reaction units, and N_m, the number of mass transfer units. The group $k\rho_b L/u_o$ is equivalent to kt for a homogeneous reaction and is called the *number of reaction units*. The group KL/u_o is the *number of mass transfer units* and is equivalent to the *NTU* in mass transfer operations such as gas absorption. Using Eq. (9.19), these terms can be combined to give N, the *overall number of units for mass transfer and reaction*:

$$\frac{1}{N} = \frac{1}{N_r} + \frac{1}{N_m} = \frac{1}{K_o L/u_o} \tag{9.22}$$

where

$$N_r = \frac{k\rho_b L}{u_o}$$

$$N_m = \frac{KL}{u_o}$$

Then Eq. (9.21) can be written as

$$x = 1 - e^{-N} \tag{9.23}$$

Model III: Plug Flow in Bubble Phase, Complete Mixing in Dense Phase

When there is plug flow in the bubble phase, complete mixing in the dense phase, and all gas flows through the bubble phase, Eq. (9.12) is integrated with $u_B = u_o$ and a constant value for x_D:

$$\ln\left(\frac{x_D}{x_D - x_B}\right) = \frac{KL}{u_o} \tag{9.24}$$

The total reactant transferred to the dense phase equals the amount consumed:

$$\int_0^L K(x_D - x_B)dl = k\rho_b L(1 - x_D) \tag{9.25}$$

Substituting from Eq. (9.24) and integrating gives

$$\frac{1}{x_D} = \frac{u_o}{k\rho_b L}\left(1 - e^{-KL/u_o} + k\rho_b \frac{L}{u_o}\right) \tag{9.26}$$

The conversion for the bubble phase, which is the same as the overall conversion, is given by an equation similar to Eq (9.21), but the exponential term includes K rather than K_0, and the limiting conversion is x_D rather than 1.0:

$$x_B = x = x_D(1 - e^{-KL/u_o}) \tag{9.27}$$

The conversions for Model II and Model III are compared in Figure 9.9.

When N_m is small—say, 0.5–2.0– and $-N_r \geq 3$, the conversion is limited by the rate of mass transfer between bubbles and the dense phase, and increasing N_r to high values has little effect on the conversion. Under these conditions, it makes little difference whether the dense phase is well mixed or not. This makes it easier when fitting data from small lab reactors to Model II or III, since the value of K obtained is almost the same. However, when predicting the performance of a tall reactor, where $N_m \geq 5$, there is a large difference in conversion between the plug-flow and perfect-mixing models. With Model III, the conversion can never exceed that for a completely mixed reactor, whereas for Model II, a large value of N_m means only slightly lower conversion than for ideal plug flow.

Model IV: Axial Dispersion in Dense Phase

This model allows for partial mixing in the dense phase using an effective axial diffusivity, D_{ea}. The model is based on plug flow for the bubble phase

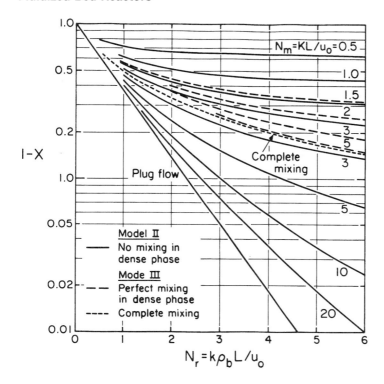

FIGURE 9.9 Conversion for first-order reaction in a fluidized bed.

and no flow through the dense phase. The equation for the bubble phase is the same as for Models II and III:

$$\frac{dx_B}{dl} = \frac{dx}{dl} = \frac{K}{u_o}(x_D - x_B) \tag{9.28}$$

The equation for the dense phase is

$$K(C_B - C_D) + D_{ea}\frac{d^2C_D}{dl^2} = k\rho_b(1 - x_D) \tag{9.29}$$

The conversion now depends on three dimensionless variables: $N_r = k\rho_b L/u_o$, $N_m = KL/uo$, and a modified Peclet number, $\text{Pe}' = u_o L/D_{ea}$. DeVries and coworkers [12] used this model to predict the conversion for the Shell Chlorine Process, the high-temperature oxidation of HCl in a fluidized bed. This is a very fast reaction that is limited by mass transfer and mixing rates, as shown in Figure 9.10. When $N_r >> \text{Pe}'$ and $\text{Pe}' \gtrsim 2N_m$,

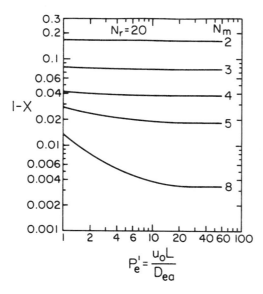

FIGURE 9.10 Effect of axial mixing and gas interchange on conversion in a fluidized bed. (From Ref. 12.)

the gas interchange rate is the limiting factor, and the conversion is about the same as for plug flow in the dense phase (Model II).

In an earlier paper by May [15], flow through the dense phase was allowed for as well as axial dispersion and gas interchange. This gives four dimensionless parameters and a complex cubic equation for the solution. However, for typical conditions with cracking catalyst, $u_o >> u_{mf}$ and the effect of dense-phase flow is negligible. A few examples were presented for moderate values of N_r, N_m, and Pe′ using mass transfer parameters obtained for large-diameter reactors. The conversions predicted for $N_m = 10$ and Pe′ = 5.3 are compared with those for ideal plug flow and a completely mixed reactor in Figure 9.11. The conversions are about the same as those predicted for Model II with $N_m = 6.0$. Although for this example the conversion is midway between the values expected for plug flow and for perfect mixing, May reported that the conversions are generally closer to those for plug flow than for perfect mixing.

THE INTERCHANGE PARAMETER *K*

In many laboratory studies of catalytic reactions in fluid beds, two-phase models have been used to obtain values of *K* from the conversion data. Test

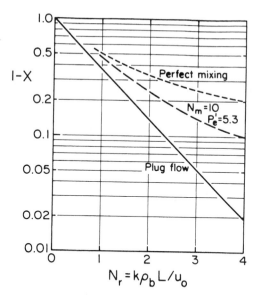

FIGURE **9.11** Conversion predicted for a typical reactor based on the two-phase model of May [15].

reactions include the decomposition of ozone [16–18] and the hydrogenation of olefins [6], which are irreversible first-order reactions. The cracking of cumene to propylene and benzene has also been studied [5,7], but this is a reversible reaction with complex kinetics, which makes interpretation of the results more difficult. Several of these studies were carried out in short, small-diameter beds with conversions less than for a completely mixed reactor. For these conditions or for other tests with less than about 50% conversion, it makes little difference which model is used to get K, since the gas interchange rate is the limiting factor. Note the small difference between the conversion plots for zero mixing and complete mixing in the dense phase in Figure 9.9 for $N_m \leqq 2$. The reported values of K in these early studies ranged from 0.05 sec^{-1} to 1 sec^{-1}, with no consistent trends, and no general correlation has been presented. Others studies of reactions in larger-diameter and taller beds gave conversion between the values for plug flow and for perfect mixing. Although the degree of mixing is now more important for the interpretation of results and the choice of model, the data are more useful for understanding the performance of industrial reactors. Data from a few studies were analyzed to obtain K values based on Model II. The values shown in Figure 9.12 are for beds at least 0.1 m in diameter with type A catalyst operated at velocities of at least 0.1 m/sec, far above the

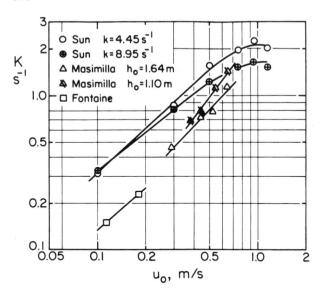

FIGURE 9.12 Gas interchange coefficients derived from kinetic tests and tracer tests.

minimum fluidization velocity. The method of calculating K is illustrated in Example 9.1

Example 9.1

The decomposition of ozone was studied in a fluidized bed 0.1 m in diameter using different samples of fluid cracking catalyst [16,19]. Some data for catalyst with a broad size distribution and a mean particle size of 60 μm are given in Table 9.1. Use Model II to calculate the values of K.

Solution.   Use Eqs. (9.21) and (9.19). For the first run

TABLE 9.1 Data for Example 9.1

u_o, m/sec	0.1	0.3	0.5	0.75	0.95	1.15
x		0.923	0.872	0.846	0.775	0.728	0.664
$\dfrac{h}{h_o}$		1.26	1.44	1.66	2.0	~2.3	~2.7

$W = 5$ kg, $\epsilon_m = 0.456$, $h_o = 0.75$ m, $k_r = 4.45$ sec^{-1} based on particle volume.

$$\ln \frac{1}{1 - 0.923} = 2.56 = \frac{K_o L}{u_o}$$

$$L = 0.75(1.26) = 0.945 \text{ m}$$

$$K_o = \frac{2.56(0.1)}{0.945} = 0.271 \text{ sec}^{-1}$$

For $k\rho_b$, the rate constant per unit volume of bed, the particle fraction in the expanded bed is needed:

$$1 - \epsilon = \frac{1 - 0.456}{1.27} = 0.432$$

$$k\rho_b = 4.45(0.432) = 1.92 \text{ sec}^{-1}$$

$$\frac{1}{K} = \frac{1}{K_o} - \frac{1}{k\rho_b} = \frac{1}{0.271} - \frac{1}{1.92}$$

$$K = 0.316 \text{ sec}^{-1}$$

Results for the other runs are shown in Table 9.2 and are plotted in Figure 9.12.

The values for the highest velocities are uncertain because the bed height data were extrapolated. In some other studies, K may have been based on the initial bed height or volume, which makes the values larger, but it is not always clear which basis was used.

Figure 9.12 shows that K increases almost in proportion to the gas velocity. In the bubbling regime, higher velocity means more bubbles and more frequent coalescence. The increase in K continues in the turbulent regime, where there is no longer a distinct bubble phase but there are concentration differences between the regions of high and low bed density. The value of K should not depend on k, and the differences shown in Figure 9.12 indicate a weakness in the simple model. The data of Massimilla [20] are for acrylonitrile synthesis in a 15.6-cm reactor using catalyst with $\overline{d}_p = 56 \, \mu m$, and they also show a strong effect of velocity. The higher K values for the shorter bed might be due to more reaction in the freeboard region, which is not accounted for in the simple model.

TABLE 9.2 Results for Example 9.1

u, m/sec	0.1	0.3	0.5	0.75	0.95	1.15
K, sec^{-1}	0.316	0.86	1.56	1.96	2.27	2.01

The particle size distribution, or PSD, has long been known to affect the performance of fluid-bed reactors. A broad PSD is better than a narrow one, and it is particularly important to have a large amount of fines, generally defined as particles smaller than 44 microns (325-mesh screen opening). The amount of fines influences the average bubble size, the gas interchange rate, and the axial dispersion. When the percentage fines is too low, the bed does not fluidize smoothly and large bubbles or slugs are more frequent. Fines are produced continuously by attrition in fluidized beds, but they are entrained at a high rate, and some escape the cyclone collection system. In catalytic cracking units, high-velocity jets are sometimes used to increase the rate of attrition and help maintain a suitable fines concentration.

In a study of acrylonitrile synthesis in a 0.5-m × 9-m reactor, Pell and Jordan [21] found a steady increase in conversion as the fines concentration was increased from 23% to 44%. The effect was more pronounced at a velocity of 0.38 m/sec than at 0.66 m/sec, and the optimum fines concentration was about 40%. A strong effect of size distribution was also demonstrated in the ozone decomposition tests of Sun and Grace [16]. For the same average particle size (60 μm), a wide PSD gave significantly higher conversion than a narrow PSD, as shown in Figure 9.13. The study covered

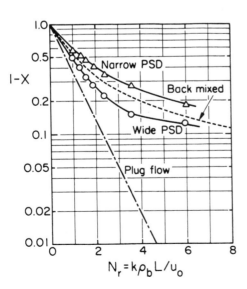

FIGURE 9.13 Effect of particle size distribution on ozone conversion in a fluidized bed. (Data from Ref. 16.)

velocities of 0.1–1.7 m/sec; the benefit of the broad PSD was greater at the intermediate velocities.

The type of gas distribution plate can have a significant effect on reactor performance, particularly for short laboratory reactors. A sintered porous metal plate gives higher conversion than a wire screen or a perforated plate [7,17,21]. With the porous plate, small bubbles are formed near the bottom of the bed, and they grow larger by coalescence as they rise. With a perforated plate, gas enters as jets, which form moderate size bubbles as they break up. The porous plates have too much pressure drop to be suitable for large reactors, and sieve plates, bubble-cap plates, or grids of pipes with multiple orifices are generally used. A satisfactory distributor has a great many holes and moderate pressure drop to ensure even gas distribution [22]. With a good distributor, the rate of gas exchange is greater near the gas inlet, though the overall effect may be small for beds that are one to several meters tall.

To allow for changes in the parameters with height, the two-phase model can be applied to sections of the reactor, as was done by Pell and Jordan [21]. They extended Model II by including a grid region, where mass transfer is rapid, the major part of the bed, and the freeboard region, where the catalyst density is low but there is better contact between the catalyst and the gas. At the high velocities used, about 20% of the catalyst was in the freeboard region. Parameters for this three-part model were determined by trial and error, but the values were not reported. Perhaps the large effect of fines concentrations is partly due to having more of the catalyst in the freeboard region.

MODEL V: SOME REACTION IN BUBBLES

A weakness of the previous two-phase models is that they are based on no reaction in the bubble phase, and they predict a limiting conversion as N_r becomes very large. Because catalyst in the cloud is contacted by gas circulating in the bubble phase, and catalyst rains down through bubbles as they split up or coalesce, there should be no limit to the conversion other than a thermodynamic limit. A model allowing for some catalyst in the bubbles was presented by Lewis, Gilliland, and Glass [6]. The fraction of the catalyst that contacts bubble gas is called a. For negligible flow and no mixing in the dense phase, the equation for the bubble phase is

$$\frac{dx_B}{dl} = \frac{K(x_D - x_B)}{u_o} + \frac{ak\rho_b(1 - x_B)}{u_o} \tag{9.30}$$

The material balance for the dense phase is

$$k\rho_b(1-a)(1-x_D) = K(x_D - x_B) \tag{9.31}$$

The solution of these equations for plug flow in the bubble phase is

$$\ln\frac{1}{1-x_B} = \ln\frac{1}{1-x} = \frac{k\rho_b L}{u_o}\left(a + \frac{K(1-a)}{K + k\rho_b(1-a)}\right) \tag{9.32}$$

or

$$\ln\frac{1}{1-x} = N_r a + N'(1-a) \tag{9.33}$$

where

$$\frac{1}{N'} = \frac{1}{N_m} + \frac{1}{N_r}(1-a)$$

This model was fitted to the data for hydrogenation of ethylene in a 2-inch fluidized bed to give values of a and K ($K = F$ in Ref. 6). The values of a ranged from 0.04 to 0.16, increasing with gas velocity. These values are higher than what would be predicted based on bubble and cloud sizes. The values of K ranged from 0.3 to 0.8 sec^{-1} and showed an unexpected decrease with increasing gas velocity. Although allowing for some reaction in the bubble phase is realistic and doesn't make the model very complicated, most workers have ignored this effect and either used the simple models with one empirical parameter, K, or made the models more complex by adding axial dispersion or by writing separate equations for different sections of the reactor.

AXIAL DISPERSION

Although axial mixing decreases the reactor conversion by lowering the average reactant concentration, it is difficult to determine D_{ea}, the dispersion coefficient, by using the results of kinetic tests, since the conversion is also dependent on the gas interchange parameter. Conversion data that fall between the limits of plug flow and perfect mixing can generally be fitted by various combinations of D_{ea} and K. It might seem that measuring the concentration profile in the dense phase of a bubbling bed would be the approach to take, since the profiles for the limiting cases are so different, as illustrated in Figure 9.6. However, the velocity in the dense phase is very low, and a sample probe may draw in gas from a wide region that includes some bubbles. Also, the probe acts as a filter, and catalyst stuck to the probe increases the conversion of the gas being sampled. This method has not yet led to reliable values of D_{ea}; but with better probes, it might be useful.

Most published values of D_{ea} have come from some type of tracer test. In early tests by May at Esso Research with FCC catalyst, a pulse of radioactive solid was injected at the top of the bed, and scintillation counters were used to monitor the solids mixing rate [15]. Curves showing the approach to equilibrium at different depths were compared with theoretical mixing curves to get the best value of D_{ea}. The gas-mixing rate in the dense phase was assumed to equal the solids-mixing rate, which seems reasonable because of the very low velocity in the dense phase. Values of D_{ea} from May's tests at $u_o = 0.8$ ft/sec (0.24 m/sec) are plotted in Figure 9.14. The effect of bed diameter on D_{ea} was attributed to an increase in average bubble size, and D_{ea} was estimated to level off at about 6 ft^2/sec (0.56 m^2/sec) for $D = 20$ ft (6 m).

De Groot also used the tagged-solids method to measure D_{ea} in beds of silica with either a broad or narrow PSD [23]. For the broad-range silica, which is similar to cracking catalyst, the values of D_{ea} at 0.2 m/sec were about the same as at 0.1 m/sec, and the average values are shown in Figure 9.14. For the largest bed, De Groot's value of D_{ea} was only half that reported by May, though there was fair agreement for the smaller diameter. For silica with a narrow size distribution, which did not fluidize

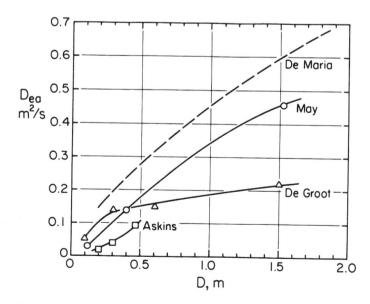

FIGURE 9.14 Effect of bed diameter on axial dispersion coefficient.

smoothly, values of D_{ea} (not shown in Figure 9.14) were several-fold lower.

The top line in Figure 9.14 is taken from the paper by De Maria and Longfield [24], who studied gas mixing in laboratory and commercial reactors. They used He and CO_2 tracers and found that D_{ea} was proportional to u_o and increased strongly with D. Their values of D_{ea} are greater than those of May and do not appear to level out at high D. However, in analyzing the tracer response data, they used a one-dimensional model, which does not account for bubble–dense bed interchange and therefore overestimates the diffusivity in the dense phase. The lowest values of D_{ea} in Figure 9.14 are from Shell Oil for FCC catalyst [8], and for small beds, D_{ea} increases rapidly with increasing D.

A few other mixing studies are reviewed by Pell [22], but most are for small beds, and no general correlation has been presented. It is clear that D_{ea} increases with bed diameter and is very dependent on particle size distribution. However, the effect of gas velocity is uncertain, since some work shows that D_{ea} goes through a maximum with increasing velocity, while others show either a proportional increase or almost no effect of velocity. This makes it hard to select an appropriate value to use in analyzing pilot studies or for design. It also provides some justification for using simpler models based on plug flow or complete mixing.

For tall beds operated at high velocity, the conversion may be close to that predicted by Model II in spite of a high value of D_{ea}. For example, if $D = 1.5$ m, $L = 6$ m, and $u_o = 0.5$ m/sec, D_{ea} is estimated as 0.3 m^2/sec from Figure 9.14. Then $Pe' = u_o L/D_{ea} = 0.5(6)/0.3 = 10$, and the importance of axial dispersion can be estimated using the solution given in Chapter 6 for reaction plus dispersion in a fixed bed. Figure 6.12 shows that for $Pe' = 10$, the conversion is only a few percent less than for plug flow and a similar error is expected if axial diffusion is neglected in modeling the fluid-bed reactor. When Model II is used to estimate the conversion, the values of K should be derived from experimental data using the same model, as was done for Figure 9.12. These effective values of K are lower than the "true" values obtained using a more complex model, such as Model IV or Model V.

If N_m, the number of mass transfer units, is quite large and there is little concentration difference between the bubbles and the dense phase, then axial mixing may be the most important factor determining the departure from ideal-flow performance. Then the conversion could be estimated from Figure 6.12 using Pe' based on the effective D_{ea} obtained from kinetic tests. The "true" D_{ea} from solids-mixing experiments is lower than the effective D_{ea}, since the latter must also account for the effect of neglecting the gas interchange parameter.

SELECTIVITY

For consecutive reactions where the intermediate is the desired product, the local selectivity and the maximum yield are decreased because of the concentration differences between the bubble phase and the dense phase. Consider the simple case of two first-order reactions in series, and assume Model II is applicable:

$$A \xrightarrow{1} B \xrightarrow{2} C$$

$$S = \frac{r_1 - r_2}{r_1} = 1 - \frac{r_2}{r_1} = 1 - \frac{k_2 C_B}{k_1 C_A} \tag{9.34}$$

The concentrations at an intermediate position in the bed are sketched in Figure 9.15. In the modeling studies, attention was focused on the concentration of reactant A, which always has a lower value in the dense phase than in the bubbles. The concentration of B is higher in the dense phase than in the bubbles, and the difference is only slightly less than for A if the selectivity is high. The combination of low C_A and high C_B decreases the selectivity. The loss in yield is greater for low values of the interchange parameters, as shown for a particular example in Figure 9.16. The effect of axial mixing on selectivity was predicted by De Maria et al. [25], who studied the partial oxidation of naphthalene. As expected, high dispersion coefficients or lower Peclet numbers decreased the maximum yield of phthalic anhydride.

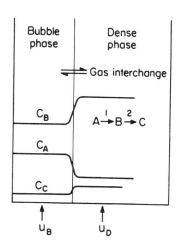

FIGURE 9.15 Concentration gradients affect selectivity with consecutive reactions.

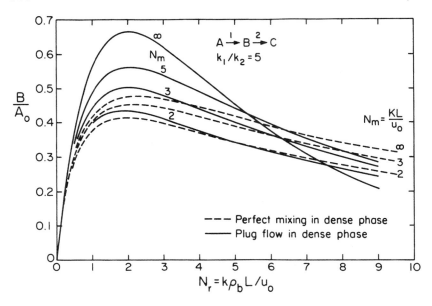

FIGURE 9.16 Effect of gas interchange on yield for consecutive reactions in a fluid bed.

HEAT TRANSFER

Heat transfer to or from a fluidized bed can be accomplished with an external jacket or with bundles of horizontal or vertical tubes immersed in the bed. The heat transfer coefficients are generally higher than for packed-tube reactors and many times greater than for gas flow at the same velocity in a reactor with no solids. The high coefficients are due to the violent and erratic motion of the solids, which is characteristic of vigorously bubbling beds or beds operating in the turbulent regime.

With an exothermic reaction in the bed, clusters or packets of hot solid come in contact with the cooler surface of the wall or the tubes, and they give up some of their heat in the short time before they are swept away. A model based on the penetration theory of unsteady-state heat transfer and some supporting data were presented by Mickley and Fairbanks [26]. The average coefficient is predicted to vary with the square root of the thermal conductivity, density, and heat capacity of the clusters and inversely with the square root of the average contact time. The fraction of the surface in contact with clusters is taken to be $(1 - \alpha)$, where α is the volume fraction bubbles in the bed. Heat transfer to the bare surface is assumed negligible because of the low conductivity of the gas. These assumptions lead to

$$h = 2(1 - \alpha) \left(\frac{k_e \rho_b c_p}{\pi \tau} \right)^{0.5}$$

(9.35)

where

k_e = effective conductivity of the clusters

ρ_b = density of the dense phase

c_p = heat capacity of the solid (contribution of gas heat capacity is negligible)

τ = contact time for clusters at the surface

The average contact time cannot be predicted, but it would be expected to depend on the particle type and size, the gas velocity, the size and arrangement of the tubes, and perhaps the bed dimensions. However, the theory does help explain some of the experimental results, and it is a guide in the development of empirical correlations. Only a few studies are reviewed here, to show the major trends and typical coefficients for fine solids.

Film coefficients reported by Beeby and Potter [27] for a 1-inch horizontal tube in beds of type A solids are shown in Figure 9.17. For fine sand

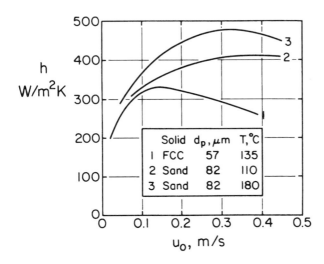

FIGURE 9.17   Heat transfer in fluidized beds.

fluidized with air, the coefficient first goes up sharply with increasing velocity and then changes more gradually, reaching a maximum at 0.3–0.4 m/sec. The initial increase is due to more and bigger bubbles, which causes more violent solids motion, but this is eventually offset by the lower bed density and the greater chance that gas will cover the surface. The lower coefficients for FCC catalyst are probably due to the lower bed density and lower conductivity of the porous solid. The fact that the maximum coefficient occurs at a relatively low velocity may be due to greater bed expansion with the FCC catalyst.

The data in Figure 9.17 for sand show about a 15% increase in coefficient for a temperature change from 110°C to 180°C. This increase is greater than predicted by Eq. (9.35), since the change in gas conductivity is only 15% and k_e changes more slowly with T than does k_g. The values of k_e can be estimated from the gas and solid conductivities using the method shown in Chapter 5 (see Figure 5.16). A better check on the penetration theory is afforded by data for beds fluidized with argon, air, or helium [26] to give a wide range of bed properties. At a given velocity, the coefficients increased with the 0.52–0.55 power of the bed conductivity, in good agreement with theory. To predict coefficients for gas mixtures at reaction conditions, test could be made with air at moderate temperature and corrected using Eq. (9.35). For temperatures of 300°C or higher, the radiation contribution to k_e is significant and can be evaluated using Eqs. (5.68) and (5.69).

There have been a great many studies of heat transfer in fluidized beds of large particles, which are important for the design of fluid-bed boilers. The maximum coefficients are lower than with type A solids, and some experiments show a gradual decrease in h with increasing d_p and a minimum h at $d_p = 2$–3 mm [28]. Extensive data are presented in books [29,30] and in the proceedings of the Engineering Foundation Conferences on Fluidization.

COMMERCIAL APPLICATIONS

The first major application of fluidization technology was the fluid-bed catalytic cracking (FCC) process introduced during World War II. Heavy oil vapors are cracked to gasoline and fuel oil plus low-molecular-weight paraffins and olefins by contact with very hot particles of fine zeolite–silica–alumina catalyst. The catalyst provides energy for vaporization of the feed and for the endothermic reactions. A few percent of the feed forms carbonaceous deposits on the catalyst, rapidly decreasing its activity, so frequent regeneration is necessary. Spent catalyst is continuously removed from the reactor and sent to the regenerator, where air is introduced to burn off the "coke," reheat the catalyst, and restore its activity.

In early versions of the FCC process, the reactor and regenerator were fluidized beds placed side by side. Catalyst from the reactor passed down by gravity through a stripper, where upflowing steam displaced the hydrocarbon vapors and maintained the solid in a fluidized state. The catalyst then flowed in a transfer line to a point below the beds, where it was picked up by the air stream and carried into the regenerator. Catalyst from the regenerator flowed through another transfer line to a tee, where it joined the oil feed and passed up into the reactor.

Other versions of FCC units had different methods of controlling the solid flow between reactors; in one "single-vessel" unit, the reactor was placed on top of the regenerator. However, the biggest change came after the introduction of very active zeolite catalysts and the realization that much of the cracking took place in the transfer line carrying catalyst into the reactor. Current designs feature a riser reactor, a tall, large-diameter pipe, where all the cracking occurs as catalyst particles are carried upward at high velocity by the oil vapors. (More details of the riser reactors are given in Chapter 10.) The gas–solid suspension is discharged through cyclones into a vessel that serves as a stripper and a feed reservoir for spent catalyst. The regenerator is a large reactor (up to 18 m in diameter) with a bed depth of 10–15 m, and it is often the largest vessel in the refinery. Figure 9.18 shows an FCC unit designed with a riser reactor. Some steam or lift gas is fed to the bottom of the riser to maintain fluidization until the feed oil is vaporized. Gas and solid exit the riser horizontally through swirl nozzles that make some of the solids drop out prior to the cyclones. Steam is introduced at several points in the multistage stripper to maximize hydrocarbon removal.

The regenerator is operated in the turbulent regime at superficial velocities of 0.3–1.0 m/sec, over 100 times the minimum fluidization velocity. Entrainment of fines is severe, and the top of the regenerator is crowded with many sets of two-stage cyclones (only one set is shown in Figure 9.18). The cyclones recover over 99.99% of the entrained solids, which are returned to the bed through dip legs discharging below the top of the bed. Most of the oxygen is consumed in the reactor, but the catalyst is not completely regenerated. Because of solids mixing, there is a wide distribution of residence times and a corresponding distribution of carbon content on the catalyst particles. Typically, over 90% of the carbon is burned off. Two-stage regenerators are used in some refineries to provide greater control over the degree of regeneration and the off-gas composition.

A variety of reactor models have been proposed for the regenerator, including two-phase models, a perfectly mixed reactor, and two or three mixers in series [31]. These models have been combined with a

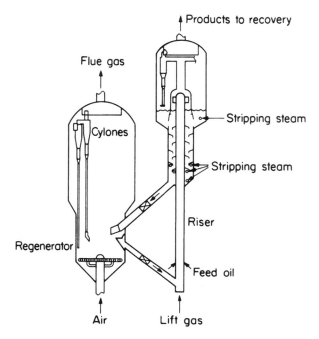

FIGURE **9.18**   Modern FCC unit.

plug-flow model for the riser reactor to predict the dynamic response of
the FCC unit and to devise control strategies. It is difficult to determine
the best model for the regenerator because of uncertainty in the kinetics
as well as in the fluidization parameters. Both CO and CO_2 are produced as the coke burns, and some CO is oxidized in the gas phase.
The rest of the CO can be burned later to generate steam. Oxidation of
CO above the bed can lead to large, undesirable temperature increases,
and some catalysts are promoted with platinum to favor CO oxidation
in the bed.

     The production of phthalic anhydride by partial oxidation of naphtha-
lene was carried out in fluid-bed reactors for many years, and performance
data for industrial scale reactors are available [25,32,33]. There are several
steps in the overall reaction, and rate constants for the following scheme
have been presented [25]. Parallel paths convert naphthalene, NA, directly
to phthalic anhydride, PA, and indirectly to PA via a naphthoquinone
intermediate, NQ. Some phthalic anhydride is further oxidized to maleic
anhydride, MA, and to CO and CO_2:

Steps 1 and 2 are first order to naphthalene and to oxygen; but with excess air, they can be treated as pseudo-first-order and combined to give a first-order rate constant for naphthalene conversion. This offers a chance to test various fluid-bed models that have been developed for first-order reactions. Johnson and coworkers [33] compared the conversions and yields predicted by three different single-bubble models with data provided by Badger Engineers, Inc. Although the predicted PA yields of 86–88% were close to the measured value of 89%, the calculated outlet naphthalene concentrations were 2.13, 1.28, and 0.14% of the feed concentration, whereas the actual concentration was too low to measure and was probably less than 0.01% of the feed. As pointed out by Bolthrunis [32], the large differences in the predicted and actual amounts of unconverted naphthalene indicate that the bubble models are not relevant for commercial reactors, at least for those operating at moderate to high velocity.

The two-phase model can be tested using the Badger data and empirical values of the interchange parameter. In the following example, Model II is chosen, since the K values in Figure 9.12 were calculated using the same model.

Example 9.2

Operating conditions for naphthalene oxidation are given. Use Model II to predict the fraction of naphthalene unconverted.

Reactor: $D = 2.13$ m, $L = 7.9$ m (expanded bed)

Catalyst: $\overline{d_p} = 53$ μm, 28% < 44 μm

$u_{mf} = 0.077$ cm/sec

$u_{mb} = 0.5$ cm/sec

$\rho_{bulk} = 770$ kg/m^3, $\quad \epsilon_m = 0.44$

$\rho_b = 350$ kg/m^3 (expanded bed)

Reaction at 636 K, 266 kPa:

$k_1 = k_2 = 1.8$ sec^{-1}, based on particle volume

$u_o = 0.43$ m/sec

$C_o = 2\%$

Solution.

$$k = k_1 + k_2 = 3.6 \ \text{sec}^{-1}$$

$$1 - \epsilon = (1 - 0.44)\left(\frac{350}{770}\right) = 0.255$$

Corrected $k = 3.6(0.255) = 0.918 \ \text{sec}^{-1}$, based on bed volume

$$N_r = \frac{kL}{u_o} = \frac{0.918(7.9)}{0.43} = 16.9$$

From Figure 9.12, at 0.43 m/sec, $K = 0.7$–$1.2 \ \text{sec}^{-1}$. Choose $K \cong 0.8 \ \text{sec}^{-1}$:

$$N_m = \frac{KL}{u_o} = \frac{0.8(7.9)}{0.43} = 14.7$$

$$\frac{1}{N} = \frac{1}{N_m} + \frac{1}{N_r} = \frac{1}{14.7} + \frac{1}{16.9}$$

$$N = 7.86$$

$$1 - x = e^{-7.86} = 3.9 \times 10^{-4}$$

If $C_O = 2\%$,

$$C_{out} = 2(3.9 \times 10^{-4}) = 7.8 \times 10^{-4}\% = 8 \ \text{ppm}$$

The bed described in Example 9.2 was probably operating in the turbulent fluidization region, since the bed height was over twice the initial value. Therefore it would have been surprising if the bubble models had been adequate. Even for beds operating at somewhat lower velocity in the bubbling bed region, the two-phase model, with an empirical correlation for K, is expected to give a better prediction of reactor performance than the single-bubble models.

The use of fluidized beds for gas-phase polymerization started in 1968 with the UNIPOL™ process, which was developed by Union Carbide to make high-density polyethylene. This process has now been adapted to produce other grades of polyethylene as well as polypropylene and various copolymers. The fluid bed is composed of porous particles, which are aggregates of polymer containing fine grains of titanium, chromium, or other metal catalyst. Polymerization takes place at the polymer–catalyst interface, and the particles grow larger over a period of several hours. Some of the polymer is withdrawn continuously or at intervals to maintain the bed inventory, and catalyst is added to make up for what is lost in the product. The catalyst content of the product is small enough (about 0.01%) so that it does not have to be removed. Typical reaction conditions are 90–110°C, a pressure of 10–30 atm, and a conversion of 2–5% per pass, with a large recycle ratio [34]. A flow diagram is shown in Figure 9.19.

The fluidization conditions and design problems for polymer production are different from those for catalytic cracking or partial oxidation. The average size of the polymer particles is 300–500 microns, and the bed is operated at gas velocities only 2–5 times u_{mf}. The particles are of the class B type and show no region of particulate fluidization. Large bubbles are formed, up to 0.7 m in diameter [35], and lab or pilot-plant reactors often operate in the slugging regime. Commercial reactors are 2–6 meters in diameter, and they operate in the bubbling or turbulent regime. In spite of these differences, scaleup is not a problem as far as conversion is concerned. A low conversion is typical, because a high gas flow is needed to keep the bed fluidized and to facilitate heat removal. With little or no diluent in the feed and less than 5% conversion, there is essentially no difference between the gas in the bubbles and the gas in the dense phase, so gas interchange and axial mixing of the gas are not important. In some reactors, hydrogen is added to the feed to control the molecular weight, but the amount used is too low to introduce mass transfer effects.

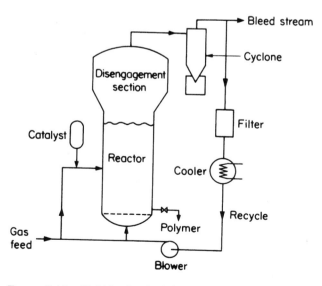

FIGURE **9.19** Fluid-bed polyolefin reactor.

The polymerization is quite exothermic, and heat must be removed to control the reactor temperature. This is usually done with an external cooler, since heat exchanger tubes in the bed might become coated with polymer. Close control of temperature is important, since local hot spots can lead to softening or melting and formation of large aggregates [36]. For large reactors, the rate of polymer production may be limited by the rate of heat removal. An increase in capacity can be achieved by cooling all or part of the recycle gas below the dew point so that the feed gas has suspended droplets of liquid [37]. The droplets evaporate on contact with the polymer particles and provide additional cooling.

NOMENCLATURE

Symbols

a	fraction of catalyst in bubbles
C	reactant concentration
C_B	reactant concentration in bubbles
C_D	reactant concentration in dense phase
C_O	reactant concentration in feed gas
D	diameter
D_b	diameter of bubble
D_{AB}	molecular diffusivity
D_{ea}	effective axial diffusivity
d_p	particle diameter
$\bar{d}_p$	average particle diameter
g	gravitational constant
h	bed height
h_o	initial height
K	interchange coefficient, $\sec^{-1}$ or ft^3 gas/sec, ft^3 bed
K_o	overall coefficient
k	reaction rate constant
L	bed length
l	distance
N	Overall number of units for mass transfer plus reaction
N_m	number of mass transfer units, KL/uo
N_r	number of reaction units, $k\rho_b L/u_o$
P	pressure
ΔP	pressure drop
Pe'	Peclet number, $u_o L/D_{ea}$
S	selectivity, cross-sectional area
u_o	superficial velocity

u_c	critical velocity for turbulent fluidization
u_{mf}	minimum fluidization velocity
u_{mb}	minimum bubbling velocity
u_B	flow carried by bubbles per unit cross section
u_D	dense-phase flow per unit cross section
v_b	bubble velocity
v_t	terminal velocity
x	fraction conversion
x_B	fraction conversion in bubbles
x_D	fraction conversion in dense phase

Greek Letters

α	volume fraction bubbles in bed
β	coefficient in Eq. (9.10)
ϵ	void fraction
ϵ_m	void fraction at minimum fluidization
μ	viscosity
ρ	density
ρ_b	density of bed
ρ_p	density of particles
ρ_s	density of solid
ρ_{bulk}	density of settled bed
ϕ	shape factor

PROBLEMS

9.1 Microspheroidal catalyst particles with a mean size of 70 microns and a particle density of 0.9 g/cm^3 are fluidized with air at 1 atm and 120°C.

 a. Predict u_{mf} and u_{mb} assuming $\epsilon_m = 0.45$.
 b. Predict the bed expansion in the region of particulate fluidization, and plot h/h_o versus u_o.
 c. What is the bed density in lb/ft^3 and in kg/m^3 at u_{mf} and at u_{mb}?

9.2 A fluidized bed for flue-gas cleaning uses 1/16-inch Al$_2$O$_3$ spheres impregnated with a catalyst. The bed is operated at 1.1 atm and 400°C, and the particle density is 1.25 g/cm^3.

 a. Predict the minimum fluidization velocity.
 b. Estimate the bed expansion at $u_o = 3\ u_{mf}$ for different assumed bubble sizes.

9.3   The ozone decomposition reaction was studied in an 18-inch-diameter fluidized reactor [17]. With 68-μm particles and a sintered-plate distributor, the conversion was approximately the same as predicted for a perfectly mixed reactor, though for the same N_r, the conversions were somewhat dependent on gas velocity, as shown in Table 9.3. The initial bed height was 2.17 feet.

    a.   Use the simple two-phase model to calculate the gas interchange coefficients and compare with values in Figure 9.12.
    b.   How would the values of K change if allowance were made for gas flow in the dense phase?
    c.   How would the values of K change if some allowance were made for axial dispersion?

9.4   The catalytic cracking of cumene to propylene and benzene was studied at 800°F using a fluidized bed 3 inches in diameter [7]. The silica-alumina catalyst had 13% Al_2O_3 and a BET surface of 490 m^2/g. The 100- to 200-mesh fraction of the catalyst was used after fluidizing for several hours to remove fines. The predicted equilibrium conversion was 0.77, and in many of the fixed bed tests this value was almost reached. With a porous-plate distributor and 8-inch initial bed height, the conversion was 62% at 0.1 ft/sec and 50% at 0.2 ft/sec. Treating the reaction as pseudo-first-order, N_r was estimated to be 8.4 for 0.1 ft/sec and 4.2 for 0.2 ft/sec.

    a.   Estimate the minimum fluidization velocity and the fraction of the gas passing through the bed as bubbles for these runs.
    b.   Use Model II or Model III to estimate the gas interchange coefficients. Compare these values with the authors' coefficients of 130 and 290 hr^{-1}, which were based on Model I.

TABLE 9.3   Data for Problem 9.3

u, ft/sec	$\dfrac{u}{u_{mf}}$	N_r	$1-x$
0.0472	4	5	0.11
0.0472	4	10	0.08
0.0944	8	5	0.16
0.0944	8	10	0.12
0.146	12	5	0.19
0.146	12	10	0.13

9.5   For the naphthalene oxidation conditions given in Example 9.2, what would be the fraction unreacted for operation at 0.2 m/sec or 0.1 m/sec, about half or one-quarter the normal velocity? Assume the bed expansion is a linear function of the gas velocity.

9.6   The partial oxidation of hydrocarbon A to product B is carried out with a large excess of air at 350°C. The pseudo-first-order rate constant $k\rho_b$ is 2.6 sec^{-1}. Product B oxidizes to CO_2, but the rate constant is only 0.2 sec^{-1}.

  a.   For a fixed-bed reactor operating at 1.0 ft/sec, what bed length gives the maximum yield of B? What is the conversion of A at this point?
  b.   For a fluidized-bed reactor, what is the maximum yield and bed length if $u_o = 1$ ft/sec and mixing in the dense phase can be neglected?
  c.   What is the maximum yield if the fluid bed is completely back-mixed?

9.7   An 8-foot-diameter fluidized reactor is to be designed for a superficial velocity of 1.5 ft/sec and a conversion of 85%. The reaction is first order and irreversible, and the rate constant based on tests in a small fixed bed ($\epsilon = 0.40$) is $k\rho_b = 0.18$ sec^{-1}. The minimum fluidization velocity is 0.03 ft/sec, and the estimated bed expansion is 40% at 1.5 ft/sec.

  a.   What bed height would be needed for an isothermal fixed-bed reactor with plug flow of gas?
  b.   Estimate the bed height for the fluidized reactor if there is negligible axial mixing in the dense phase, no reaction in the bubbles, and a constant value of the interchange parameter.
  c.   Estimate the required bed height based on the axial diffusion model and published values of D_{ax}.

REFERENCES

1.   S Simone, P Harriott. Powder Technology 26:161, 1980.
2.   D Geldart. Powder Technology 7:285, 1973.
3.   JF Davidson, D Harrison. Fluidized Particles. Cambridge, UK: Cambridge University Press, 1963.
4.   R Cliff. In: Gas Fluidization Technology. D Geldhart, ed. New York: Wiley, 1986.
5.   JF Mathis, CC Watson. AIChEJ 2:518, 1956.
6.   WK Lewis, ER Gilliland, W Glass. AIChEJ 5:419, 1959.

7. A Gomezplata, WW Shuster. AIChEJ 6:454, 1960.
8. JW Askins, GP Hinds Jr, F Kunreuther. Chem Eng Progr 47:401, 1951.
9. PN Rowe, BA Partridge, E Lyall. Chem Eng Sci 19:973, 1964.
10. D Kunii. O Levenspiel. Fluidization Engineering. 2nd ed. Boston: Butterworth-Heineman, 1991.
11. RW Fontaine, P Harriott. Chem Eng Sci 27:2189, 1972.
12. RJ de Vries, WPM van Swaaij, C Mantovani, A Heijkoop. 2nd Int Symp Chem Reaction Eng, 1972, pp 39–59.
13. T Chiba, H Kobayashi. Chem Eng Sci 25:1375, 1970.
14. CY Shen, HF Johnstone. AIChEJ 6:454, 1960.
15. WG May. Chem Eng Progr 55(12):49, 1959.
16. G Sun, JR Grace. Chem Eng Sci 45:2187, 1990.
17. FD Toor, PH Calderbank. In: AAH Drinkenburg, ed. Proceedings of the Symposium on Fluidization. Amsterdam: Netherlands Univ. Press, 1967.
18. WPM Van Swaaij, FJ Zuiderweg. 2nd Int Symp Chem Reaction Eng, 1972, pp 9–25.
19. G Sun. PhD dissertation. Univ. of British Columbia, Vancouver, 1991.
20. L Massimilla. AIChE Symp Ser 69(128):11, 1973.
21. M Pell, SP Jordan. Paper presented at AIChE meeting, New York, Nov. 1987.
22. M Pell. Gas Fluidization. Handbook of Powder Technology. Vol 8. New York: Elsevier, 1990.
23. JH de Groot. Proceedings Int Symp on Fluidization, Amsterdam, Netherlands Univ. Press, 1967, p 348.
24. F De Maria, JE Longfield. Chem Eng Prog Symp Ser 58(38):16, 1962.
25. F de Maria, JE Longfield, G Butler. Ind Eng Chem 53:259, 1961.
26. HS Mickley, DF Fairbanks. AIChEJ 1:374, 1955.
27. C Beeby, OE Potter. AIChEJ 30:977, 1984.
28. D Kunii, O Levenspiel. Ind Eng Chem Res 30:136, 1991.
29. JSM Botterill. Fluid-Bed Heat Transfer. New York: Academic Press, 1975.
30. D Kunii, O Levenspiel. Fluidization Engineering. 2nd ed. Boston: Butterworth-Heineman, 1991.
31. A Arbel, Z Huang, IH Rinard, R Shinnar, AV Sapre. Ind Eng Chem Res 34:1228, 1995.
32. CO Bolthrunis. Chem Eng Progr 85(5):51, May 1989.
33. JE Johnson, JR Grace, JJ Graham. AIChEJ 33:619, 1987.
34. T Xie, KB McAuley, JCC Hsu, DW Bacon. Ind Eng Chem Res 33:449, 1994.
35. H Koda, T Kurisaka. In: M Kwauk, D Kunii, ed. Fluidization 1985. Beijing: Science Press, 1985, p 402.
36. EM Ali, AE Abasaeed, SM Al-Zahrani. Ind Eng Chem Res 37:3414, 1998.
37. ID Burdett, RS Eisinger, P Cai, KH Lee. Paper presented at Miami meeting of AIChemE, 1998.

10

Novel Reactors

RISER REACTORS

In modern FCC units, the cracking reaction takes place in a vertical pipe, or riser, where hot catalyst introduced at the bottom is carried upward at high velocity by lift gas plus the hydrocarbon vapors formed from the oil feed. The riser may be 1–2 m in diameter and up to 40 m tall, and the exit vapor velocity is generally 15–20 m/sec [1], giving a gas residence time of a few seconds. At the top of the riser, catalyst is recovered by cyclones, stripped with steam, and sent to the regenerator. A sketch of an FCC unit was shown in Figure 9.18. The mass flow rate of catalyst in the riser is usually about six times that of the gas, but the volume fraction of solids in the reactor is quite small. Over much of the length, the average suspension density is about 50 kg/m^3, corresponding to a solids fraction $(1 - \epsilon) \cong 0.04$, but near the inlet, where the particles are accelerated, the suspension density may be three to four times greater [2].

A major advantage of the riser reactor for catalytic cracking is that the gas and solid move in nearly plug flow, which gives more uniform catalytic activity and better selectivity than with a bubbling or turbulent fluidized bed. A riser reactor can be used for other rapid catalytic reactions, such as the production of acrolein from propylene [3] or the partial oxidation of n-butane to make maleic anhydride. In Du Pont's butane oxidation process

[4], some or all of the required oxygen is supplied by lattice oxygen, and the catalyst is partially reduced in the process. The entrained catalyst is regenerated with air in a separate vessel before being returned to the reactor. The advantages of this process over a single fluidized bed are better control of the two steps in the process and freedom to use a higher butane concentration without risk of explosion.

The terms *fast fluidized bed* (FFB) and *circulating fluid bed* (CFB) are also used for upflow reactors where the gas velocity is high and the solids concentration quite low. However, unlike the riser reactors for catalytic cracking or butane oxidation, most CFB units return all the entrained solids to the base of the reactor without any processing step. Particles from the cyclones are collected in a standpipe, which is kept fluidized at low velocity, and the solids are sent at a controlled rate to the bottom of the reactor. The CFB units are used mainly for gas–solid reactions such as the combustion of coal or other solid fuels and processing of metal ores or other inorganic compounds. Usually the particles are type B solids, with sizes up to a few millimeters in diameter, and they may pass through the reactor and cyclone many times before being consumed or discharged. The gas velocities range from 2–10 m/sec, somewhat lower than in FCC risers. A great many experimental and modeling studies of CFB reactors are discussed in recent reviews [5,6], but the fluid dynamics are still not completely understood. Models and empirical correlations derived for one reactor may give a poor fit when tested for others solids or flow conditions. A few studies are cited here to illustrate the general characteristics of riser or CFB reactors, but detailed models and design procedures are not discussed. The focus is on riser reactors operating with type A catalysts with once-through flow of solid and gas.

Suspension Density

In a riser reactor, the reaction rate depends on the suspension density, usually expressed in kg/m^3. Predicting the density is difficult, because there are axial and radial density gradients, which may cover a several-fold range of values. The suspension density is much greater than if the particles acted independently and had a slip velocity equal to the terminal velocity. For example, with 60–μm FCC catalyst, $v_t \cong 0.1$ m/sec, and if $u_g =$ 15 m/sec, the particle velocity for the ideal case would be 14.9 m/s, almost equal to the gas velocity. However, based on tracer studies and density measurements, FCC risers operate with particle velocities that are much less than the gas velocity. Experimental results are often expressed as a slip factor, ψ, the ratio of actual gas and particle velocities:

$$\psi = \frac{u_g}{\epsilon v_p} \tag{10.1}$$

A slip factor of 2 is considered typical, but values of ψ may range from 1.2 to 4, and no reliable correlation is available. For the preceding example, if $\psi = 2.0$, $v_p \cong 8$ m/sec, and $u_g - v_p \cong 7$ m/sec, which is 70 times v_t. This large difference is due to the formation of loose aggregates or clusters of particles, which have much greater terminal velocities than single particles. Theory shows that a uniform suspension of fine particles in a gas is unstable, and the particles will tend to form clusters, but the degree of aggregation is not yet predictable.

The average catalyst density in the suspension is proportional to the mass flow rate G_s and varies inversely with the particle velocity:

$$\rho_p(1 - \epsilon) = \rho_s = \frac{G_s}{v_p} \tag{10.2}$$

In an FCC unit with $G_s = 400$ kg/sec, m^2 and $v_p \cong 8$ m/sec, $\rho_s \cong 50$ kg/m^3. Since the particle density is about 1200 kg/m^3, the average volume fraction catalyst in the riser would be 50/1200, or 0.042. This is an order of magnitude less than the catalyst concentration in a bubbling bed.

In a large riser, the bed density can be determined from pressure measurements, since the wall friction is small. Above the inlet region, the pressure drop is proportional to the suspension density:

$$\frac{dP}{dL} = g\rho_p(1 - \epsilon) \tag{10.3}$$

Near the inlet, the pressure gradient is higher because of the energy needed to accelerate the solids. The suspension density is also higher, though Eq. (10.3) is no longer valid when the solids are accelerating [4]. The entry region can be less than 1 meter or up to several meters in length, depending on the solids flow rate, the gas rate, and the gas density [7]. For FCC units, the entrance region is a small fraction of the riser length, but for laboratory reactors it may be relatively large. Near the top of the riser, the suspension density gradually increases if the flow direction changes abruptly at the exit [8,9], as shown in Figure 10.1. The use of a blind tee reduces erosion, since a pad of solids forms at the top of the tee. Clusters of solids disengage from the gas and fall back, increasing the density. With a large-radius bend, there is no change in density near the exit [10].

Radial gradients in solids concentration and solids velocity have been measured in small laboratory reactors and in a few commercial FCC risers. Solids move downward at the wall, since the gas velocity is zero, but even at moderate distance from the wall the solids still move down or are nearly stationary. The riser has a core where gas and solid move up at high velocity and an annular region where the solids concentration is high and the solids are nearly stationary or moving downward. Measurements show a core

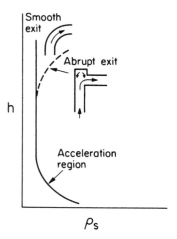

FIGURE 10.1   Effect of riser exit geometry on suspension density. (After Ref. 10.)

radius, r_c, of $0.9R–0.95R$, so the core occupies 80–90% of the riser area [9]. In the annulus, the average suspension density is a few times higher than in the core, but the density is less than in the dense phase of a bubbling bed. The solids move in strands or clusters and rapidly interchange with solids in the core, though there is no definite boundary between the core and annulus. The radial density profile in a commercial FCC riser is shown in Figure 10.2 [11]. The unsymmetrical profile is caused by solids entering at one side of the riser. It is difficult to get uniform distribution of the feed even where two or more feed points are used.

Kinetic Models

A number of workers have used core–annulus models for riser reactors [3,4,8,12]. In the simpler models, all the gas is assumed to flow in the central core, as shown in Figure 10.3. The core gas velocity is higher than the superficial velocity by the ratio of total area to core area. The solids in the core are in clusters that are carried upward at about half the gas velocity. In the annular region, which contains one-third to one-half the total solids, there is no net flow of gas or solids, though there may be slight downflow very near the wall and upflow further away from the wall. There is interchange of gas and solid between the annulus and the core. In some models, this interchange is described using an area-based mass transfer coefficient, and correlations similar to those for wetted-wall columns have been proposed [3,4]. However, considering the erratic movement of the solids and the

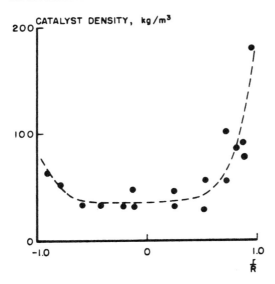

FIGURE **10.2**   Typical radial catalyst density profile in feed riser (From Ref. 11.)

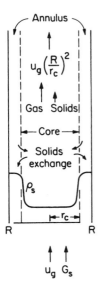

FIGURE **10.3**   Core–annulus model for a riser reactor.

absence of a definite boundary between the core and annulus, it seems likely that the interchange process is dominated by turbulent movements and not by molecular diffusion.

The interchange coefficient, k_c, has been estimated from the response to pulse inputs of tracer gas [8,13]. Patience and Chaouki [8] used sand in a 0.083-m riser and reported k_c values (their k) that ranged from 0.03 to 0.08 m/sec. The values increased with gas velocity, but with considerable scatter in the data. White and coworkers [13] used sand and FCC catalyst in a 0.09-m riser and found k_c (their k_{ca}) values of 0.05–0.02 m/sec that decreased with gas velocity. The interchange coefficient k_c is based on a unit core–annulus area and can be converted to a volumetric coefficient K for comparison with other models. For $k_c = 0.03$ m/sec, $K \cong 1.2$ sec^{-1}, which seems the right order of magnitude based on the values shown in Figure 9.12. However, if k_c is independent of diameter, D, the volumetric coefficient K will vary inversely with D if r_c/R is constant.

More direct evaluation of the interchange coefficient from conversion measurements would be helpful, but analysis of FCC data is difficult because of the complex kinetics and large changes in velocity and temperature in the riser. The ozone decomposition reaction was used as a test reaction by Ouyang and Potter [6]. In a 0.25-m riser with gas velocities of 2–8 m/sec, the overall ozone conversion was only 5–30%, and never was more than half the conversion predicted for an ideal reactor. Such low conversions are surprising, since a core–annulus model with half the catalyst in the core would predict considerably higher conversions for any reasonable value of k_c or K. The recommended model to use if more data become available is the two-phase model of Glass, which was presented in Chapter 9 [Ref. 6 and Eq. (9.29)]. Parameter a was the fraction of the catalyst in the bubbles and now becomes the fraction of catalyst in the core region. For a first-order reaction with plug flow in both regions, the conversion is [Eq. (9.32)].

$$\ln\left(\frac{1}{1-x}\right) = \frac{k\rho_b L}{u_G}\left(a + \frac{K(1-a)}{K + k\rho_b(1-a)}\right)$$

The effect of high catalyst concentrations near the inlet and exit of the reactor could be allowed for by using separate equations for each section of the reactor, where a and K might be different.

Axial mixing in commercial FCC risers has been measured using radioactive argon as a tracer gas and irradiated catalyst particles as a solids tracer. For a 1.30-m riser and $u_g \cong 10$ m/sec, Viitanen [2] reported D_{ea} was 9–23 m^2/sec for the gas phase, with higher values near the top of the riser. Comparable values for the catalyst were 3–15 m^2/sec, but the average particle velocity was only half the gas velocity, and the residence time distribution of the catalyst was slightly broader than for the gas. Tracer studies by

Martin et al. [9] in a 0.94-m riser gave axial dispersion coefficients for the catalyst of $10-18$ m^2/sec, in reasonable agreement with other studies. Although these values of D_{ea} are more than an order of magnitude greater than those shown in Figure 9.14 for fluidized beds, the high gas velocities and reactor lengths make the effect of axial dispersion in risers fairly small. For example, if $D_{ea} = 20$ m^2/sec, $L = 30$ m, and $u_g = 10$ m/sec, then $Pe = 10 \times 30/20 = 15$. As shown in Figure 6.12, for $Pe = 10-20$, the conversion is much closer to that for plug flow than for perfect mixing. Considering the uncertainty in predicting the radial catalyst distribution and the gas interchange parameter, it seems reasonable to neglect axial dispersion when using a core–annulus model for the reactor.

MONOLITHIC CATALYSTS

Monolithic catalysts present an alternative to the use of small particles of catalyst for very rapid reactions of gases. The monolith may be a ceramic or metal support similar to a honeycomb, with square, hexagonal, triangular, or sinusoidal-shaped cells that are separated by thin walls to give a large number of parallel channels. The catalyst is deposited on the cell walls or impregnated into a thin coating of Al_2O_3 or other porous support. The cross section of an extruded ceramic monolith with a catalyzed wash coat is shown in Figure 10.4. This monolith has 40 cells/cm^2, or 260 cells/in.^{2} (cpsi), and monoliths with up to 400 cpsi are commercially available. The advantages of monolithic catalysts are low pressure drop, high external surface area, and minimal loss of catalyst by attrition or erosion.

The major use of monoliths is in catalytic converters for gasoline-fueled vehicles. All new cars need a converter to control emissions of carbon monoxide, hydrocarbons, and nitric oxide, and most use a monolith impregnated with Pt or Pd and Rh. Even though the catalysts have 0.1% or less of noble metal, the annual cost of these units exceeds that of all other catalysts sold for the petroleum and chemical industries.

Catalyst monoliths are also effective in the control of air pollution from stationary sources. They have been used for many years to oxidize hydrocarbon vapors in the vent streams from chemical plants and to reduce solvent emissions from printing and cleaning processes. More recent applications include CO removal from gas turbine exhaust and the selective catalytic reduction of NO in flue gas. Performance curves for the oxidation of various compounds over a Pt/Al$_2$O$_3$ catalyst are shown in Figure 10.5, where the conversion is plotted against the feed temperature. The reactors operate adiabatically, and the exit temperature may be 10–100°F above the feed temperature. At first, the conversion increases exponentially with temperature, as expected from the Arrhenius relationship. The decrease in slope

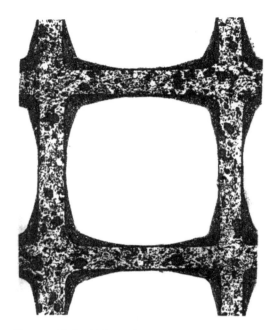

FIGURE 10.4 Enlarged cross section of a ceramic monolith. (From Engelhard Company.)

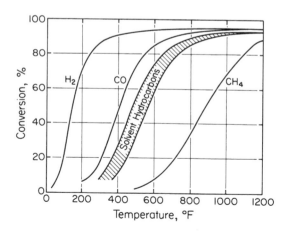

FIGURE 10.5 Typical conversion curve for a Pt/Al$_2$O$_3$ catalyst. (From Ref. 14. Reproduced with permission of the American Institute of Chemical Engineers. Copyright 1974 AIChE. All rights reserved.)

at moderate conversion is due to the change in reactant concentration and to the increasing importance of mass transfer. If the reaction is first order, the intrinsic rate constant k and the mass transfer coefficient k_c can be combined to give an overall rate constant K_o:

$$r = K_o C \tag{10.4}$$

$$\frac{1}{K_o} = \frac{1}{k} + \frac{1}{k_c} \tag{10.5}$$

The mass transfer coefficient increases only slightly with temperature, so above a certain temperature the reaction becomes mass transfer controlled. Further increases in temperature give almost no change in conversion. The transition to mass transfer control occurs at a lower temperature for very reactive species, such as H_2 and CO, than for hydrocarbons, but the kinetics of oxidation are often not known. The design temperature and flow rate are based on lab tests or experience with similar materials. The reactor is usually operated in the mass transfer control regime, where the conversion depends on the rate of mass transfer and the gas flow rate.

Mass Transfer Coefficients

Gas flow through the small channels of a honeycomb matrix is nearly always laminar, and analytical solutions are available for heat and mass transfer for fully developed laminar flow in smooth tubes. In the inlet region, where the boundary layers are developing, the coefficients are higher, and numerical solutions were combined with the analytical solution for fully developed flow and fitted to a semitheoretical equation [14]:

$$\mathrm{Sh} = 3.66\left(1 + 0.078\,\mathrm{ReSc}\frac{d}{L}\right)^{0.45} \tag{10.6}$$

Experimental data gave somewhat higher coefficients, and the correlation was modified to reflect the increase [14]:

$$\mathrm{Sh} = 3.66\left(1 + 0.095\,\mathrm{ReSc}\frac{d}{L}\right)^{0.45} \tag{10.7}$$

The limiting value of 3.66 is for tubes of circular cross section. Other values would be used for triangular or square cross sections.

There are more data for heat transfer in laminar flow than for mass transfer, and the correlations should be similar, with Pr and Nu replacing Sc and Sh. An empirical equation for heat transfer at Graetz numbers greater than 20 is [15]

$$\mathrm{Nu} = 2.0\mathrm{Gz}^{1/3} = 1.85(\mathrm{RePr}\frac{d}{L})^{1/3} \tag{10.8}$$

$$\mathrm{Gz} \equiv \frac{\dot{m}c_p}{kL} = \frac{\pi}{4}\mathrm{RePr}\frac{d}{L}$$

The exponents and the forms of the equations are different, but, as shown by Figure 10.6, the Nussselt numbers predicted by Eq. (10.8) are only 15–20% less than the Sherwood numbers given by Eq. (10.7). Considering that both equations are modifications of theoretical equations and may include effects of surface roughness and natural convection, the agreement is reasonably good.

Other studies of mass transfer in monoliths have given coefficients much lower than those predicted by Hawthorn's correlation, Eq. (10.7). Votruba and coworkers [16] reported heat and mass transfer coefficients for evaporation of liquids from a wetted ceramic monolith. Their empirical equations include $\mathrm{Re}^{0.43}$ and $\mathrm{Sc}^{0.56}$, and the predicted Sherwood numbers for $\mathrm{Sc} = 2$ are shown in Figure 10.6. Much of the data fall below the minimum Sherwood number of 3.66, and the calculated Nusselt and Sherwood numbers were quite different, which could be due to errors in temperature measurement. In the work of Ullah et al. [17], CO was oxidized with stochiometric O_2 in ceramic monoliths at temperatures high enough to ensure mass transfer control. High conversions were obtained, but most of the Sherwood numbers were below 3.66. It would have been interesting to repeat some of the test with excess O_2 so that gas blending would be less important. The results of Heck and Farrauto [18] for oxidation of CO and C_3H_6 on different supports fall close to those of Hawthorn at high values of ReScd/L, but there is a lot of scatter in the data, and some of the Sherwood numbers are less than 3.66.

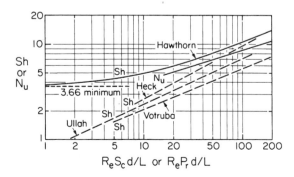

FIGURE 10.6 Mass and heat transfer in catalyst monoliths.

Faced with the widely different results shown in Figure 10.6, how can the appropriate correlation for design be selected? Sherwood numbers less than the theoretical minimum cast doubt on the accuracy of the measurements, since factors such as surface roughness and entrance effects would only increase the average coefficient. A wide distribution of channel sizes would lower the average coefficient, since a disproportionate fraction of the total flow would go through the larger openings. The ceramic monoliths appear very uniform in cell size, and flow maldistribution is unlikely. However, if a few cells do not receive a catalyst coating, complete conversion of reactant would not be possible, and low apparent mass transfer coefficients would result. Pending further studies, it is recommended that coefficients be predicted using Eq (10.7) or the upper part of the Heck–Farrauto plot in Figure 10.6. Suppliers of catalytic monoliths should be able to provide test data to confirm the predicted performance.

Design Equations

When the temperature is high enough to make mass transfer the controlling step, the conversion can be predicted from the flow rate, the external area, and the mass transfer coefficient. For an incremental length of monolith and a surface concentration equal to zero, the mass balance is

$$u_o \, dC = -k_c a C \, dl \tag{10.9}$$

where

u_o = superficial velocity

a = external area per unit volume of monolith

With an exothermic reaction, the temperature increases with conversion, but for integration, k_c is assumed constant at the average value:

$$\ln \frac{C_o}{C} = \ln \frac{1}{1-x} = \frac{k_c a L}{u_o} \tag{10.10}$$

The effects of flow rate and length can also be expressed using the space velocity (see Chapter 5):

$$SV = \frac{F}{V} = \frac{u_o S}{SL} = \frac{u_o}{L} \tag{10.11}$$

$$\ln \frac{1}{1-x} = \frac{k_c a}{SV} \tag{10.12}$$

The space velocity is often given in terms of the gas flow rate at standard conditions, so for Eq. (10.12), the space velocity is corrected to reaction conditions:

$$\text{SV} = \text{SV}_{\text{STP}}\left(\frac{T}{273}\right)\left(\frac{1}{P}\right) \tag{10.13}$$

The reciprocal of SV is a time, but it is not the residence time in the reactor, and it has no fundamental significance.

For catalytic combustion, the space velocities range from 10,000 hr^{-1} to 200,000 h^{-1} (STP), with larger values for the monoliths with high cell densities. The effects of space velocity and temperature on CO conversion in a medium-density monolith are shown in Figure 10.7. The almost-level plots show that mass transfer is controlling for temperatures above 250°C. Because of the increase in diffusivity with temperature ($D\alpha T^{\sim1.7}$), there should be some effect of temperature on conversion. However, the superficial velocity is proportional to T for tests at constant SV_{STP}, and, as shown by Eq. (10.10), the increase in u_o cancels much of the effect of increased k_c. For a change in temperature from 300 to 450°C at $\text{SV}_{\text{STP}} = 120,000$, the CO conversion should increase by about 2%.

Example 10.1

A monolithic converter with 100 cpsi and a length of 6 inches is used to oxidize CO in the exhaust stream from a gas turbine. The catalyst is 0.2%

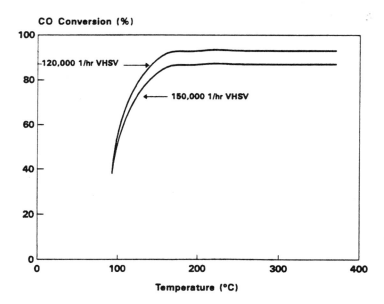

FIGURE 10.7 Oxidation of CO in a 100 cpsi monolith. (From Ref. 18.)

Pt/Al_2O_3, and the feed temperature is 300°C, high enough to make mass transfer the controlling step. What is the expected conversion for space velocities of 60,000 hr^{-1} and 120,000 hr^{-1} based on STP?

Solution. Assume the uncoated monolith has square cells with a wall thickness of 0.04 cm.

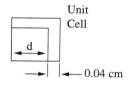

Unit
Cell

d

0.04 cm

$$100 \, \text{cpsi} = \frac{100}{(2.54)^2} = 15.5 \text{ cells/cm}^2$$

$$(d + 0.04)^2 = \frac{1}{15.5} = 0.0645 \text{ cm}^2$$

$$d = 0.214 \text{ cm}$$

$$\epsilon = \frac{d^2}{(d + 0.04)^2} = 0.71$$

Assume the wash coating lowers d to 0.21 cm and ϵ to 0.68:

$$a = \frac{4}{0.21} \times 0.68 = 12.9 \text{ cm}^2/\text{cm}^3$$

$SV_{STP} = 60,000 \, hr^{-1}$ so

$$SV = \frac{60,000}{3600} \frac{573}{273} = 35.0 \text{ sec}^{-1} \text{ at } 300°C$$

$$L = 6 \times 2.54 = 15.2 \text{ cm}$$

Since $SV = u_o/L$,

$$u_o = 35(15.2) = 532 \text{ cm/sec}$$

$$u = \frac{u_o}{\epsilon} = \frac{532}{0.68} = 782 \text{ cm/sec}$$

For air, $\mu = 0.0284$ cp, $\rho = 6.17 \times 10^{-4} \text{g/cm}^3$, so

$$\frac{\mu}{\rho} = 0.460 \text{ cm}^2/\text{sec}$$

$$Re = \frac{0.21(782)}{0.46} = 357$$

From the table of diffusion coefficients for binary gas mixtures, p. ix, $D_{CO-N_2} = 0.192 \text{ cm}^2/\text{sec}$ at 288 K. At 573 K,

$$D_{CO-N_2} \cong 0.192\left(\frac{573}{288}\right)^{1.7} = 0.618 \text{ cm}^2/\text{sec}$$

$$Sc = \frac{0.46}{0.618} = 0.744$$

$$ReSc\frac{d}{L} = 357(0.744)\left(\frac{0.21}{15.2}\right) = 3.67$$

From Figure 10.6 or Eq (10.7), $Sh = 4.19$, so

$$k_c = \frac{4.19(0.618)}{0.21} = 12.3 \text{ cm/sec}$$

$$\ln\frac{1}{1-x} = \frac{k_c aL}{u_o} = \frac{12.3(12.9)15.2}{532} = 4.53$$

$$1 - x = e^{-4.53} = 0.0108, \qquad 98.9\% \text{ conversion}$$

At 120,000 hr^{-1}, $u_o = 2 \times 532 = 1064$ cm/sec, so

$$Re = 2 \times 357 = 714$$

$$ReSc\frac{d}{L} = 3.67 \times 2 = 7.34$$

$$Sh = 4.64$$

$$k_c = \frac{4.64(0.618)}{0.21} = 13.65 \text{ cm/sec}$$

$$\ln\frac{1}{1-x} = \frac{13.65(12.9)15.2}{1064} = 2.52,$$

$$1 - x = 0.080, \qquad 92\% \text{ conversion}$$

This is in good agreement with the results shown in Figure 10.7.

The cell density has a strong effect on the conversion in catalytic monoliths. Going from 100 to 400 cpsi means a twofold decrease in cell diameter and a twofold increase in a, the area per unit volume. If the Sherwood number increases with about $Re^{1/3}$, as in the intermediate-flow region, a twofold decrease in d and Re gives about a 60% increase in k_c and a 3.2-fold increase in $k_c a$. For the second part of Example 10.1, a 400-cpsi monolith would raise the predicted conversion from 92% to 99.9%. If the Sherwood number is nearly constant (near the minimum value), there would be a fourfold increase in $k_c a$ on going from 100 to 400 cpsi. For clean gases, which do not foul the catalyst, monoliths with 400 cpsi are widely used. For gases with suspended particulate, coarser monoliths are recommended. For selective catalytic reduction of NO (SCR) in flue gas from coal-fired boilers, catalyst monoliths with 6 to 7-mm channels are generally used because of the fly ash. If SCR was used after ash removal by precipitation, much smaller channels could be used, but the gas would have to be reheated to reaction temperature.

As shown by Figure 10.6 and Eq. 10.7, higher values of k_c can be obtained using shorter lengths. Using several short monoliths separated by gaps gives a higher conversion, since the mass transfer rate is high near the inlet, where the boundary layer is not fully developed. Using four 1-inch sections might give as high conversion as a single 6-inch length. However, the savings in the amount of catalyst might be offset by the extra cost of installing and supporting several separate catalyst layers.

WIRE-SCREEN CATALYSTS

When a metal-catalyzed reaction is so fast that external mass transfer controls, several layers of fine wire screen can be used as the catalyst bed. The catalytic oxidation of ammonia to nitric oxide, which is the first step in nitric acid production, is carried out with screens (called *gauzes*) of Pt/Rh alloy, and very high ammonia conversions are obtained. Similar gauzes are used in the Andrussov process for manufacture of HCN from CH_4, NH_3, and O_2. Wire screens are also used for catalytic incineration of pollutants and in improving combustion efficiency in gas burners.

There have been several studies of mass and heat transfer to wire screens, and the work by Shah and Roberts [19] covers the range of Reynolds numbers typical of commercial ammonia oxidation. They studied the decomposition of H_2O_2 on Ag or Pt screens of different mesh size and presented an empirical correlation for the j_D factor: For $5 < Re, \gamma < 245$,

$$j_{D,\gamma} = 0.644(Re, \gamma)^{-0.57} \tag{10.14}$$

where

$$j_{D,\gamma} = \frac{k_c}{u_o/\gamma} Sc^{2/3}$$

$$Re, \gamma = \frac{du_o\rho}{\gamma\mu}$$

γ = minimum fractional opening of a screen

The wire diameter d was chosen for the length parameter in Re, rather than a hydraulic diameter, since this permits direct comparison with correlations for flow normal to cylinders. The use of u_o/γ for the velocity is arbitrary, but it is simpler than trying to calculate an average velocity in the bed of screens. For a simple woven screen with square openings and N wires per unit length, γ is:

$$\gamma = (1 - Nd)^2 \tag{10.15}$$

Equation (10.14) gives coefficients 10–20% less than predicted using data for heat transfer to cylinders and the analogy between heat and mass transfer [19]. This small difference could be due to poorer mass transfer where the wires cross or to the use of u_o/γ rather than an average velocity for Re. Satterfield and Cortez [20] showed that a single screen gave slightly higher coefficients than multiple screens, but the spacing between screens did not affect the coefficients.

The mass transfer area per unit mass of metal varies inversely with d, so fine screens are preferred. However, very fine screens are fragile and are more likely to be damaged during installation or operation. One supplier provides Pt/Rh gauzes with 32 wires per cm, or 1024 openings per cm^2, and wire diameters of 0.076 or 0.06 mm.

When mass transfer of one reactant to the wire surface is the controlling step, the conversion is related to the number of gauzes rather than to the length (thickness) of the bed, since the spacing between gauzes is not important.

For a first-order reaction,

$$-u_o dC = k_c C da \tag{10.16}$$
$$da = a' dn \tag{10.17}$$

where

n = number of gauzes
a' = external area of one gauze per unit cross section

For a gauze with square openings,

$$a' = 2\pi Nd \tag{10.18}$$

where N = the number of wires per unit length. Integration of Eq. (10.16)
gives

$$\ln\left(\frac{C_o}{C}\right) = \ln\left(\frac{1}{1-x}\right) = \frac{k_c a' n}{u_o} \tag{10.19}$$

Ammonia Oxidation

The catalytic oxidation of ammonia is carried out at 1–10 atm and temperatures of 850–950°C in adiabatic reactors up to 6 m in diameter. The reactor
has a double-cone shape, with the gauzes supported at the middle. For
operation at 1 atm, only four or five gauzes are needed, but up to 30 are
used for high-pressure operation. Even with 30 gauzes, the bed is less than
1 cm thick, and the gas residence time is only a few milliseconds.

The feed temperature for the NH_3/air mixture is 250–350°C, but once
ignited the gauzes are at a much higher temperature that is close to the feed
temperature plus the adiabatic temperature rise. Intermediate temperatures
are unstable, and the reactor operates at an upper stable point where the
reaction is mass transfer controlled. The temperature and conversion profiles are sketched in Figure 10.8.

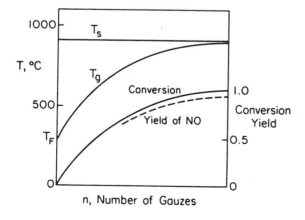

FIGURE **10.8** Temperature and conversion in an ammonia converter.

Example 10.2

Air with 10% NH_3 at 100 psig and 300°C is passed over a pack of Pt/Rh gauzes that have a surface temperature of about 900°C. The superficial velocity at inlet conditions is 1.8 m/sec. The gauzes are made of 0.076-mm wire and have 32 wires per cm.

    a.   Predict the NH_3 conversion as a function of the number of gauzes.

    b.   Assume that the rate of the main byproduct reaction, the decomposition of NO, is proportional to the surface concentration of NO and that the first gauze converts 0.25% of the NH_3 fed to N_2. Calculate the yield of NO as a function of n.

Solution.

    a.   The gas enters at 300°C and leaves at about 900°C, but for simplicity a temperature of 700°C is used to calculate the gas properties.

Main reaction: $NH_3 + 1.25 O_2 \rightarrow NO + 1.5\ H_2O$
By product reaction: $2NO \rightarrow N_2 + O_2$
Feed gas: 10 NH_3, 90 Air (71.1% N_2, 18.9% O_2)

$$M_{ave} = 0.1(17) + 0.9(29) = 27.8$$

$$100\ \text{psig} = \frac{114.7}{14.7} = 7.80\ \text{atm}$$

Assume $\mu = \mu_{air} = 0.0435$ cp at 700°C:

$$\rho = \left(\frac{27.8}{22,400}\right)\left(\frac{273}{973}\right)\left(\frac{7.8}{1}\right) = 2.72 \times 10^{-3}\ \text{g/cm}^3$$

$$u_o \text{ at } 700°C = 1.8\left(\frac{973}{573}\right) = 3.06\ \text{m/sec}$$

$$Re = \frac{7.6 \times 10^{-3}\ \text{cm}(306\ \text{cm/sec})(2.72 \times 10^{-3})}{4.35 \times 10^{-4}} = 14.5$$

From Eq. (10.15),

$$\gamma = [1 - 32(7.6 \times 10^{-3})]^2 = 0.573$$
$$Re, \gamma = \frac{14.5}{0.573} = 25.3$$

$$D_{NH_3}/N_2 = 0.23 \text{ cm}^2/\text{sec at } 298 \text{ K, 1 atm}$$
$$\text{(table, p. ix)}$$

$$D_{NH_3} \cong 0.23 \left(\frac{973}{298}\right)^{1.7} \left(\frac{1}{7.8}\right)$$

$$= 0.22 \text{ cm}^2/\text{sec at } 7.8 \text{ atm, } 700°C$$

$$Sc = \left(\frac{4.35 \times 10^{-4}}{2.72 \times 10^{-3}}\right)\left(\frac{1}{0.22}\right) = 0.727$$

$$Sc^{2/3} = 0.808$$

From Eq. (10.14), $j_{D,\gamma} = 0.644(25.3)^{-0.57} = 0.102$, so

$$k_c = j_{D,\gamma}\left(\frac{u_o}{\gamma}\right)\left(\frac{1}{Sc^{2/3}}\right) = 0.102\left(\frac{306}{0.573}\right)\frac{1}{0.808}$$

$$= 67.4 \text{ cm/sec}$$

$$a' = 2\pi(0.0076)32 = 1.53 \text{ cm}^2/\text{cm}^2$$

$$\ln\left(\frac{1}{1-x}\right) = \frac{k_c a' n}{u_o} = \frac{67.4(1.53)n}{306} = 0.337n$$

The predicted conversions are given in Table 10.1.

b. If $k_{c,NO} \cong k_{c,NH_3}$, the gradient for NO diffusing from the surface will be about equal and opposite to that for NH_3. For the first element of surface, NH_3 is 10% in the gas and 0% at the surface, and NO is about 10% at the surface and 0% in the gas. As C_{NH_3} decreases and C_{NO} increases, the driving forces for reactant and product diffusion are both reduced, as shown in Figure 10.9, and the concentration of NO at the surface is nearly constant. Thus

TABLE 10.1 Ammonia Conversion and Yield

Gauzes (n)	Conversion (x)	Yield (x − 0.0025n)
1	0.286	0.284
2	0.490	0.485
5	0.815	0.802
10	0.966	0.941
15	0.994	0.956
20	0.999	0.949

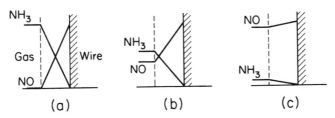

FIGURE 10.9 Gradients of NH$_3$ and NO near the catalyst surface: (a) at inlet; (b) 40% conversion; (c) near exit.

at each gauze, the same amount of NO is converted to N$_2$ by side reactions. The net yield of NO is $x - 0.0025n$, which goes through a maximum of 95.6% at $n \cong 15$, as shown in Table 10.1. This is close to the yield reported for commercial reactors, though some of the yield loss is due to N$_2$O formation, and the detailed kinetics of N$_2$ and N$_2$O formation are uncertain [21].

The simple approach in Example 10.2 does not take into account the increase in viscosity and diffusivity as the gas becomes hotter, but these effects are less important than the rearrangement of the catalyst surface. Photomicrographs of used Pt/Rh gauze show a very rough surface and an increase in the apparent diameter of the wires [28], which might increase $k_c a$ by as much as 50%. No data are available on the mass transfer characteristics of these roughened gauzes.

With a very thin bed of screens in a large converter, it is difficult to get uniform gas distribution, since the velocity in the inlet pipe is high and the pressure drop across the gauze pack is low. To improve gas distribution, the support plate for the gauzes should be designed to have a pressures drop as great as or greater than that due to the gauzes. Gas distribution is improved when several "getter" screens of Pd/Au alloy are placed below the Pt/Rh screens to capture volatilized platinum oxide [22], since they increase the overall pressure drop.

Catalytic Combustion

The oxidation of carbon monoxide and hydrocarbons in vent streams or in auto exhaust can be carried out at moderate temperatures (300–500°C) using screens or an array of wires or ribbons coated with metal or metal oxide catalysts. The main advantage of using wires rather than a ceramic monolith for the catalyst support is that the concentration boundary layer around the wires reforms at each layer of screen instead of gradually getting thicker, as it does with laminar flow in a honeycomb-type monolith. The mass transfer coefficient is about the same for each screen, and the value is comparable to that for the first few millimeters of a ceramic monolith. Some workers [23,24] have used the term *short metal monolith* or *Microlith*^{TM} to describe these catalysts, though they are more like the screens used for ammonia oxidation than the ceramic or metal monoliths that have a large number of parallel channels. The mass transfer coefficient for the wires can be estimated using Eq. (10.14).

Another advantage of wire-screen units for treating auto exhaust is the low mass, which leads to more rapid warm-up. With typical catalytic converters, 60–80% of the total emissions in the test cycle occur during the 2- to 3-minute warm-up period following a cold start [25]. The emissions can be greatly decreased by reducing the warm-up period to 1/2 to 1 minute.

A promising application of wire-supported catalysts is for combustion in gas turbines. Using a lot of excess air lowers the peak combustion temperature and reduces NO_x formation, but the homogeneous reactions are slowed, leading to incomplete conversion. With a wire-screen catalyst to start the reactions, combustion can be completed at a much lower temperature so that CO and NO_x emissions are minimized [26].

REACTIVE DISTILLATION

When a liquid-phase reaction can proceed to only a moderate conversion because of an equilibrium limitation, the traditional approach has been to use distillation to recover the products and then to recycle the unused reactants. A large excess of one reactant can be used to get a high conversion of the other reactant, but this increases the cost of separating and recycling the excess reactant. In some cases the separations are very difficult because of close-boiling materials or azeotrope formation, and the cost of the distillation columns and accessories may be several times the cost of the reactor. A promising option for such cases is to carry out reaction and separation simultaneously in a reactive distillation column. The concept of reactive distillation was first applied to homogeneous reactions such as esterfication with an acid catalyst [27,28]. The reaction takes place in the pools of aerated liquid on the distillation trays, while vapor passing through the trays is enriched in the low-boiling components, just as in normal distillation. Continuous removal of one or both reaction products shifts the equilibrium so that the reaction continues, and nearly complete conversion of the limiting reactant can be obtained. Another advantage for exothermic reactions is that heat released by reaction generates additional vapor in the column, so less energy needs to be supplied in the reboiler, and no heat transfer surface is required to remove the heat of reaction. Finally, the gradual changes in liquid composition due to reaction may prevent the formation of azeotropes that would otherwise limit the separation in conventional distillation columns.

Reactive distillation columns generally have a reactive section placed between a stripping section and a rectifying section, as shown in Figure 10.10. The reactants can be introduced at the same point or fed separately, as in Figure 10.10. Sometimes a prereactor is used to bring the system nearly to chemical equilibrium, and the mixture of reactants and products is then sent to the reactive section of the column. If the feed has a slight excess of one reactant or contains inert compounds, the distillate or bottoms product would be sent to other columns for final purification. With pure reactants and a stochiometric feed, it may be possible to have a high conversion of both reactants and nearly pure products from the reactive distillation column.

Predicting the performance of a reactive distillation system is a challenging problem, because the liquids often form nonideal mixtures, and the addition of chemical reactions to the normal distillation programs may lead to convergence problems unless special algorithms are used [29]. When the reactions are very rapid, it may be satisfactory to assume chemical equilibrium as well as vapor–liquid equilibrium at each stage and then to determine the number of ideal stages needed. Rigorous models

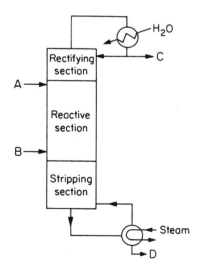

FIGURE 10.10 Reactive distillation column for A + B $\rightleftharpoons$ C + D with a stochiometric feed.

have been developed to include the kinetics of reaction; the liquid holdup on each stage is a variable that can be adjusted to provide a certain approach to chemical equilibrium. The complexity of the models needed for an accurate simulation probably delayed the industrial applications of reactive distillation.

The production of high-purity methyl acetate was the first large-scale application of reactive distillation; the process was developed by Agreda and coworkers at Eastman Chemical Company [27]. The reaction of methanol with acetic acid takes place in solution with an acid catalyst:

$$CH_3OH + CH_3COOH \rightleftharpoons CH_3COOCH_3 + H_2O \qquad (10.20)$$

The equilibrium constant for this reaction is 5.2, so only moderate conversion could be obtained in a plug-flow or batch reactor. Purification of the product mixture would be very difficult, because there are two azeotropes, with boiling points close to that of methyl acetate. With reactive distillation, the higher boiling reactant, acetic acid, is fed near the top of the column, as shown in Figure 10.11, and methanol is fed near the bottom. This counter-

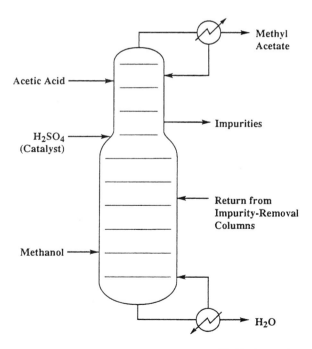

FIGURE 10.11 Methyl acetate reactive distillation column. (From Ref. 30.)

flow arrangement gives a high ratio of acetic acid to methanol in the upper part of the reactive section, so a high conversion of methanol is reached. In the lower part of the reactive section, nearly all of the acetic acid is converted by reaction with excess methanol. Sulfuric acid catalyst is added near the top and is removed with the water. A side stream is withdrawn and distilled in separate columns to remove impurities, such as other esters. A rigorous computer simulation that included kinetics, mass transfer, and tray hydraulics was used to guide the scaleup and optimization of the process [27].

Catalytic Distillation

Reactive distillation can also be carried out with heterogeneous catalysts by using screens or baskets to keep the catalyst particles confined between the trays of the column. Another option is to use a packed bed, with catalyst prepared as rings, saddles, or other standard packing shapes. Small particles are desired, to get a high effectiveness factor, but beds of small particles have low capacity with countercurrent flow because of flooding limits. A promising solution to this problem is to use structured packing that has small particles of catalyst held between layers of crimped screen. Katamax structured packing, developed by Koch Engineering Co., has almost the same gas–liquid mass transfer coefficient as standard structured packing and permits good access of liquid to the small catalyst particles [28].

Structured packing is used in the production of methyl *tert*-butyl ether (MTBE) from methanol and isobutene by catalytic distillation [28,30]. In some plants, a primary converter is employed, as shown in Figure 10.12, to bring the feed mixture nearly to equilibrium. A small excess of methanol is generally used to ensure a high conversion of isobutene in the reactive section. The stripping section removes most of the methanol and butenes from the MTBE product, and the rectifying section removes MTBE from the unreacted methanol and the inerts, which include *n*-butene and butanes in the feed. Although the reaction prevents formation of the usual azeotropes, a reactive azeotrope may form near the MTBE end of the column [30].

Potential applications of reactive distillation to other systems, including those with unfavorable thermodynamics, are discussed in recent articles [31,32]. Other methods of combining reaction with separations, such as extraction, crystallization, and adsorption, are being explored, but none have been used on a large scale. Using reactors with membranes that selectively remove a reaction product is a very promising development, but improvements in membrane permeability, selectivity, and high-temperature stability are needed for practical processes.

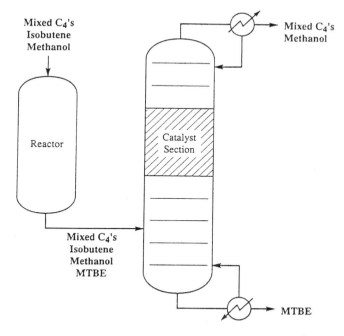

Primary Reactor   Catalytic Distillation

FIGURE 10.12 Catalytic distillation process for MTBE. (From Ref. 30.)

NOMENCLATURE

Symbols

a	external area per unit reactor volume
a'	external area of one gauze per unit cross section
C	reactant concentration
c_p	heat capacity
D	molecular diffusivity, diameter
D_{ea}	axial dispersion coefficient
d	channel diameter, wire diameter
G_s	solids mass flux
Gz	Graetz number
g	gravitational constant
$j_{D,\gamma}$	j-factor for mass transfer based on u_o/γ
K	volumetric mass transfer coefficient
K_o	overall coefficient for mass transfer plus reaction

k	reaction rate constant, thermal conductivity
k_c	mass transfer coefficient per unit area
L, l	bed length or channel length
M	molecular weight
$\dot{m}$	mass flow rate
N	number of wires per unit length of screen
n	number of screens
P	pressure
Pe	Peclet number
Pr	Prandtl number
R	radius of riser reactor
r	reaction rate
r_c	core radius
Re	Reynolds number, Re, γ based on u_o/γ
S	cross-sectional area
Sc	Schmidt number
SV	space velocity
SV_{STP}	space velocity at standard conditions
T	absolute temperature
u_g, u_o	superficial gas velocity
V	reactor volume
v_p	particle velocity
v_t	terminal velocity
x	fraction conversion

Greek Letters

ϵ	void fraction
μ	viscosity
ρ	fluid Density
ρ_p	particle density
ρ_s	suspension density
ψ	slip factor

PROBLEMS

10.1 Compare the limiting conversions for air oxidation of CO, C_3H_8, and $n\text{-}C_6H_{14}$ in a 100-cpsi monolith operating at 400°C and $SV_{STP} = 150,000\ hr^{-1}$.

10.2 For oxidation of CO under the conditions of Example 10.1 with $SV_{STP} = 120,000\ hr^{-1}$, calculate the conversion after each section if four 1.5-inch sections are placed in series instead of a single 6-inch monolith.

10.3 For ammonia oxidation under the conditions of Example 10.2, what is the effect on the average conversion if three-fourths of the gas goes through the central 50% of the gauze pack and one-fourth goes through the outer 50%? Assume 15 gauzes, and neglect the effect of roughening. Is the selectivity also affected by the flow maldistribution?

10.4 Given that external mass transfer is the controlling step in a catalyzed exothermic reaction, derive the equation for the surface temperature and relate the temperature difference to the adiabatic temperature rise. Show that under some conditions the surface temperature could exceed the adiabatic reaction temperature.

10.5 The catalytic oxidation of propylene was studied in short metal monoliths (SMM) at temperatures of 150–500°C [24]. With a single SMM element, the conversion was 65% at 400°C and 75% at 500°C. With three elements in series, the propylene conversion was 86% at 400°C and 93% at 500°C.

    a.   Calculate the apparent activation energy, and compare with the value expected for a diffusion-controlled reaction.

    b.   Are the results for three elements in series consistent with those for a single element and a first-order reaction?

10.6 A bed of screens formed from 0.2-mm wire and coated with a supported noble metal catalyst is used to remove solvents from a vent stream. There are 15 wires per centimeter in each direction, and the superficial gas velocity is 4.8 m/sec at 1 atm and 450°C. If external mass transfer is controlling, how many screens are needed to get 99% removal? Assume $M = 80$ to predict the diffusivity.

REFERENCES

1. D King. In: OE Potter, DJ Nicklin, eds. Fluidization VII. New York: Engineering Foundation, 1992, p 15.
2. PI Viitanen. Ind Eng Chem Res 32:577, 1993.
3. GS Patience, PL Mills. In: VC Corberan, SV Bellon, eds. New Developments in Selective Oxidation II. Amsterdam: Elsevier, 1994.
4. TS Pugsley, GS Patience, F Berruti, J Chaouki. Ind Eng Chem Res 31:2652, 1992.
5. L Godfroy, GS Patience, J Chaouki. Ind Eng Chem Res 38:8, 1999.
6. S Ouyang, OE Potter. Ind Eng Chem Res 32:1041, 1993.
7. MY Louge, V Bricout, S Martin-Letellier. Chem Eng Sci 54:1811, 1999.
8. GS Patience, J Chaouki. Chem Eng Sci 48:3195, 1993.
9. MP Martin, P Turlier, JR Bernard, G Wild. Powder Technology 70:249, 1992.
10. JR Grace. Chem Eng Sci 45:1953, 1990.

11. HJA Schuurmans. Ind Eng Chem Process Des Dev 19:267, 1980.
12. MJ Rhodes. Powder Technology 61:27, 1990.
13. CC White, RJ Dry, OE Potter. In: OE Potter, DJ Nicklin, eds. Fluidization VII. New York: Engineering Foundation, 1992, p 265.
14. RD Hawthorn. AIChE Symposium Ser 70(137):428, 1974.
15. WL McCabe, JC Smith, P Harriott. Unit Operations of Chemical Engineering. 6th ed. New York: McGraw-Hill, 2001, p 344.
16. J Votruba, O Mikus, K Nguen, V Hlavacek, J Skrivanek. Chem Eng Sci 30:201, 1975.
17. U Ullah, SP Waldram, CJ Bennett, T Truex. Chem Eng Sci 47:2413, 1992.
18. RM Heck, RJ Farrauto. Catalytic Air Pollution Control—Commercial Technology. New York: Van Nostrand, 1996, p 55.
19. MA Shah, D Roberts. Chemical Reaction Eng II. ACS Adv in Chem Ser 133:259, 1974.
20. CN Satterfield, DH Cortez. Ind Eng Chem Fund 9:613, 1970.
21. RM Heck, JC Bonacci, WR Hatfield, TH Hsung. Ind Eng Chem Proc Des Dev 21:73, 1982.
22. GR Gillespie, RE Kenson. Chem Tech (October) 1971, p 627.
23. S Roychoudhury, G Muench, JF Bianchi, WC Pfefferle. SAE Technical Paper No. 971023, 1997.
24. RN Carter, P Menacherry, WC Pfefferle, G Muench, S Roychoudhury. SAE Technical Paper No. 980672, 1998.
25. RP Hesketh, D Bosak, L Kline. Chem Eng Education (Summer) 2000, p 240.
26. RN Carter, LL Smith, H Karim, M Castaldi, S Etemad, G Muench, RS Boorse, P Menacherry, WC Pfefferle. Mat Res Soc Symp Proc 549:93, 1999.
27. VH Agreda, LR Partin, WH Heise. Chem Eng Prog 86(2):40, February 1990.
28. JL DeGarmo, VN Parulekar, V Pinjala. Chem Eng Prog 88(3):42, March 1992.
29. S Venkataraman, WK Chan, JF Boston. Chem Eng Prog 86(8):45, August 1990.
30. MF Doherty, G Buzad. I Chem E Symp Ser 128:A51, 1992.
31. MJ Okasinaki, MF Doherty. I Chem E Symp Ser 142:625, 1997.
32. AI Stankiewicz, JA Moulin. Chem Eng Prog 96(1):22, January 2000.

Index

Absorption plus reaction:
  of CO, 281
  of CO_2, 274
  of O_2, 260, 300, 304
  of SO_2, 317
  theories for, 257–281
Acrylamide, polymerization of, 37
Activation energy, 10, 97
  apparent value of, 149
Adiabatic reactors, 106
  kinetic data from, 22
  with quench cooling, 110, 111
  radial flow type, 108
  for SO_2 conversion, 115, 124
  two-stage type, 108
Adiabatic temperature rise, 109, 187, 196
Adsorption:
  chemisorption (see also Langmuir isotherm), 44
  nitrogen adsorption, 129, 137

[Adsorption]
  as rate-controlling step, 58
Agglomeration:
  of catalyst in liquid, 319
  of particles in riser, 399
Agitation:
  by nozzles in jacket, 177
  in stirred-tank reactors, 232, 297, 323
Air pollution control:
  with catalyst monoliths, 403
  by incineration of pollutants, 411
  by selective catalytic reduction (SCR), 411
Alcohol dehydrogenation, 62
Alkylation of isobutane, 227
Arrhenius relationship, 10, 149
  derivative of, 11, 182
Association parameter, 142
Axial diffusivity (see Axial dispersion)

RETURN TO: CHEMISTRY LIBRARY

100 Hildebrand Hall • 510-642-3753

LOAN PERIOD	1	2 _1 Month_	3
4		5	6

ALL BOOKS MAY BE RECALLED AFTER 7 DAYS.

Renewals may be requested by phone or, using GLADIS,
type **inv** followed by your patron ID number.

DUE AS STAMPED BELOW.

NON-CIRCULATING UNTIL: ~~MAY 14 2003~~	
AUG 1 5 2003	
DEC 1 8 2003	
DEC 2 0 2005	
DEC 0 3 2005	
DEC 2 0	
SEP 0 8	
NOV 1 7	

FORM NO. DD 10
2M 5-01

UNIVERSITY OF CALIFORNIA, BERKELEY
Berkeley, California 94720–6000